[JR HIGH – YEAR 1 STUDENT]

BOOK 1

PRINCIPLES OF MATHEMATICS

BIBLICAL WORLDVIEW CURRICULUM

Katherine A. [Loop] Hannon

Author: Katherine A. [Loop] Hannon

Master Books Creative Team:

Editor: Craig Froman

Design: Jennifer Bauer

Cover Design: Diana Bogardus

Copy Editors:
Judy Lewis
Willow Meek

Curriculum Review:
Kristen Pratt
Laura Welch
Diana Bogardus

First printing: April 2015
Ninth printing: October 2021

Copyright © 2015 by Katherine A. Hannon and Master Books®. All rights reserved. No part of this book may be reproduced, copied, broadcast, stored, or shared in any form whatsoever without written permission from the publisher, except in the case of brief quotations in articles and reviews. For information write:

Master Books®, P.O. Box 726, Green Forest, AR 72638
Master Books® is a division of the New Leaf Publishing Group, Inc.

ISBN: 978-0-89051-875-5
ISBN: 978-1-61458-453-7 (digital)
Library of Congress Number: 2015937807

Unless otherwise noted, Scripture quotations are from the King James Version of the Bible.

Some sections are heavily adapted with permission from Katherine A. [Loop] Hannon, *Revealing Arithmetic: Math Concepts from a Biblical Worldview* (Fairfax, VA: Christian Perspective, 2010).

Printed in the United States of America

Please visit our website for other great titles:
www.masterbooks.com

Katherine A. (Loop) Hannon, is a homeschool graduate who had her own view of math transformed. Understanding the biblical worldview in math made a tremendous difference in her life and started her on a journey of researching and sharing on the topic. For more than a decade, she has been researching, writing, and speaking on math, as well as other topics. Her previous books on math and a biblical worldview have been used by various Christian colleges, homeschool groups, and individuals.

Acknowledgments

This curriculum was a major part of my life for several years, but I don't believe you would be holding it in your hands today if it were not for some very special people. I'd like to acknowledge and thank

my mom (Cris Loop) for being my right-hand helper throughout this project, solving hundreds of math problems and painstakingly working with me to make the material understandable.

my dad (Bob Loop) for his overall support (and specifically for all his input into the statistics chapter).

my brother (Brian Loop) for putting up with math questions at all hours of the day and helping me think through concepts and flow.

my cousin Hannah and my aunt Marie Ferreira for testing the initial draft, giving me a weekly deadline and providing invaluable feedback.

Joy Dubbs for being my technical sounding board on so many concepts, for reading drafts for me, and for never tiring of a math question.

Abbey Ryan for reading the material, cheering me onward, and reminding me why I was writing this.

the people at Master Books for believing in the project and moving it into publication.

everyone else who has prayed for and supported me during this project in one way or another.

Above all, I'd like to thank the Lord, without whom all our labors are in vain.

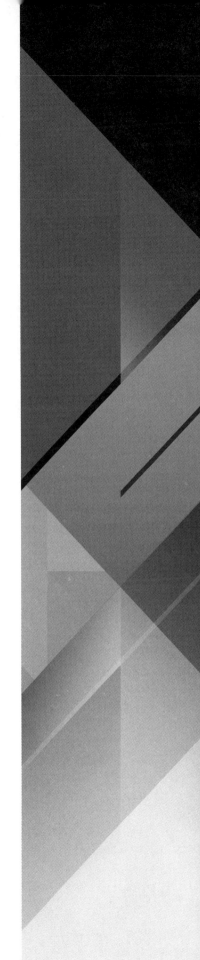

Once you're done with Book 1 ... you'll be ready for Book 2!

Get ready to continue exploring principles of mathematics! Now that you know the core principles of arithmetic and geometry, you're ready to move on to learning even more skills that will allow you to explore more aspects of God's creation.

In Book 2, we'll focus on the core principles of algebra, coordinate graphing, probability, statistics, functions, and other important areas of mathematics. The topics may sound intimidating, but you'll discover that they are simply useful techniques that serve a wide range of practical uses. As we do in this book, we'll continue to discover that all of math boils down to a way of describing God's creation and is a useful tool we can use to serve God, all while worshiping Him!

About This Curriculum

What Is This Curriculum?

This is Year 1 of a two-year math course designed to give students a firm mathematical foundation, both academically and spiritually. Not only does the curriculum build mathematical thinking and problem-solving skills, it also shows students how a biblical worldview affects our approach to math's various concepts. Students learn to see math not as an academic exercise, but as a way of exploring and describing consistencies God created and sustains. The worldview is not just an addition to the curriculum, but the starting point. Science, history, and real life are integrated throughout.

How Does a Biblical Worldview Apply to Math… and Why Does It Matter?

Please see lessons 1.1–1.3 and 2.7 for a brief introduction to how a biblical worldview applies to math and why it matters.

Whom Is This Curriculum For?

This curriculum is aimed at **grades 7-8**, fitting into most math approaches the **year or two years prior to starting high school algebra**. If following traditional grade levels, Year 1 can be completed in grade 6 or 7, and Year 2 in grade 7 or 8. Students needing a slower pace can complete both books over 3 years (grades 6, 7, and 8).

The curriculum also works well for **high school students** looking to firm up math's foundational concepts and grasp how a biblical worldview applies to math. High school students may want to follow the alternate accelerated schedule in the *Student Workbook* and complete each year of the program in a semester, or use the material alongside a high school course.

Where Do I Go Upon Completion?

Upon completion of Year 1, students will be ready to move on to Year 2. Upon completion of both years, **students should be prepared to begin or return to any high school algebra course.**

Are There Any Prerequisites?

Year 1: Students should have a **basic knowledge of arithmetic** (basic arithmetic will be reviewed, but at a fast pace and while teaching problem-solving skills and a biblical worldview of math) and **sufficient mental development** to think through the concepts and examples given. Typically, anyone in 6th grade or higher should be prepared to begin.

Teacher Guide

Year 2: It is strongly recommended that students complete Year 1 before beginning Year 2, as math builds on itself.

What Are the Curriculum's Components?

The curriculum consists of the **Student Textbook** and the **Student Workbook**. The *Student Textbook* contains the lessons, and the *Student Workbook* contains all the worksheets, quizzes, and tests, along with an answer key and suggested schedule. An optional eCourse is available on the author's website (ChristianPerspective.net).

How Do I Use This Curriculum?

General Structure — This curriculum is designed to be self-taught, so students should be able to read the material and complete assignments on their own, with a parent or teacher available for questions. This student book is divided into chapters and then into lessons. The number system used to label the lessons expresses this order. The first lesson is labeled as 1.1 because it is Chapter 1, Lesson 1.

Worksheets, Quizzes, and Tests — The accompanying *Student Workbook* includes worksheets, quizzes, and tests to go along with the material in this book, along with a suggested schedule and answer key.

Answer Key — A complete answer key is located in the *Student Workbook*.

Schedule — A suggested schedule for completing the material in 1 year, along with an accelerated 1-semester schedule, are located in the *Student Workbook*. A 3-year schedule can be found online at https://www.christianperspective.net/math/pom1#3-year.

[CONTENTS]

Preface 11

Chapter 1: Introduction and Place Value 13
    1.1 Math Misconceptions 13
    1.2 What Is Math? 15
    1.3 The Spiritual Battle in Math 18
    1.4 Numbers, Place Value, and Comparisons 22
    1.5 Different Number Systems 27
    1.6 Binary and Hexadecimal Place-value Systems 31

Chapter 2: Operations, Algorithms, and Problem Solving 37
    2.1 Addition and Subtraction 37
    2.2 Applying Basic Addition and Subtraction to Time 42
    2.3 Multi-digit Addition and Subtraction 46
    2.4 Keeping and Balancing a Checkbook Register 52
    2.5 Different Addition and Subtraction Methods 56
    2.6 Problem Solving: An Introduction 58
    2.7 Chapter 1 and 2 Synopsis 63

Chapter 3: Mental Math and More Operations 65
    3.1 Mental Math 65
    3.2 Rounding — Approximate Answers 67
    3.3 Multiplication 68
    3.4 Division 73
    3.5 Properties 76
    3.6 Conventions — Order of Operations 81
    3.7 Chapter Synopsis 83

Chapter 4: Multi-digit Multiplication and Division 85
    4.1 Multi-digit Multiplication 85
    4.2 The Distributive Property — Understanding Why 92
    4.3 Multi-digit Division 95
    4.4 More Multi-digit Division, Remainders, and Operations with Zero 99
    4.5 More Problem-Solving Practice/Enter the Calculator 102
    4.6 Chapter Synopsis 105

Chapter 5: Fractions and Factoring 109
    5.1 Understanding Fractions 109
    5.2 Mixed Numbers 114
    5.3 Equivalent Fractions and Simplifying Fractions 117
    5.4 Understanding Factoring 121
    5.5 Simplifying Fractions, Common Factors, and the Greatest Common Factor 124
    5.6 Adding and Subtracting Fractions 125
    5.7 Least Common Denominator/Multiple 130
    5.8 Chapter Synopsis and Expanding Our Worldview 131

Chapter 6: More with Fractions...135
6.1 Multiplying Fractions...135
6.2 Working with Mixed Numbers 138
6.3 Simplifying While Multiplying 142
6.4 Reciprocal/Multiplicative Inverse and More..146
6.5 Dividing Fractions...148
6.6 Chapter Synopsis..152

Chapter 7: Decimals...153
7.1 Introducing Decimals 153
7.2 Adding and Subtracting Decimals 158
7.3 Multiplying Decimals..162
7.4 Dividing and Rounding with Decimals..164
7.5 Conversion and More with Decimals 166
7.6 Chapter Synopsis..168

Chapter 8: Ratios and Proportions...169
8.1 Ratios..169
8.2 Proportions..171
8.3 Ratios and Proportions Containing Decimals...174
8.4 Scale Drawings and Models 176
8.5 Mental Math and Decimals 179
8.6 Chapter Synopsis..181

Chapter 9: Percents...183
9.1 Introducing Percents 183
9.2 Finding Percentages 186
9.3 More Finding Percentages/Multiplying and Dividing Percents 187
9.4 Adding and Subtracting Percents/Finding Totals 189
9.5 Mental Percents..192
9.6 Chapter Synopsis..193

Chapter 10: Negative Numbers...195
10.1 Understanding Negative Numbers..195
10.2 Adding Negative Numbers...197
10.3 Subtracting Negative Numbers...201
10.4 Temperature and Negative Numbers...203
10.5 Absolute Value 206
10.6 Multiplying and Dividing Negative Numbers 208
10.7 Negative Mixed Numbers...210
10.8 Negative Fractions...212
10.9 Chapter Synopsis..213

Chapter 11: Sets...215
11.1 Sets and Venn Diagrams...215
11.2 Number Sets...219
11.3 More on Sets...222

11.4 Ordered Sets (i.e., Sequences) 224

11.5 Chapter Synopsis 226

Chapter 12: Statistics and Graphing 227

12.1 Introduction to Statistics 227

12.2 Collecting Data — Sampling..229

12.3 Organizing Data — Understanding Graphs 232

12.4 Organizing Data — Drawing Bar/Column Graphs 236

12.5 Coordinates 238

12.6 Organizing Data — Line Graphs 241

12.7 Organizing Data — Averages 244

12.8 Organizing Data — More on Averages 249

12.9 Chapter Synopsis 250

Chapter 13: Naming Shapes: Introducing Geometry 253

13.1 Understanding Geometry...253

13.2 Lines and Angles 256

13.3 Polygons...260

13.4 Circles, Triangles, and Three-Dimensional Shapes...................263

13.5 Fun with Shapes...267

13.6 Chapter Synopsis 272

Chapter 14: Measuring Distance 275

14.1 Chapters for Measuring Distance...275

14.2 Conversions via Proportions..281

14.3 Different Conversion Methods 283

14.4 Currency Conversions 286

14.5 Metric Conversions 288

14.6 Multistep Conversions...291

14.7 Conversions Between U.S. Customary and Metric 292

14.8 Time Conversions..294

14.9 Chapter Synopsis 297

Chapter 15: Perimeter and Area of Polygons................................299

15.1 Perimeter 299

15.2 Formulas...303

15.3 Area — Rectangles and Squares 305

15.4 Area — Parallelograms 309

15.5 Chapter Synopsis and the Basis of Math and Truth...................311

Chapter 16: Exponents, Square Roots, and Scientific Notation 315

16.1 Introducing Exponents 315

16.2 Understanding Square Roots 319

16.3 Square Unit Conversions...322

16.4 Scientific Notation...324

16.5 More Scientific Notation 328

16.6 Chapter Synopsis 331

Chapter 17: More Measuring: Triangles, Irregular Polygons, and Circles............333

17.1 Area — Triangles..333

17.2 Area — More Polygons...336

17.3 Measuring Circles..338

17.4 Irrational Numbers..343

17.5 Chapter Synopsis and Pi in the Bible...344

Chapter 18: Solid Objects and Volume...347

18.1 Surface Area..347

18.2 Volume...350

18.3 Cubic Unit Conversion...356

18.4 Measuring Capacity in the U.S. Customary System...........................358

18.5 Conversion to and from Cubic Units..361

18.6 Measuring Capacity in the Metric System 362

18.7 Measuring Weight and Mass...364

18.8 Chapter Synopsis...365

Chapter 19: Angles...367

19.1 Measuring Angles..367

19.2 More with Angles..370

19.3 Angles in Pie Graphs..374

19.4 Expanding Beyond..376

19.5 Chapter Synopsis and Faulty Assumptions..379

Chapter 20: Congruent and Similar...381

20.1 Congruent..381

20.2 Corresponding Parts — Applying Congruency 384

20.3 Exploring Similar Shapes..390

20.4 Angles in Triangles and AA Similarity 392

20.5 Similar Triangles in Action — Finding the Height of a Tree.................395

20.6 Chapter Synopsis and a Peek Ahead..396

Chapter 21: Review..399

21.1 Arithmetic Synopsis..399

21.2 Geometry Synopsis...400

21.3 Life of Johannes Kepler..400

21.4 Course Review...403

Appendix A 407

Endnotes...409

Index 415

Preface

Growing up, I never pictured myself writing a math curriculum. While I was good at math, I never grasped its importance nor understood how to apply some of the more advanced concepts outside of a textbook. It was a subject of rules to be memorized, applied, and forgotten. After all, would I ever really use most of it in life?

Nor did I see how math could be viewed from a biblical worldview. I was blessed to grow up in a Christian homeschool family and to see the worldview battle in my other subjects, but not in math. I subconsciously delegated math to some sort of "neutral" category.

Obviously, something changed. During my senior year, my mom had me read a book by James D. Nickel titled *Mathematics: Is God Silent?* As the book delved in depth into the history and philosophy of math, I realized that math wasn't neutral at all — and I saw all of math began to transform.

Excited, I desperately wanted to share what I'd learned with everyone I could. My heart ached for the many young people growing up without seeing God in math or really understanding how it served as a tool to describe His creation. Yet there simply weren't many resources available on the topic.

The need prompted me to begin speaking and writing on math. I left library after library with stacks of math books. I read and read and read. The more I learned, the more amazed I grew. Around two years later, I wrote my first book, *Beyond Numbers: A Practical Guide to Teaching Math Biblically*. After much more research, I wrote my second book, *Revealing Arithmetic: Math Concepts from a Biblical Worldview*.

Both these books gave parents tools and information on how to teach math from a biblical worldview. *Revealing Arithmetic* added details and specific ideas for arithmetic concepts. Still, though, everyone kept asking for an actual curriculum — something I insisted I'd never write. The mere thought was overwhelming.

It eventually became clear that writing a curriculum was exactly what I was supposed to do. After unsuccessfully trying my hardest to get out of the task, I succumbed and began writing.

The journey proved tougher than I imagined, but also more amazing as I got to watch the Lord provide despite (or rather through) many obstacles, including a concussion that took more than a year and a half to recover from. I discovered over and over again that it's truly only by God's enabling that we can do anything. He's the One who causes our brains to work and who sustains the universe in a consistent fashion, making math possible. God's provisions also included many precious people who helped, as you'll see from the Acknowledgments.

It's my earnest prayer that as you use this curriculum and study math from a biblical worldview, you'll be reminded of God's greatness and encouraged that you can trust Him completely.

By His grace,

Katherine [Loop] Hannon

[CHAPTER 1]

Introduction and Place Value

1.1 Math Misconceptions

Math — what does the word bring to mind? Numbers in a textbook? Lists of multiplication and division facts? Problems to solve?

That about sums up the typical view of math, doesn't it? Yet while math does have numbers, facts, and problems, there's much more to math than typically presented.

But before we look at what math is, let's start by examining what it is not. Specifically, let's take a look at three common — but dangerous — misconceptions about math.

Misconception 1: Math Is Neutral

Most math books approach math as a neutral subject. And at first glance, math certainly appears neutral. Neutral means "not engaged on either side; not aligned with a political or ideological grouping."[1] Christians and atheists all can agree that "1 + 1 = 2." This makes math neutral, right?

To answer this, consider a tree. A tree seems pretty neutral too, doesn't it? People of all religions can see a tree, touch a tree, smell a tree, and study a tree, agreeing on a tree's basic features. But this does not mean a tree is neutral. A tree's very existence begs for an explanation. Where did trees originate? Why does a tree have intricate parts that all work and grow together?

Our underlying perspective regarding a tree is determined by what we would call a **worldview**. In *Understanding the Times*, David Noebel (founder of Summit Ministries) defines a worldview this way: "A worldview is like a pair of glasses — it is something through which you view everything. And the fact is, everyone has a worldview, a way he or she looks at the world."[2] In other words, a worldview is a set of truths (or falsehoods we believe to be true) through which we interpret life.

Those with a biblical worldview — those looking at life in light of what the Bible teaches — would see a tree as part of God's originally perfect but now fallen creation, while those with a naturalistic worldview might say a tree evolved from a cosmic bang. When we look at the essential questions of a tree — where it came from, how we should use it, etc. — we see a tree is not really neutral.

In a similar way, math facts may seem neutral. People of all religions can use math and agree that "1 + 1 = 2." But this does not mean math is neutral. Where did math originate? Why does math work the way it does? Why are we able to use math?

Just as it does in the case of a tree, the Bible gives us a framework from which we can answer these questions and build our understanding of math. As we'll discover, only the biblical explanation for math's very existence makes sense out of math and transforms math from a dry list of facts to an exciting exploration.

The point here is simply that math cannot be neutral. The Bible teaches Jesus is Lord of all — the Creator and Sustainer of *all* things (Colossians 1:16–17). He doesn't exempt math from that. Math cannot be separated into a "neutral" box.

Misconception 2: A Biblical Math Curriculum Is the Same as Any Other, with a Bible Verse or Problem Thrown In Now and Then

If you're wondering if we're just going to add a Bible verse to the top of the page, mention God dividing the Red Sea when we discuss division, and have you solve Bible-based word problems, let me assure that this is *not* what this curriculum aims to do. Although thinking about how God divided the Red Sea might be helpful in turning our eyes to the Lord, it does nothing for helping us understand how to view *division itself* in light of biblical principles. In this course, we're aiming to let the Bible's principles transform our view of *math itself.*

Misconception 3: Math Is a Textbook Exercise

Quite often, math comes across as a textbook exercise. We memorize this and solve that. There are so many seemingly arbitrary rules to follow that it's easy to scratch your head and wonder who invented this complex system in the first place.

If your view of math is confined to rules and problems — or even if you know there's more to math but are not sure why it feels so dry — there is good news! Math is *not* a mere textbook exercise. Math helps us observe the design throughout creation, design instruments, draw, build boats, operate a business, work with computers, cook, sew, and more. In this course, we'll incorporate history, science, and real-life applications as we go, exploring math both inside *and* outside of a textbook.

1.2 What Is Math?

If you were to try to work in nearly any field of science — be it chemistry, engineering, or anatomy — you would need to study and use math. Why? *Because math is the tool scientists use to explore creation.*

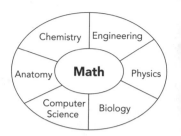

Not only do scientists use math, but artists, pilots, musicians, business managers, clerks, sailors, and homemakers do too. All occupations use math to one extent or another!

Math also shows up in everyday life. Every time you go shopping, you use math — math helps you know how much an item costs (price tags use numbers!), find the unit price of an item, estimate your total, etc. You use math in the kitchen when you measure ingredients. You use math to count the number of silverware to put on the table, to balance a checkbook and track your finances, to understand loans and car payments, to figure out how many bags of bark you need to landscape a flowerbed or how many square feet of carpet to cover a room — the list of math's everyday uses goes on and on.

Math is clearly more than intellectual rules and techniques found in a textbook. Which brings us to the question: what *is* math?

When we **start with the Bible** — God's revealed Word to man — as our source of truth, it makes sense out of every area of life, including math. It gives us a framework for answering not only *what* math is, but also *where* it came from and *why* it works. Take a look at just a few truths with me.

■ **All things were created and are sustained by the eternal, triune God of the Bible.**

> *In the beginning God created the heaven and the earth. And the earth was without form, and void; and darkness was upon the face of the deep. And the Spirit of God moved upon the face of the waters (Genesis 1:1–2).*

> *In the beginning was the Word, and the Word was with God, and the Word was God. The same was in the beginning with God. All things were made by him; and without him was not any thing made that was made. . . . And the Word was made flesh, and dwelt among us, (and we beheld his glory, the glory as of the only begotten of the Father,) full of grace and truth (John 1:1–3, 14).*

> *Jesus answered them. . . . "I and my Father are one" (John 10:25, 30).*

> *For by him [Jesus] were all things created, that are in heaven, and that are in earth, visible and invisible, whether they be thrones, or dominions, or principalities, or powers: all things were created by him, and for him: And he is before all things, and by him all things consist (Colossians 1:16–17).*

*[referring to Jesus] . . . upholding all things by the word of his power
. . . (Hebrews 1:3).*

■ **God is a consistent God who never changes — and who has appointed the ordinances of heaven and earth.**

> *For I am the LORD, I change not; therefore ye sons of Jacob are not consumed (Malachi 3:6).*

> *Thus saith the LORD; If my covenant be not with day and night, and if I have not appointed the ordinances of heaven and earth; Then will I cast away the seed of Jacob and David my servant, so that I will not take any of his seed to be rulers over the seed of Abraham, Isaac, and Jacob: for I will cause their captivity to return, and have mercy on them (Jeremiah 33:25–26).*

■ **God created man in His own image and gave him the task of subduing the earth.**

> *So God created man in his own image, in the image of God created he him; male and female created he them. And God blessed them, and God said unto them, Be fruitful, and multiply, and replenish the earth, and subdue it: and have dominion over the fish of the sea, and over the fowl of the air, and over every living thing that moveth upon the earth (Genesis 1:27–28).*

Let's look at how these truths apply to math. A never-changing God is holding *all* things together and has appointed the ordinances — or the decrees — by which heaven and earth operate. God created and sustains a consistent universe. God also created man in His image, capable of subduing and ruling over the earth.

We already established that math is the tool scientists use to describe creation. In other words, math is a way of describing the consistencies God created and sustains! Man is able to use math to, in a very limited way, think "God's thoughts after Him" (Johannes Kepler) because God made us in His image and gave us the ability to subdue the earth.

The Bible teaches that God created all things — and math is no exception. The symbols and techniques we think of as math describe on paper the ordinances by which God governs all things. Men might develop different symbols (people have used many different numerals and techniques throughout history, as we'll see throughout this course), but men have no control over the principles. No matter what symbols we use to describe it, one plus one consistently equals two because God both decided it would and, day in and day out, keeps this ordinance in place!

Math, in essence, is **a way of describing the consistent way this universe operates.** It is the language, so to speak, we use to express the quantities and consistencies around us — quantities and consistencies God created and sustains.

PRINCIPLES OF
MATHEMATICS 1

Math works outside a textbook *because* God is faithful to uphold all things. Math facts never change *because* God never changes. We can rely on math *because* we can rely on God. Math is complex and complicated *because* God created a complex universe and it takes a lot of different rules and methods to even begin to describe it! Math applies universally *because* God's rule is universal — He's present everywhere. Math helps us see the incredible wisdom and care displayed throughout creation — an order, wisdom, and care that is there *because* we have a wise and caring Creator! At the same time, math reveals the effects of sin that mar God's original design — effects that are there *because* of man's sin, but which remind us of the mercy found in Jesus.

> *Mathematics transfigures the fortuitous concourse of atoms into the tracery of the finger of God.* — Herbert Westren Turnbull[3]

Do you catch how this understanding could revolutionize our view of math? Math doesn't have to be a dry subject of mere numbers and techniques. Numbers and techniques are tools to describe God's creation and help us with the real-life tasks God's given us to do.

> *Mathematics is like a chest of tools.* — Walter W. Sawyer[4]

I love that imagery. Picture a chest of tools for a moment. Some tools — such as a screwdriver — are easy to use and apply in thousands of situations. Other tools — such as a router — take more time and dedication to master and serve a more limited, although just as necessary, purpose.

In a similar fashion, some math concepts — such as addition — are fairly easy to grasp and frequent in their applications. Others — such as some aspects of algebra or calculus — require more dedication to grasp. These higher-level concepts, while they might not come in handy as often as addition, have *very* powerful applications.

From basic to advanced, *all* of math is a tool that points us to God and can be used to complete the tasks He gives us to do!

Ready to Begin?

Some of you reading this course probably dislike math or find it an incredibly challenging task. Others of you may love it and be gifted in it.

Whatever your current view of math, I invite you to take a journey with me. While we'll be seeking to approach concepts from a biblical worldview throughout the course, these first two chapters will be extra-intensive in that department, as we want to lay a firm foundation upon which to build the rest. So please bear with the extra amount of reading.

My prayer is that you'll acquire a deeper appreciation for God's greatness and faithfulness and be encouraged in your walk with Him as you delve into the world of mathematics.

1.3 The Spiritual Battle in Math

The Bible gives us a solid foundation for why math works. Math is a tremendous testimony to God's faithfulness and power. Yet math has been sadly twisted.

Let's take a deeper look at the spiritual battle within math, at how men try to explain math apart from God, and at how ultimately only a biblical worldview makes sense out of math.

Naturalism in Math

Consider the following quote:

> *One cannot escape the feeling that these mathematical formulae have an independent existence and an intelligence of their own, that they are wiser than we are, wiser even than their discoverers, that we get more out of them than was originally put into them.*
> — Heinrich Hertz (German physicist)[5]

Notice that Mr. Hertz is claiming that mathematical formulae themselves are wise and have an independent existence. Rather than acknowledging God, he is giving *math itself* the credit for math's amazing ability to work. This is a very naturalistic view of math — an attempt to explain math from only natural causes, apart from God.

Let's think about this claim for a moment. Can math itself explain its own existence? Remember, math goes hand in hand with creation. Things don't just "happen." We live in a universe consistent enough that we can describe gravity using math and call it a law. If the universe were run by random processes, why would we see such order, design, and consistency?

Besides this, viewing math as a self-existent truth still doesn't answer the fundamental question of how we *know* it's true in the first place. Is the foundation for truth our experience? Do we know that one plus one equals two because we experience that it does, and therefore assume that it always will?

Our experience in itself is not a solid foundation for truth. For one, we can never experience everything, so therefore could never truly know anything for sure! Math is so useful because it helps us solve problems we have not experienced. It allows us to calculate the force needed to get a brand-new rocket into the air — and to predict how a bridge will hold weight before we build it. Much of math deals with things that we can never actually experience, but which help us solve a variety of real-life problems. In order to use math, we *have* to assume it works consistently in areas we have never — and never can — experience.

Humanism in Math

Now consider this quote:

> The German mathematician Leopold Kronecker (1823–91) once said, "God made the integers, all else is the work of man." First causes, this comment suggests, are divine, while the complexities, minutiae, and refinements of mathematics are a human creation. For Kronecker's contemporary Dedekind, however, the integers too were the "free creations of the human mind." . . . For him, as for many modern mathematicians and theorists, mathematics stood as an independent and secular discipline. — Denis Guedj[6]

Who did Kronecker and Dedekind give credit for math? *Man.* Both these men viewed math as the product of the human mind. Rather than giving God the credit for math's ability to work, they gave it to man. This is a humanistic view of math — a view that focuses on *man* and *his* achievements, ignoring his Creator.

Let's think about the problem with basing truth on human reasoning. Time and again, math concepts men think up using mathematical reasoning end up applying in creation. Why is this? Why do men's thoughts line up with reality? Why do we find such an orderly, mathematical world all around us?

Albert Einstein expressed the problem this way — and admits there's something miraculous in the world that can't be explained by reasoning alone.

> Even if the axioms of the theory are posited by man, the success of such a procedure supposes in the objective world a high degree of order which we are in no way entitled to expect a priori [based on man's reasoning]. Therein lies the "miracle" which becomes more and more evident as our knowledge develops. . . . And here is the weak point of positivists and of professional atheists, who feel happy because they think that they have not only pre-empted the world of the divine, but also of the miraculous.[7]

Also, why are there universal laws of logic we rely on to be true? Why can't one person decide that 1 plus 1 will equal 2 and another that it will equal 3 and they both be right? This sort of thinking, if applied consistently, would completely make math, as well as logic itself, meaningless and useless!

The Battle Defined

The spiritual battle over math is the same as the battle we find in other areas of life. Will we recognize our *dependency* on God, or claim *independence* from Him?

Our view of any area of life — including math — is either going to stem from a dependent perspective on life (one that recognizes our dependency on God and His Word) or an independent one. When we get down to the fundamental level, there is no such thing as neutrality. Even a tree is not neutral — as we saw in the first lesson, the tree was either created by God or got here some other way.

While it is true that man developed math symbols and techniques, it makes no sense why those symbols and techniques mean anything in real life if they truly are the "free creations of the human mind" as Dedekind stated.

For more information on different worldviews on math and how the biblical worldview makes sense of math, see James D. Nickel, *Mathematics: Is God Silent?* rev. ed. (Vallecito, CA: Ross House Books, 2001). For more information on how logic itself can't be explained apart from God, see Dr. Jason Lisle, *The Ultimate Proof of Creation: Resolving the Origins Debate* (Green Forest, AR: Master Books, 2009).

Likewise, math is either dependent on God or it is not. If God does not receive the glory for math's ability to work, that glory goes somewhere else. As R.J. Rushdoony points out,

> . . . mathematics is not the means of denying the idea of God's pre-established world in order to play god and create our own cosmos, but rather is a means whereby we can think God's thoughts after Him. It is a means towards furthering our knowledge of God's creation and towards establishing our dominion over it under God. The issue in mathematics today is root and branch a religious one.[8]

The Bible is clear: we are to trust and worship God; He gives us our every breath, He controls each aspect of life, and He determines truth — apart from Him, we are nothing. If man ignores this truth, he does so to his own demise.

> For the wrath of God is revealed from heaven against all ungodliness and unrighteousness of men, who hold the truth in unrighteousness; Because that which may be known of God is manifest in them; for God hath shewed it unto them. For the invisible things of him from the creation of the world are clearly seen, being understood by the things that are made, even his eternal power and Godhead; so that they are without excuse: Because that, when they knew God, they glorified him not as God, neither were thankful; but became vain in their imaginations, and their foolish heart was darkened. Professing themselves to be wise, they became fools, And changed the glory of the uncorruptible God into an image made like to corruptible man, and to birds, and fourfooted beasts, and creeping things (Romans 1:18–23).

The Depth of the Battle

The battle over math is much more than a theological squabble over numbers. It ultimately affects our entire approach to truth.

If we look at math as something spiritually neutral — a self-existent or man-made fact — then math becomes an independent source of truth. We find ourselves viewing math as the ultimate standard rather than God's Word.

Millions of people have embraced the lie of evolution because they believe it has been scientifically proven to be true. At the root of their belief is the false notion that deductive reasoning or mathematical principles are the ultimate standard ruling the universe.

Yet, apart from God, it does not even make sense why we can reason or why the universe is consistent! Science and math would be impossible in a universe without God. The very tool skeptics try to use to disprove God cannot be explained apart from God. Even honest unbelievers acknowledge their inability to explain math in their worldview. Most simply ignore the *why*.

In this article I shall not attempt any deep philosophical discussion of the reasons why mathematics supplies so much power to physics. . . . The vast majority of working scientists, myself included, find comfort in the words of the French mathematician Henri Lebesgue: "In my opinion a mathematician, in so far as he is a mathematician, need not preoccupy himself with philosophy — an opinion, moreover, which has been expressed by many philosophers."
— Freeman J. Dyson[9]

But when we look at math from a biblical perspective, we understand that math is not a source of truth; it is a description of the consistencies of God. God is the source of truth. We can only rely on math to work because we can trust God. Thus, as we study math in this course, we will not approach it as a means to determine truth or as the source of truth, but rather as a tool to help us understand the trustworthy principles our trustworthy God created and sustains.

Math and the Gospel

Although we might try our hardest, we cannot change math. We can change the symbols or names we use in math, but we cannot change what the names and symbols represent — 1 of something plus 1 of something else will consistently equal 2. Math is not relative. Why?

Because God is God and we are not! He, not us, decides how things will be. He set and keeps certain principles in place, and if we want a math that will actually work, we *have* to conform to those principles.

Math reminds us that God decides truth, not us. We need to be careful in every area that we take heed to the truth He has revealed to us in His Word, the Bible, and that we don't try to change those truths to fit what we think or want. For example, the Bible tells us that salvation is by faith in Jesus, and not by works or any other way (Ephesians 2:8; James 14:6), and that hell is real (Revelation 21:8).

It's tempting to try to change this truth, thinking there must be some good in ourselves or that God would not really send people to hell (especially those whom we love and think are nice), but God's truth is not open to negotiation. He's God, not us. If we want salvation, we have to conform to what *God* says will save us.

Over and over again, the Bible, God's Word, urges us to trust in God's way of salvation — Jesus. Only He could pay the penalty for sin. Only by believing upon Him — admitting our own helplessness — can we be saved. Just as God is faithful to hold this universe together consistently, He will be faithful to everything else He says in His Word. You can rely on what God says.

If you've not responded to God's gift of salvation, today is the day to do so! He will keep His Word — both to save and to punish.

If you're not sure if you have trusted God's way of salvation, don't delay in making sure. If you are sure, then take tremendous comfort in the knowledge that God is faithful and will complete what He began in you.

For more information on God's plan of salvation, please see www.biblicalperspective.net.

Keeping Perspective

The battle we face in math is ultimately a battle to remember our complete dependency on God. Even our ability to count comes from Him! Each math concept works only *because* of His faithfulness. Apart from Him, we truly can do *nothing*.

Ever since the Garden of Eden, Satan has been trying to distort the truth and get men to trust themselves instead of God. He has done this very thing in math — turning what should be a testimony to God into a testimony to man and math.

We can all see that math works. Someone or something has to be responsible for math's ability to work. If we're not giving the glory for math to God, then we're ending up giving it to man or to math. If math is not encouraging us that we can depend on our faithful, all-powerful God, then it is subtly telling us we can live independently from Him and determine our own truth.

Yes, indeed, there is a spiritual battle in math — and it's the same battle we face in every area of life.

1.4 Numbers, Place Value, and Comparisons

Now that we've seen the overall foundation the Bible gives us and explored a little about the spiritual battle in math, let's begin applying what we've discussed to specific aspects of math. In order to build our understanding of math from the foundation up, we'll be exploring simple review concepts for these first few chapters. As we do, though, we'll be learning important principles that apply to more advanced concepts.

An Overview of Mathematical Symbols and Terms

Math is filled with symbols and terms. Just as it is helpful if we use the same words to refer to objects (a book, sink, couch, etc.), it's helpful to use standardized symbols and terms in math.

Before we jump into looking at specific symbols and terms, though, let's take a minute to look at the big picture. Much of math is a naming process — a way of describing quantities and consistencies God created. So let's take a look at the first "naming" process the Bible describes: Adam naming the animals.

> *And out of the ground the LORD God formed every beast of the field, and every fowl of the air; and brought them unto Adam to see what he would call them: and whatsoever Adam called every living creature, that was the name thereof (Genesis 2:19).*

In naming the animals, Adam
1. observed God's creation (the animals) and
2. assigned names to describe the different animals.

In describing quantities, we
1. observe God's creation (the quantities around us) and
2. assign names (or symbols) to different quantities.

So what can we learn from Adam naming the animals? Well, notice that God brought the animals to Adam for naming — Adam was in God's presence while observing and naming. While sin separated man from his Creator, through Jesus, we can again **know God and worship Him while using math to describe His creation.** This holds true not just for basic math, but for *every* area of math we'll explore. In fact, the Bible urges us to do "whatsoever" we do "as to the Lord"!

And whatsoever ye do, do it heartily, as to the Lord, and not unto men (Colossians 3:23).

Number Systems: Beyond Quantities

Number systems prove useful in other ways besides recording quantities. For example, house numbers and telephone numbers don't record quantities — instead, they give us a way of "naming" homes and telephone lines. As another example, numbers and math are used in cryptography ("the art of writing or solving codes")[10] to help code messages. And before you picture coding as only wartime messages across enemy lines, did you realize that computers use a code to translate the letters or symbols you type on a keyboard? There's a number assigned to every letter or symbol that can be typed on a keyboard!

Reviewing Numerals and Place Value

Undoubtedly, you already know how to count (use words — like "one" — to describe quantities) and write numerals (use symbols — like "1" — to describe quantities). Below is just a quick review.

"Zero" is the name we mainly use in English to describe having nothing (you may also sometimes hear other names, such as "nought," "oh," or "nil," used to mean nothing). "One" is the name for a single unit — a single pen, dollar, CD, etc. "Two" is the name for a group of 2 units of anything.

Rather than words, we commonly use symbols. It's a lot easier to write "1" than to spell out "one" all the time! At the same time, though, it would be impossible to have a different symbol for *every* number. Instead, we use a concept known as place value.

Different Words for Quantities — The Tower of Babel

Different cultures use different words to describe quantities. For example, a single quantity is called *one* in English, *uno* in Spanish, *eine* in German, and *один* in Russian. Once again, the Bible tells us why.

Genesis 11:1–9 tell us about an event that changed the world — the Tower of Babel. Prior to the Tower of Babel, "the whole earth was of one language, and of one speech" (Genesis 11:1). Thus, men would have used the same words to describe quantities.

At the Tower of Babel, men misused the ability God had given them to communicate and sought to unite against God and make a name for themselves. The project stopped abruptly when God came down and confused their languages. The Tower of Babel accounts for the many different language systems we find, including the different words used to describe numbers.

In place value, the place, or location, of a number determines its value. So we use the symbols 0, 1, 2, 3, 4, 5, 6, 7, 8, 9 to represent up to nine. Once we reach ten, we move to the next "place" over, using the same digits, but knowing that each one represents a set, or group, of ten. So "10" represents 1 set of ten and 0 sets of one, "20" represents 2 sets of ten and 0 sets of one, and "21" represents 2 sets of ten and 1 set of one.

This place-value concept can be extended as far as necessary to represent numbers. Once we have ten tens, we move on to hundreds, then thousands, then ten thousands, then hundred thousands, and so forth.

Hundred trillion	Ten trillion	One trillion	Hundred billion	Ten billion	One billion	Hundred million	Ten million	One million	Hundred thousand	Ten thousand	One thousand	Hundred	Ten	One

Using our place-value system, we would represent the approximate world population in 2011 as 6,946,043,989 (or six billion, nine hundred forty-six million, forty-three thousand, nine hundred eighty-nine).[11]

Hundred trillion	Ten trillion	One trillion	Hundred billion	Ten billion	One billion	Hundred million	Ten million	One million	Hundred thousand	Ten thousand	One thousand	Hundred	Ten	One
					6	9	4	6	0	4	3	9	8	9

Notice how the commas every three places help our eyes count the places and determine the value.

4444444   4,444,444

In other countries, decimal points (4.444.444) or other separators are used instead of commas. (An important thing to keep in mind if ordering something online from another country!) Spaces (4 444 444) are also a recognized way of separating the places.

Reading Numbers

Notice how when reading numbers, we recycle terms. We start with ones (our basic units), tens, and hundreds. Then we have thousands (our new unit), followed by *ten* thousands and *hundred* thousands. We repeat this for millions, billions, etc.

Hundred trillion	Ten trillion	One trillion	Hundred billion	Ten billion	One billion	Hundred million	Ten million	One million	Hundred thousand	Ten thousand	One thousand	Hundred	Ten	One
☐	☐	☐	☐	☐	☐	☐	☐	☐	☐	☐	☐	☐	☐	☐

Notice also that in writing, we use commas every three digits, thereby separating the "thousands," "millions," etc.

To read a number, we read the number from left to right. If a digit has a zero, we don't read that place, as there's nothing to "report" there (as in the 0 in the hundred's place in 123,456,567,087).

123,456,567,087 would be read "one hundred twenty-three billion, four hundred fifty-six million, five hundred sixty-seven thousand, eighty-seven."

Now, I am sure you already know how to read numbers in English, but did you realize that there are variations in how to read numbers? The British often add an "and" (example: "one hundred *and* twenty-one" instead of "one hundred twenty-one"). 1,325 could also be read as "thirteen twenty-five" instead of as "one thousand three hundred twenty-five." This might make sense for dates ("the year thirteen twenty-five") or even house numbers ("I live at thirteen twenty-five Pleasant Lane"). When reading a street address over the phone, you might even just read each digit by itself, as in "one, three, two, five Pleasant Lane" to avoid confusion. These variations remind us that **names are a tool to help us communicate**, so clearly communicating is the most important thing.

When asked to write the words you would use to read a number in this course, use the traditional American method ("one thousand, three hundred twenty-five" for 1,325).

Reviewing Basic Comparison Terms and Symbols

If a number is larger, or has more, than another number, we say it **is greater than** the other number. If it is smaller/has less, we say it is **less than** the other number. If two quantities are the same, we say they are **equal**. If they are not the same and we do not want to make a specific comparison as to which one is greater, we say they are **not equal**. (Any number that is greater than or less than another number is also not equal to it — it's just a matter of what point we want to make.)

The symbols <, >, =, and ≠ are merely "shortcuts" for describing how numbers compare. They save our fingers from having to write the word out every time. It's a lot easier to write < than "less than." It also makes equations easier to read.

> Notice that the "less than" and "greater than" signs are the same, but pointing the opposite directions. You can remember which direction to put the symbol by remembering that the **larger side goes with the larger number.**

5	5 is less than or does not equal 6. < or ≠	6	5	5 equals 5 =	5
6	6 is greater than or does not equal 5. > or ≠	5	6	6 equals 6 =	6

Would it surprise you to learn that >, <, =, and ≠ are algebraic symbols? Any time we use a non-numerical symbol in math, we are actually using algebra. So >, <, =, and ≠ are actually part of algebra! Algebra is nothing to fear — it's just a way of using symbols to describe God's creation. In the case of >, <, =, and ≠, we're using symbols to describe how numbers compare.

Different Ways to Compare Numbers

Much as symbols for writing numbers have varied, so have symbols for comparing them. While we're used to using the "=" sign to mean "equal to," other symbols have been used throughout history — the box shows just a few. Instead of symbols, many cultures also used words or contractions to describe equality (*pha, equantur, aequales, gleich,* etc.).[12] Once again, history helps us see that the symbols we study in math are just an agreed-upon language system we use today to describe the quantities and consistencies God created and sustains.

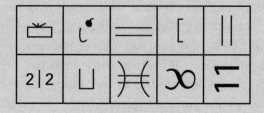

PRINCIPLES OF MATHEMATICS 1

> # Keeping Perspective
>
> We looked today at a few names (one, two, three, etc.) and symbols (1, 2, 3, =, >, <, etc.) used in math. As we continue our study of math, we're going to learn various names and symbols men have adopted to describe different consistencies or operations. Keep in mind that **terms and symbols are like a language** — agreed-upon ways of communicating about the quantities and consistencies around us.

1.5 Different Number Systems

It's all too easy to start viewing the terms, symbols, and methods we learn in math as math itself, thereby subtly thinking of math as a man-made system. A look at history, however, reveals many other approaches to representing quantities. Let's take a look at a few of them and at how they compare with our place-value system.

Place-Value Systems

In the last lesson, we reviewed how the number system we're mainly familiar with uses the place, or location, of a digit to determine its value. This is known as a **place-value system.**

Perhaps place value is easiest to picture using a device used extensively throughout the Middle Ages: an abacus. Each bead on the bottom wire of an abacus represents one; on the next, ten; on the third, one hundred; and on the fourth, one thousand. To represent a quantity on an abacus, we move the appropriate number of beads from each wire to the right. In the abacus shown, the 1 bead to the right on the thousands wire represents 1 thousand, the 4 beads to the right on the hundreds wire represent 4 hundred, the 9 beads to the right on the tens wire represent 9 tens, and the 1 bead on the ones wire represents 1. Altogether, that makes 1,491.

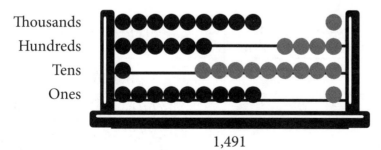

1,491

Just as the place, or line, of a bead changes its value, the place, or location, of a symbol in a place-value system changes its value. The number system commonly used today is called the **Hindu-Arabic decimal system** (or just the "**decimal system**" for short). This system came from the Hindu system, which the Arabs adopted and brought to Europe.

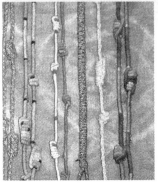

The Quipu — An Intriguing Approach

The Incas — an extensive empire in South America spanning more than 15,000 miles — had a fun approach to recording quantities. They tied knots on a device called a quipu (kē pōō).[13] The quipu system was extremely complicated, and only special quipu makers, called quipucamayocs, were able to interpret them. Although we do not know a lot about quipus, we do know they used place value. The location of the knot, along with some other factors, determined its value.

Apparently, the Incas were very successful with this innovative approach. Not only did they operate a huge empire, but the Incas baffled the Spanish conquerors by their ability to record the tiniest details as well as the largest ones on their quipus.[14]

Fixed-Value Systems

A different approach to recording quantities is to *repeat* symbols to represent other numbers. For example, here are some symbols in Egyptian numerals (hieroglyphic style).[15]

The next figure shows two different quantities represented using Egyptian numerals and our decimal place-value system. Notice how when writing twenty-two, the Egyptians repeated their symbol for one and their symbol for ten twice. They put the smaller values on the left and the larger values on the right. Thus the symbol for ten is to the right of the symbol for one.

Decimal System a place-value system	Egyptian System a fixed-value system
22	
1,491	

We'll refer to number systems that use repeated symbols like this as **fixed-value systems**.

A Deeper Look at the Egyptian System

Notice that in the Egyptian version of 1,491, the symbols representing "ninety" are stacked on top of each other.

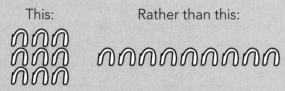

This: Rather than this:

While there were many variations within the Egyptian system over time, in general, when writing more than four of each symbol, the Egyptians **spaced, stacked, or grouped the symbols in sets (groups) of four or less**, with the larger set on top or first.[16] This practice made it easier to count the symbols (and thus to read the number!) at a glance.

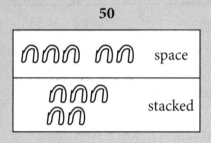

50

Let's compare our decimal place-value system with the Egyptian system. To record forty-nine objects in the Egyptian system, we would repeat the symbol for "one" nine times to show we had nine ones, and then repeat our symbol for "ten" four times to show we had four sets of ten. In the decimal system, we would instead use our symbols for four and nine, putting the 4 in the tens column so it represents four sets of ten and 9 in the ones column, representing nine sets of one.

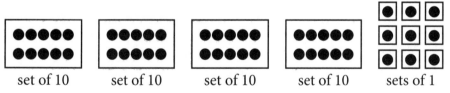

"Forty-nine" = four sets of ten and nine ones

Decimal System a place-value system	Egyptian System a fixed-value system
49	〇〇〇 〇〇〇 ∩∩∩∩ 〇〇〇

When we compare forty-nine in both systems, we see it takes significantly fewer symbols to represent the number in the decimal system. Place value saves a lot of extra writing!

To represent a number like forty in the decimal system, we would again use a 4, adding a zero (0) to represent that we have no (0) sets of one. Notice the importance of a zero (0) in a place-value system; without it, we would have no way of showing that the 4 represents 4 sets of ten instead of 4 sets of one.

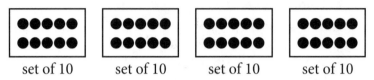

set of 10   set of 10   set of 10   set of 10

"Forty" = four sets of ten and no ones

Decimal System a place-value system	Egyptian System a fixed-value system
40	∩∩∩∩

Ordered Fixed-Value Systems

Another approach to recording quantities is to again use a limited number of symbols and repeat those symbols, but to add rules regarding their order that change the symbols' meaning. Roman numerals are an example of an ordered fixed-value system.

Take a look at these symbols used for quantities in Roman numerals:

I	1
V	5
X	10
L	50
C	100
D	500
M	1,000

As with the Egyptians, quantities in Roman numerals are represented by repeating symbols, although this time with the larger quantities on the left.

22 is written XXII in Roman numerals.

But unlike in the Egyptian system, the same symbol is generally not repeated more than three times. Instead, it is assumed that whenever a symbol representing a smaller quantity is to the *left* of a symbol representing a larger quantity, one should *subtract* the value of the smaller quantity from the value of larger quantity to get the value the two symbols represent.

I	1	XI	11
II	2	XII	12
III	3	XIII	13
IV	4	XIV	14
V	5	XV	15
VI	6	XVI	16
VII	7	XVII	17
VIII	8	XVIII	18
IX	9	XIX	19
X	10	XX	20

Notice that the smaller quantity is to the *left* of the larger — this means to subtract I from V, giving us 5 – 1, or 4.

Notice that the smaller quantity is to the *right* of the larger — this means to add I to V, giving us 5 + 1, or 6.

There was a time when "four" was written IIII instead of IV. But IV is easier to read, as there are fewer symbols involved.

Now let's take a look at the same number we looked at with the Egyptians: 1,491.

Decimal System a place-value system	Roman Numeral System an ordered fixed-value system
1,491	**MCDXCI** M = 1,000 CD = 500 – 100 = 400 XC = 100 – 10 = 90 I = 1 1,000 + 400 + 90 + 1 = 1,491

Notice that Roman numerals would not lend themselves well to quickly adding or subtracting on paper! There is a reason we use the decimal place-value system for most purposes.

Keeping Perspective

While you may use only our current decimal place-value system on a regular basis, being aware of other systems will help you learn to better see our place-value system as just one system to help us describe quantities.

1.6 Binary and Hexadecimal Place-value Systems (Optional Lesson)

See https://www.christianperspective.net/math/pom1 for a hands-on idea of how to explain these different systems.

Before we move on, we're going to take one more look at the concept of place value, as it's a pretty important concept. While I'm sure you're quite familiar with our current place-value system, did you realize computers use place-value systems based on a value besides ten?

Well, they do! They use what's known as a binary place-value system. Exploring this system, along with the hexadecimal place-value system, is not only cool, but it can also help provide an even firmer grasp of the decimal place-value system. Let's take a look.

Unwrapping Place-Value Systems

The value we choose for each place in the system is called our **base**. You can picture a base like a container — the size of the container determines how much it can hold. In the decimal system, each place, or container, can hold *ten* digits (0, 1, 2, 3, 4, 5, 6, 7, 8, 9); once we reach *ten* of a unit, we move to the next place over, using the same digits, but knowing that each one represents ten of the previous place's value.

We write forty-two as "42" to represent 4 sets of ten (or 40) plus 2.

Binary System

Computers actually operate off a base-two place-value system called the **binary system** (*bi* means *two*). In a binary system, instead of allowing *ten* values (0, 1, 2, 3, 4, 5, 6, 7, 8, 9) in each place, we only allow *two* (0, 1). It's as if each place, or container, can only hold the *two* digits: 0 and 1. Once we reach two, we move to the next place over. While in the decimal system, each place is worth ten times the previous place, and each place in the binary system is worth two times the previous place.

In binary, the number "10" represents 1 set of *two* and 0 sets of *one*, or two!

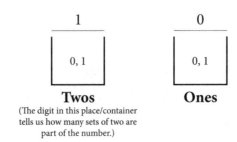

To make things clearer, take a look at the first four places, or containers, for both systems.

Decimal (base 10)

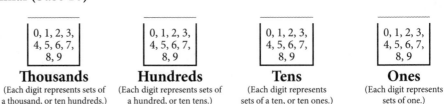

Binary (base 2)

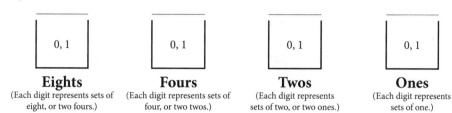

Let's take a look at how this plays out with a few numbers.

Example: Find the decimal value of the binary number 1010.

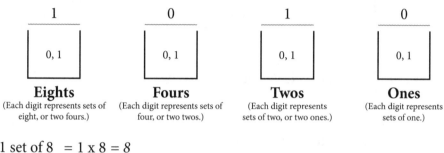

1 set of 8 = 1 x 8 = *8*
0 sets of 4 = 0 x 4 = *0*
1 set of 2 = 1 x 2 = *2*
0 sets of 1 = 0 x 1 = *0*
8 + 0 + 2 + 0 = 10

1010 in binary is the same as the decimal number 10.

Example: Find the decimal value of the binary number 1111.

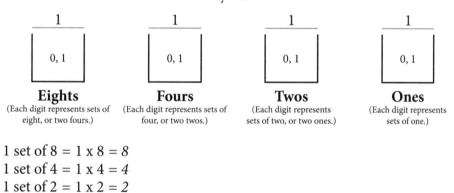

1 set of 8 = 1 x 8 = *8*
1 set of 4 = 1 x 4 = *4*
1 set of 2 = 1 x 2 = *2*
1 set of 1 = 1 x 1 = *1*
8 + 4 + 2 + 1 = 15

1111 in binary is the same as the decimal number 15.

Computer Circuits

Because computer circuits run on electricity, the 0 and 1 used in binary numbers can easily describe the "off" and "on" flows of electricity controlled by an open or closed switch. Whenever there's electricity, the computer interprets it as a 1. When there's no electricity, it interprets it as a 0.

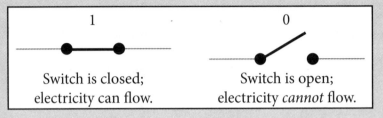

Making Computer Talk More Concise: Hexadecimal Numbers

Although binary numbers translate well to electrical pulses, they tend to get long quickly (eight is written 1000), making them difficult for us to read. To help make numbers more readable, computer programs often use hexadecimal numbers (a place-value system based on 16) to represent binary numbers. Because it has a larger base (i.e., a container that can hold more digits), the hexadecimal system can represent very large numbers with fewer digits.

Decimal: 1,200
Binary: 10010110000
Hexadecimal: 4B0

Base 16-Hexadecimal System

16 Symbols: 0, 1, 2, 3, 4, 5, 6, 7, 8, 9, A, B, C, D, E, F

A represents the decimal value of 10.
B represents the decimal value of 11.
C represents the decimal value of 12.
D represents the decimal value of 13.
E represents the decimal value of 14.
F represents the decimal value of 15.

0, 1, 2, 3, 4, 5, 6, 7, 8, 9, A, B, C, D, E, F	0, 1, 2, 3, 4, 5, 6, 7, 8, 9, A, B, C, D, E, F	0, 1, 2, 3, 4, 5, 6, 7, 8, 9, A, B, C, D, E, F	0, 1, 2, 3, 4, 5, 6, 7, 8, 9, A, B, C, D, E, F
Four thousand ninety-sixes (Each digit represents sets of four thousand ninety-six, or sixteen two hundred fifty-sixes.)	**Two hundred fifty-sixes** (Each digit represents sets of two hundred fifty-six, or sixteen sixteens.)	**Sixteens** (Each digit represents sets of sixteen, or sixteen ones.)	**Ones** (Each digit represents sets of one.)

Example: Find the decimal value of the hexadecimal number 4B0.

4	B	0
0, 1, 2, 3, 4, 5, 6, 7, 8, 9, A, B, C, D, E, F	0, 1, 2, 3, 4, 5, 6, 7, 8, 9, A, B, C, D, E, F	0, 1, 2, 3, 4, 5, 6, 7, 8, 9, A, B, C, D, E, F
Two hundred fifty-sixes (Each digit represents sets of two hundred fifty-six, or sixteen sixteens.)	**Sixteens** (Each digit represents sets of sixteen, or sixteen ones.)	**Ones** (Each digit represents sets of one.)

4 sets of 256 = 4 x 256 = *1,024*
11 sets of 16 = 11 x 16 = *176*
0 sets of 1   = 0 x 1   = *0*
1,024 + 176 = 1,200

4B0 in hexadecimal is the same as the decimal number 1,200.

Keeping Perspective

Place-value systems can be based off *any* quantity — and other systems besides the decimal one are in common use today! Each system is a tool to help us describe quantities . . . and each works best in different situations.

While it's not necessary for you to learn the binary or hexadecimal systems (unless you plan to go into computer programming), take some time to explore them a little. Thinking outside the box this way will help you develop your mathematical skills and grow in your ability to use math as a tool.

[CHAPTER 2]

Operations, Algorithms, and Problem Solving

2.1 Addition and Subtraction

A large portion of math deals with recording operations/relationships. If you add $5 to $9, you get $14. If you add 5 cookies to 9 cookies, you get 14 cookies. This relationship holds true so consistently that we can call it a fact that if we take 5 and 9, we'll get 14.

Likewise, if we went to the store with $14 and spent $5, we'd have $9 left. If we started with 14 cookies and then ate 5, we'd have 9 cookies left. We can call it a fact that if we take 5 from 14, we'll end up with 9.

You've no doubt been adding and subtracting for years. But stop and ponder for a moment the amazing fact that all those addition and subtraction facts you memorized as a young child actually apply in real life. No matter where you are, you can take one of something and one of something else and end up with two.

Math is consistent because Jesus is consistently upholding all things by the Word of His power.

> *Who [Jesus] being the brightness of his glory, and the express image of his person, and upholding all things by the word of his power . . . (Hebrews 1:3).*

Jesus governs this universe so predictably and reliably that we can memorize how objects will add and subtract and rely on them to work! Wow.

In Jeremiah, God draws His people's attention to the consistencies around them ("ordinances" are "that which is established or defined . . . law"[1]) to remind them He keeps His covenants.

Thus saith the LORD; If my covenant be not with day and night, and if I have not appointed the ordinances of heaven and earth; Then will I cast away the seed of Jacob and David my servant, so that I will not take any of his seed to be rulers over the seed of Abraham, Isaac, and Jacob: for I will cause their captivity to return, and have mercy on them (Jeremiah 33:25–26).

Every time we add and subtract (or do any other operation in math) and see that it works — that God is still consistently governing all things — it's shouting out at us that we serve a covenant-keeping God whom we can trust! Just as He is faithful to the day and the night and the "ordinances," or laws, He has placed all around us (including addition and subtraction), He will be faithful to everything else He says in His Word.

What About Miracles?

What about when Jesus fed 5,000 with five loaves of bread and two fish (Matthew 14:13–21)? Those five loaves of bread certainly did not divide in a typical manner!

We need to understand that God is consistent to His nature. In His mercy and faithfulness, He governs this universe in a predictable pattern. Out of His same mercy and faithfulness, He sometimes chooses to perform a miracle. Miracles help us see God's sovereignty. Miracles show us that God is the One in charge all along.

Addition Facts

$1 + 1 = 2$	$2 + 1 = 3$	$3 + 1 = 4$	$4 + 1 = 5$	$5 + 1 = 6$
$1 + 2 = 3$	$2 + 2 = 4$	$3 + 2 = 5$	$4 + 2 = 6$	$5 + 2 = 7$
$1 + 3 = 4$	$2 + 3 = 5$	$3 + 3 = 6$	$4 + 3 = 7$	$5 + 3 = 8$
$1 + 4 = 5$	$2 + 4 = 6$	$3 + 4 = 7$	$4 + 4 = 8$	$5 + 4 = 9$
$1 + 5 = 6$	$2 + 5 = 7$	$3 + 5 = 8$	$4 + 5 = 9$	$5 + 5 = 10$
$1 + 6 = 7$	$2 + 6 = 8$	$3 + 6 = 9$	$4 + 6 = 10$	$5 + 6 = 11$
$1 + 7 = 8$	$2 + 7 = 9$	$3 + 7 = 10$	$4 + 7 = 11$	$5 + 7 = 12$
$1 + 8 = 9$	$2 + 8 = 10$	$3 + 8 = 11$	$4 + 8 = 12$	$5 + 8 = 13$
$1 + 9 = 10$	$2 + 9 = 11$	$3 + 9 = 12$	$4 + 9 = 13$	$5 + 9 = 14$
$6 + 1 = 7$	$7 + 1 = 8$	$8 + 1 = 9$	$9 + 1 = 10$	
$6 + 2 = 8$	$7 + 2 = 9$	$8 + 2 = 10$	$9 + 2 = 11$	
$6 + 3 = 9$	$7 + 3 = 10$	$8 + 3 = 11$	$9 + 3 = 12$	
$6 + 4 = 10$	$7 + 4 = 11$	$8 + 4 = 12$	$9 + 4 = 13$	
$6 + 5 = 11$	$7 + 5 = 12$	$8 + 5 = 13$	$9 + 5 = 14$	
$6 + 6 = 12$	$7 + 6 = 13$	$8 + 6 = 14$	$9 + 6 = 15$	
$6 + 7 = 13$	$7 + 7 = 14$	$8 + 7 = 15$	$9 + 7 = 16$	
$6 + 8 = 14$	$7 + 8 = 15$	$8 + 8 = 16$	$9 + 8 = 17$	
$6 + 9 = 15$	$7 + 9 = 16$	$8 + 9 = 17$	$9 + 9 = 18$	

Subtraction Facts

10 – 1 = 9	11 – 2 = 9	12 – 3 = 9	13 – 4 = 9	14 – 5 = 9
9 – 1 = 8	10 – 2 = 8	11 – 3 = 8	12 – 4 = 8	13 – 5 = 8
8 – 1 = 7	9 – 2 = 7	10 – 3 = 7	11 – 4 = 7	12 – 5 = 7
7 – 1 = 6	8 – 2 = 6	9 – 3 = 6	10 – 4 = 6	11 – 5 = 6
6 – 1 = 5	7 – 2 = 5	8 – 3 = 5	9 – 4 = 5	10 – 5 = 5
5 – 1 = 4	6 – 2 = 4	7 – 3 = 4	8 – 4 = 4	9 – 5 = 4
4 – 1 = 3	5 – 2 = 3	6 – 3 = 3	7 – 4 = 3	8 – 5 = 3
3 – 1 = 2	4 – 2 = 2	5 – 3 = 2	6 – 4 = 2	7 – 5 = 2
2 – 1 = 1	3 – 2 = 1	4 – 3 = 1	5 – 4 = 1	6 – 5 = 1
1 – 1 = 0	2 – 2 = 0	3 – 3 = 0	4 – 4 = 0	5 – 5 = 0

15 – 6 = 9	16 – 7 = 9	17 – 8 = 9	18 – 9 = 9
14 – 6 = 8	15 – 7 = 8	16 – 8 = 8	17 – 9 = 8
13 – 6 = 7	14 – 7 = 7	15 – 8 = 7	16 – 9 = 7
12 – 6 = 6	13 – 7 = 6	14 – 8 = 6	15 – 9 = 6
11 – 6 = 5	12 – 7 = 5	13 – 8 = 5	14 – 9 = 5
10 – 6 = 4	11 – 7 = 4	12 – 8 = 4	13 – 9 = 4
9 – 6 = 3	10 – 7 = 3	11 – 8 = 3	12 – 9 = 3
8 – 6 = 2	9 – 7 = 2	10 – 8 = 2	11 – 9 = 2
7 – 6 = 1	8 – 7 = 1	9 – 8 = 1	10 – 9 = 1
6 – 6 = 0	7 – 7 = 0	8 – 8 = 0	9 – 9 = 0

Reviewing the Concept

Simple addition is the putting together of two or more numbers, of the same denomination, so as to make them one whole or total number, called the sum, or amount. — Daniel Adams[2]

"Addition" is a name we use to describe putting quantities together. If you add one crayon to another crayon, you will end up with two crayons.

We would describe this using the language of mathematics as "one plus one equals two." To make it easier to write and read, we can use symbols instead of words. The **"+" sign**, called the **plus sign**, is the symbol we use today to represent addition. As we have seen, the equal sign is a way of showing that the quantities on both sides of it ("1 + 1" and "2") are equivalent. So in symbols, we would write 1 + 1 = 2. We could also write the problem vertically like this:

$$\begin{array}{r} 1 \\ +\ 1 \\ \hline 2 \end{array}$$

Subtraction is just the opposite of addition. It is the name we use for taking a quantity away. If we start with two crayons in our hand, and take one away, we would end up with one crayon left.

We would describe this using the language of mathematics as "two minus one equals one." Once again, to make it easier to write and read, we can use symbols instead of words. The **"–" sign**, called the **minus sign**, is the symbol we use today to represent subtraction. As with addition, we can represent subtraction either horizontally or vertically.

$$2 - 1 = 1 \qquad \begin{array}{r} 2 \\ -\,1 \\ \hline 1 \end{array}$$

$\wedge$	Egyptian
/	Some Greek Papyri
yu	Bakhshali
P, P̄, P̃, or *p*	English

Different ways to show addition

Variations of the plus sign

Did You Know . . .

People have not always used our current plus sign (+) to represent addition. Many cultures throughout history did not use a sign at all. They used other means to signify addition, such as a number's location in relationship to another number, or words (example: Add 2 to 4). Others used very different signs from what we use today. The Egyptian Ahmes papyrus used a picture of legs walking forward,[3] the Greeks sometimes used a line,[4] and some mathematicians used a *p* or a variation of p.[5] The Bakhshali *Arithmetic* uses a *yu* to represent addition and uses the + sign to signify a quantity you would subtract![6]

Our current plus sign (+) seems to have begun in Germany sometime between A.D. 1450–1500. It probably came from the word *et*, which meant *and*.[7]

Even after our current plus sign arrived on the scene, it took a long time for everyone to adopt it. Along the way, people used different variations.[8]

God's principles do not change, although the way of recording them can (and does!). Addition is addition, no matter what symbol we use to describe it. Symbols only *represent the reality of addition God created and sustains.*

Addition of Liquids

Dip your hand in a cup of water and let a droplet fall onto an empty plate. Dip your hand again and drop another droplet of water on top of the first one. What happened?

Note that at the atomic level, the starting atoms plus the final atoms would add together following the rules of regular addition — if we started with 1 atom and added 1 more atom, the ending larger droplet would have 2 atoms. But at the visible level, we end up with one larger droplet.

The droplets should have merged together into a single droplet of water. Although one water molecule plus another equals two, one droplet of water plus one droplet of water does *not* equal two droplets — they merge together to form one larger droplet. This seems like a huge inconsistency. Why doesn't 1 water droplet + 1 water droplet = 2 water droplets?

Because God has different, though equally consistent, principles for governing liquids at the visible level (the water molecules follow the addition we would expect) than He does solids. The apparent contradiction in how liquids combine reminds us that, no matter how well we think we have things figured out, God's laws and universe are more complex than we can imagine.

40 | PRINCIPLES OF MATHEMATICS 1

On the other hand, the way raindrops combine presents a huge problem if we believe math is a product of human reasoning. After all, human reasoning says $1 + 1 = 2$. Obviously, human reasoning cannot be trusted when dealing with raindrops. And if not here, can we trust any of math? Or is it all just a delusion?

Whenever we abandon a biblical worldview, we end up with a host of difficulties. Thankfully, though, God has given us His truth to live upon. If we take God at His word, we can use addition confidently, knowing it does not rest on man's reasoning, but on God's power. At the same time, we can recognize God's greatness and expect to see Him holding this universe together in a marvelous way that no one set of math facts could ever fully describe.

On page 92 of *Mathematics: The Loss of Certainty*, Morris Kline shares how the raindrop quandary, along with many others, was brought up by Hermann von Helmholtz in 1887 and explores the problem it posed to those who had enthroned human reason and math.[9]

Term Reminder

Wouldn't it be hard if you had to define friends by a lengthy description every time you wanted to refer to them? What if, rather than saying "Fred Smith," you had to use a descriptive phrase such as "That boy with the red hair and the freckles who lives on top of the hill by the shopping center"? Names make life SO much simpler!

Names help in math too. Below are a few names for the different numbers in an addition and subtraction problem. You will need to be familiar with these terms.

Addend — The name for numbers (quantities) we are adding together.

Sum — The name for the total of the numbers added (what they equal).

Minuend — The name for the total we are subtracting (taking away) from.

Subtrahend — The name for the amount we are subtracting (taking away).

Difference — The name for the amount left after we have finished subtracting.

$$9 \quad + \quad 7 \quad = \quad 16$$
addend   addend   sum

$$16 \quad - \quad 7 \quad = \quad 9$$
minuend   subtrahend   difference

Understanding the Connection

Because addition and subtraction are opposite operations, we will often work with them in combination. For example, we can also use addition to check if we did subtraction correctly . . . and vice versa. If we know $4 + 6 = 10$, we also know $10 - 6 = 4$.

Or say we have collected $10 total from two people, and we know one person gave us $4. How much did the other person give us?

We can think of this in terms of subtraction ($10 – $4 = $6), or in terms of addition in which we need to find a missing addend ($4 + ___ = $10) . . . which we would do by subtracting $4 from $10, giving us $6. Either way, we end up with $6. If we get $10 from two people and one person gave us $4, the other one *had* to have given us $6. Much as we might sometimes wish otherwise, we know $4 won't suddenly become $5 while sitting in our pocket. We live in a consistent universe held together by a consistent God.

Comparison Signs and Operations

Now that we've covered addition and subtraction, you'll be given comparison problems that include operations, such as deciding if a greater than, less than, or equal sign belongs in this:

$$5 + 6 \qquad\qquad 8 + 2$$

Be sure to perform the operations first and then compare the end results.

$$11 \quad > \quad 10$$

Keeping Perspective

As you review addition and subtraction, and then later as we study different operations in math, keep in mind that each one's very existence is shouting out at us that we serve a faithful, powerful God whom we can trust.

2.2 Applying Basic Addition and Subtraction to Time

We all apply basic addition and subtraction throughout the day when dealing with time. For example, if a 2-hour meeting starts at 1 p.m., we know it will finish at 1 + 2, or 3 p.m. Notice the addition! So before we move on from basic addition and subtraction, let's spend some time on time.

As we do, keep in mind that God gave us time. Genesis tells us that He created the heavens and earth "in the beginning" (Genesis 1:1). At that moment, time began. Knowing we'd need to keep track of time, He gave us lights that mark the seasons and days and years.

> *And God said, Let there be lights in the firmament of the heaven to divide the day from the night; and let them be for signs, and for seasons, and for days, and years (Genesis 1:14).*

Every morning, the sun rises, signifying a new day. Every night, it sets, and we know a day has gone by. As the earth rotates around the sun, we experience seasons and see the stars from different views. Even if we had no calendar, we'd know of time's passage. The calendars and time-keeping devices we have merely describe the passage of time God set in place.

Adding and Subtracting Time

As you probably already know, when counting hours using the 12-hour clock common in the United States, we only count the hours up to 12. Once we hit 12, or *noon*, we start our counting over again, using *a.m.* to indicate a time *before* noon, and *p.m.* a time noon or after. So 10 *a.m.* means 10 hours into the day (we haven't gotten to noon yet), while 10 *p.m.* means 10 hours *after* noon. A new day begins when we get to 12 again — that is, to what we call *midnight*.

To learn more about timekeeping and time zones, see John Hudson Tiner, *Exploring the World of Mathematics: From Ancient Record Keeping to the Latest Advances in Computers* (Green Forest, AR: Master Books, 2004), chapter 2.

12:00 a.m. *Midnight*	12:00 p.m. *Noon*
1:00 a.m.	1:00 p.m.
2:00 a.m.	2:00 p.m.
3:00 a.m.	3:00 p.m.
4:00 a.m.	4:00 p.m.
5:00 a.m.	5:00 p.m.
6:00 a.m.	6:00 p.m.
7:00 a.m.	7:00 p.m.
8:00 a.m.	8:00 p.m.
9:00 a.m.	9:00 p.m.
10:00 a.m.	10:00 p.m.
11:00 a.m.	11:00 p.m.
12:00 p.m. *Noon*	12:00 a.m. *Midnight*

When working with time using a 12-hour clock, it's helpful to think in terms of how many hours you are from a reference point, such as noon or midnight.

Example: Find 3 hours before 1 p.m.

We know 1 p.m. is 1 hour past noon. Since we want to go back 3 hours, we have to go back 2 more hours (2 + 1 = 3), putting us at 10 a.m. (12 – 2 = 10). Therefore, 10 a.m. is 3 hours before 1 p.m.

2. OPERATIONS, ALGORITHMS, AND PROBLEM SOLVING

Example: Find 10 hours before 7 p.m.

Seven hours before 7 p.m. would have been noon. Since we want to go back 10 hours, we have to go back an additional 3 hours (7 + 3 = 10), putting us at 9 a.m. (12 – 3 = 9).

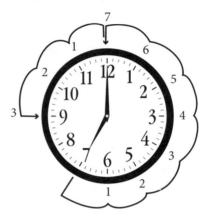

Alternately, we could have thought about the fact that 12 hours before 7 p.m. would have been 7 a.m. Since we want to find 10 hours before, we know it would be 2 hours later than 7 a.m. . . . which gives us 9 a.m. (2 + 7 = 9).

Hopefully, you're already used to doing time problems in your head — we're including it as an example of how basic addition and subtraction is often done mentally.

Time Zones

Another common (and related) application of basic addition and subtraction is that of converting between time zones. While we will learn more about time zones in Book 2, they're basically divisions across the earth's surface, each one being about the distance the sun takes an hour to travel. (In time zones, we're trying to describe the movement of the sun that God put in place.) In general, when it's 11 a.m. in one zone, it's 10 a.m. in the zone to its west, and 12 p.m. (noon) in the zone to its east. In this way, it's noon in each zone at approximately the time the sun is directly overhead in that zone.

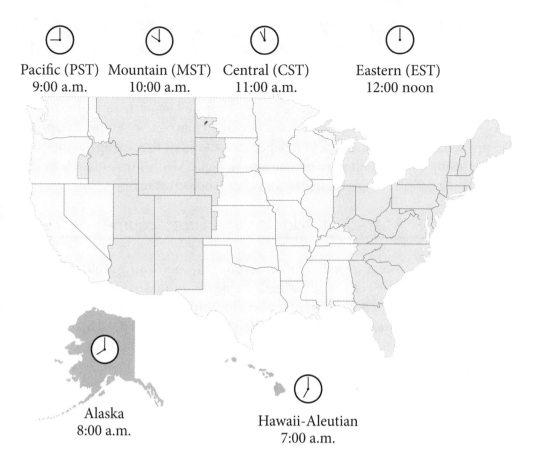

These times assume Daylight Savings Time (DST) is not in effect. During DST, those portions of the U.S. that do not follow DST (such as Hawaii) are an additional hour behind.

To find out what time it is in a different time zone within the United States, all you have to do is figure out how many hours earlier or later it is in that time zone and subtract or add that number of hours, being careful to take into account the a.m. and p.m.

For example, you can see that at 12 p.m. Eastern Standard Time (EST or ET) it's 9 a.m. Pacific Standard Time (PST or PT). So there's a 3-hour difference (12 − 9 = 3).

Thus, at 8 p.m. EST it's 3 hours earlier (5 p.m. PST, as 8 − 3 = 5), and at 2 p.m. PST it's 5 p.m. EST (2 + 3 = 5).

Again, notice how we just used basic addition and subtraction!

Keeping Perspective

Since we know math is a way of describing God's creation, we want to learn how to use it practically. Keeping track of time is one example of how math applies practically. As you review time, remember that God is the One who gave us the lights in the sky to mark the passage of time.

2.3 Multi-digit Addition and Subtraction

A large portion of math focuses on step-by-step methods, or **algorithms**, that we use to solve problems on paper. For example, since we obviously cannot memorize every single quantity we might ever want to add or subtract, we follow a step-by-step process for adding quantities on paper that allows us to solve any addition problem using a limited number of memorized facts. Algorithms are tools to automate the process of describing an aspect of God's creation.

Exploring Our Multi-digit Addition and Subtraction Algorithm

See https://www.christianperspective.net/math/pom1 for a free video on adding and subtracting on an abacus.

The step-by-step process we follow when adding and subtracting multi-digit numbers automatically keeps track of place value for us, making it easy to add or subtract any number if we just know our addition and subtraction facts from 0–9. We can see this easiest on an abacus. Remember that each bead on the bottom row of an abacus represents one set of *one*, each bead on the next (second from bottom) represents one set of *ten*, each bead on the third from bottom row represents one set of a *hundred*, etc. (refer back to 1.5 if needed).

Adding

To add on an abacus, we simply form our first quantity, and then add to it the quantity we want to add. Once we reach ten on one row, we exchange those ten beads for one bead on the next row.

Example: Add 12 + 19 on an abacus.

Step 1: *Form the starting quantity of 12 (1 ten and 2 ones).*

Step 2: *Add the next addend, in this case, 19 (1 ten and 9 ones).*
Add the 9 ones first.
Notice we run out of beads after adding 8 of the 9 ones. When we have used all 10 beads on the ones row, we...

...exchange 10 beads on the ones row for 1 bead on the tens row.

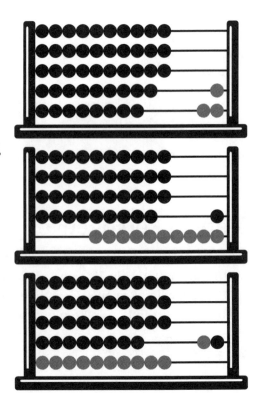

46 PRINCIPLES OF MATHEMATICS 1

Step 3: *Now we can add the remaining 1.*

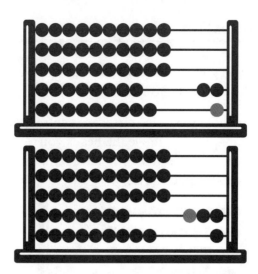

Step 4: *Add the 1 ten in 19.*

Notice how the abacus kept track of place value for us, making addition easy. When we used 10 beads in the ones row, we exchanged all those beads for one bead in the tens row.

When we "carry" or rename digits on paper, we're really doing the same thing we did on the abacus when we exchanged 10 beads from one row for 1 bead on the next row. Our algorithm, or method, keeps each digit in the correct place so we don't have to think about it as we add.

Example: Add 12 + 19 on paper using our typical method.

> **Step 1:** First, we line up the numbers so that the ones columns are on top of each other. This keeps us from accidentally adding digits that represent different values (such as a digit in a ones column with a digit from the tens column).
>
> ```
>    1 2
>  + 1 9
> ```
>
> **Step 2:** We next add the ones column much as we would add the beads on the ones row of an abacus. We end up with 11. If adding on an abacus, we would then replace 10 beads with 1 bead on the tens row; here, we write a 1 at the top of the tens column.
>
> ```
>   1
>    1 2
>  + 1 9
>      1
> ```
>
> **Step 3:** We then add up the tens column, and get 3. Now we have our answer: 31.
>
> ```
>   1
>    1 2
>  + 1 9
>    3 1
> ```

Notice how this process works for any number of digits — we can exchange 10 tens for 1 hundred (10 sets of 10 equals 100), 10 hundreds for 1 thousand (10 sets of 100 equals 1,000), etc.

Subtracting

Can you guess how to subtract using an abacus? Give it a try on your own with the equation 27 – 4. Don't look below until you've tried yourself!

As you hopefully discovered, to subtract on an abacus, you simply form your starting quantity; move the beads you're subtracting, or taking away, to the left; and read what number you have remaining. (In the figure, the grey beads show the beads you would move.)

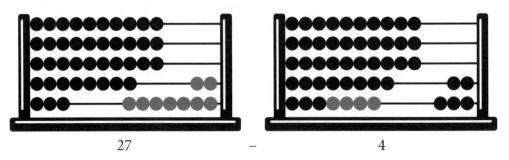

27 – 4

Now try solving 24 – 9 on an abacus. Notice that 24 has only 4 beads in the ones row, but you need to take away, or subtract, 9. Any ideas what you can do?

Since each bead on the tens row represents one set of 10, you can exchange it for 10 beads on the ones row. Since you know you want to take away 9 of those beads, you would be left adding just 1 bead to the ones row, leaving the answer: 15.

Example: Solve 24 – 9 on an abacus.

Step 1: *Form the starting quantity of 24 (2 tens and 4 ones).*

Step 2: *Subtract 9. Rename one ten to ten ones. Mentally subtract the 9 from that 10. Move the remaining 1 over to the right in the ones column.*

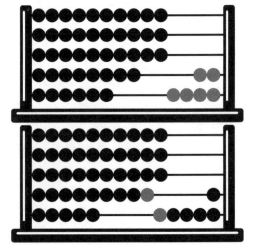

When we add on paper, we do a similar thing. Notice how we represent exchanging one set of ten from our tens column for ten in our ones column.

Example: Solve 24 – 9 on paper using our typical method.

Step 1: Write the two numbers on top of each other so the ones columns line up and we don't accidentally subtract the wrong columns.

$$\begin{array}{r} 2\,4 \\ -\ \ 9 \\ \hline \end{array}$$

Step 2: Subtract the ones column. Since we can't subtract 9 from 4, we rename 1 ten from 24 as 10 ones, showing this by crossing out the 2 in 24 and writing a 1 next to the 4 and a 1 next to the 2 to show the 1 we have left. We then subtract 9 from the 14, giving us 5, which we put below the line.

$$\begin{array}{r} {\scriptstyle 1\ 1} \\ \cancel{2}\,4 \\ -\ \ 9 \\ \hline 5 \end{array}$$

Step 3: Subtract the tens column. Since we renamed 1 ten, we only have 1 left. 1 – 0 = 1. We put a 1 below the line.

$$\begin{array}{r} {\scriptstyle 1\ 1} \\ \cancel{2}\,4 \\ -\ \ 9 \\ \hline 1\,5 \end{array}$$

> Some textbooks call exchanging quantities like this "borrowing." Renaming is a more accurate term, as we're really expressing the quantity a different way so we can subtract from it.

This method works for numbers with numerous digits as well as numbers with fewer digits — since each place is ten times the previous one, we can continue to rename from the column to the left as needed.

In subtracting 378 from 1,467, we have to rename twice — we first change the 6 in 1,467 to a 5 to add a ten in our ones column, giving us 17 – 8, and then we rename the 4 in 1,467 to get us a ten in our tens column, giving us 15 – 7.

$$\begin{array}{r} 1,4\,6\,7 \\ -\ \ 3\,7\,8 \\ \hline \end{array} \qquad \begin{array}{r} {\scriptstyle 5\ 1} \\ 1,4\cancel{6}\,7 \\ -\ \ 3\,7\,8 \\ \hline 9 \end{array} \qquad \begin{array}{r} {\scriptstyle 3\ \ 15\,1} \\ 1,\cancel{4}\cancel{6}\,7 \\ -\ \ \ 3\,7\,8 \\ \hline 1,0\,8\,9 \end{array}$$

Now take a look at this same process on an abacus.

Example: Solve 1,467 – 378 on an abacus.

Step 1: *Form the starting quantity of 1,467 (1 thousand, 4 hundreds, 6 tens, 7 ones).*

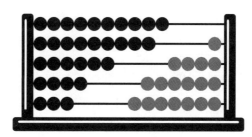

Step 2: *Subtract 378.*
Rename one ten to ten ones.
Mentally subtract the 8 from the 10, leaving 2.
Move 2 beads over to the right in the ones column.

Step 3: *Rename 1 hundred to ten tens.*
Mentally subtract 7 from the 10, leaving 3.
Move 3 beads over to the right in the tens column.

Step 4: *Subtract the 3 from the hundreds, leaving our answer: 1,089.*

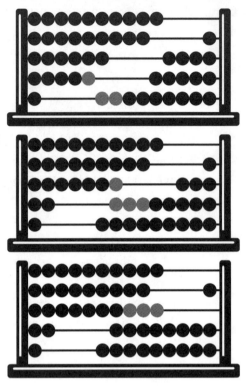

Applying Addition and Subtraction

Rather than dividing a day into two 12-hour sections, the military (and many other countries) use a 24-hour clock. As with a 12-hour clock, the clock starts ticking at midnight, but with a 24-hour clock, the clock keeps going all day instead of restarting after 12. The chart shows a general comparison between the two systems.

12-Hour Clock	24-Hour Clock		12-Hour Clock	24-Hour Clock	
12:00 a.m.	00:00	*Midnight*	12:00 p.m.	12:00	*Noon*
1:00 a.m.	01:00		1:00 p.m.	13:00	
2:00 a.m.	02:00		2:00 p.m.	14:00	
3:00 a.m.	03:00		3:00 p.m.	15:00	
4:00 a.m.	04:00		4:00 p.m.	16:00	
5:00 a.m.	05:00		5:00 p.m.	17:00	
6:00 a.m.	06:00		6:00 p.m.	18:00	
7:00 a.m.	07:00		7:00 p.m.	19:00	
8:00 a.m.	08:00		8:00 p.m.	20:00	
9:00 a.m.	09:00		9:00 p.m.	21:00	
10:00 a.m.	10:00		10:00 p.m.	22:00	
11:00 a.m.	11:00		11:00 p.m.	23:00	
12:00 p.m.	12:00	*Noon*	12:00 a.m.	24:00 / 00:00	*Midnight*

Although the military uses a 24-hour clock, the military doesn't put the ":" in the time. For example, 09:00 is written as 0900 in military time.

Converting between the two clocks uses addition and subtraction. Converting time before noon is very easy — both systems are basically the same (although they are pronounced slightly differently: 02:00 is pronounced "oh two hundred" or "zero two zero zero" or some variation along those lines in the 24-hour clock, and 2 a.m. in the 12-hour one). But notice in the afternoon that to convert between the

24-hour clock and the 12-hour clock, we have only to *subtract 12* to the time listed. And to convert from the 12-hour clock to the 24-hour clock, we need only to *add 12* to the number listed. That's because at 12 (noon), the 12-hour clock starts over again, while the 24-hour clock continues on until it reaches 24.

Example: Convert 16:00 to the 12-hour clock.

16 – 12 = 4. 16:00 is 4 p.m. in the 12-hour clock.

Example: Convert 4 p.m. to the 24-hour clock.

4 + 12 = 16. 4 p.m. is 16:00 in the 24-hour clock.

The 24-hour clock reminds us that there are different systems to describe the passage of time, but no matter how we describe time's passing, God is the one who causes the sun to faithfully rise and set in a consistent, predictable fashion.

> *Thus saith the LORD, which giveth the sun for a light by day, and the ordinances of the moon and of the stars for a light by night, which divideth the sea when the waves thereof roar; The LORD of hosts is his name: (Jeremiah 31:35).*

Keeping Perspective

No doubt, you've been adding numbers on paper for so long you can do it mechanically without thinking. But it's important to understand that **all we're really doing is keeping track of place value** so that we can use the basic facts we have memorized to add any two numbers, regardless of the number of digits.

The methods we use to add and subtract on paper are not some sort of truth in themselves — they're just a way of describing the consistencies God created and sustains.

After all, in order to use any addition or subtraction method outside of a textbook, we have to make these two presuppositions:

1. quantities will add and subtract in a consistent manner, and

2. man can observe and record this consistency.

As we have discussed in previous lessons, the Bible gives us a logical basis for these presuppositions. *Because* Jesus is "upholding all things by the word of his power" (Hebrews 1:3), the universe teams with consistency and order. *Because* God made man in His image, man has the ability to observe and record this consistency and order.

As we explore various other algorithms later, these same two principles will continue to apply. All of math builds on itself — and all of math ultimately rests on God.

2.4 Keeping and Balancing a Checkbook Register

Since we've been looking at addition and subtraction this chapter, let's take a look at one common application for these operations: working with a checking account.

Banks offer checking accounts that allow you to store money and dispense it through various methods, such as writing checks, paying bills online, and withdrawing money at the ATM. By recording what you spend and deposit into your account in your checkbook and then making sure your record matches the bank's record, you can make sure that you don't accidentally spend more than you have or go below the bank's required minimum balance.

Keeping a Checkbook Register

Take a look at the sample checkbook register shown and the explanations of the different columns.

Check Number	Date	Memo	Payment Amount	Deposit Amount	$ Balance

Check number

Check Number: Every check you write has a number in the top right corner. You can use this column to record that number.

Date: This is the date the transaction occurred.

Memo: This column is for you to record something that will help you remember what the payment or deposit was.

Payment Amount: If you wrote a check or took money out of your account in some way, you'd put the amount you took out here.

Deposit Amount: If you deposited or put money into your account in some way, you'd put that amount here.

Balance: The balance column is for you to keep a total of how much you have in the account after subtracting whatever you just took out (if a payment) or adding whatever you just put in (if a deposit).

Notice that there are two columns or boxes under payment, deposit, and balance; the one on the left is for recording dollars, and the one on the right is for recording cents.

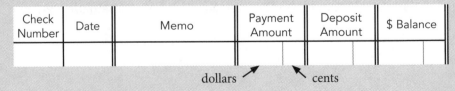

52 | PRINCIPLES OF MATHEMATICS 1

Let's walk through recording a few transactions. Say you start on March 1 with $2,500 in the bank. You would record that in the checkbook like this:

Check Number	Date	Memo	Payment Amount	Deposit Amount	$ Balance
	3/1	**Opening Balance**			**2,500**

Now let's say on March 2 you write check number 172 to your cell phone company for $45. You would record that on the next line.

Check Number	Date	Memo	Payment Amount	Deposit Amount	$ Balance
	3/1	Opening Balance			2,500
172	3/2	**Cell Phone Company**	**45**		

Now, since you *spent* $45, you now have $45 less in the bank than you did before. So you would need to *subtract* $45 from your previous balance and record the difference in the balance column in order to know how much is still in your bank account.

Check Number	Date	Memo	Payment Amount	Deposit Amount	$ Balance
	3/1	Opening Balance			2,500
172	3/2	Cell Phone Company	45		**2,455**

Now let's say you deposit a paycheck of $1,000 on March 8. You'd record this in the *deposit* column.

Check Number	Date	Memo	Payment Amount	Deposit Amount	$ Balance
	3/1	Opening Balance			2,500
172	3/2	Cell Phone Company	45		2,455
	3/8	**Paycheck**		**1,000**	

This time, you would *add* $1,000 to the previous balance, since you put money *into* the account.

Check Number	Date	Memo	Payment Amount	Deposit Amount	$ Balance
	3/1	Opening Balance			2,500
172	3/2	Cell Phone Company	45		2,455
	3/8	Paycheck		1,000	**3,455**

Balancing a Checkbook Register

Now that you know how to keep a checkbook register, it's time to take a look at a process called **balancing the checkbook** or **reconciling a statement.** Periodically (often once a month), banks send statements that show what has been deposited into your account or withdrawn from your account (payments). You can double-check your checkbook against the bank's statement to make sure that your balance agrees with theirs. This process is known as balancing the checkbook or reconciling the statement.

Most statements include instructions on how to balance a checkbook, but let's walk through some basic steps you can follow.

Your Hometown Bank
Anytown, USA

Statement
Dates: 3/1–3/31
Beginning Balance: $2,500
Ending Balance: $3,476

Payments

| 3/2 | Check 172 | Cell Phone Company | $45 |
| 3/18 | Withdrawal | | $30 |

Deposits

3/8	Auto Deposit		$1,000
3/19	Counter Deposit		$50
3/31	Interest		$1

---

1. **Make sure that everything the bank statement shows is in your checkbook.** If it's not, add it (unless you think the bank made a mistake, in which case, research it). An easy way to do this is to use a highlighter to highlight each transaction in your checkbook that's on the statement, or to put a checkmark next to it.

Check Number	Date	Memo	Payment Amount	Deposit Amount	$ Balance
	3/1	Opening Balance			2,500
172	✓ 3/2	Cell Phone Company	45		2,455
	✓ 3/8	Paycheck		1,000	3,455
173	3/9	My Favorite Charity	400		3,055
	✓ 3/18	Withdrawal	30		3,025
	✓ 3/19	Deposit–gift		50	3,075

Check 173 wasn't on the statement, so we didn't put a checkmark next to it. Sometimes, checks will not have cashed yet and thus will not be on the statement even though they are in your register.

If your account earns interest, you'll usually have to add the interest to your checkbook.

In this case, we find that everything is included except the interest. So we need to record the interest.

	✓ 3/18	Withdrawal	30		3,025
	✓ 3/19	Deposit–gift		50	3,075
	✓ 3/31	**Interest**		1	

2. **Find the balance of your checkbook.** Don't be surprised if this balance is not the same as that on the statement — it rarely is. In an active checking account, there are usually transactions that are in your checkbook that are not on the statement. You may have withdrawn money after the statement date . . . or perhaps there's a check that you wrote that the recipient has not yet cashed.

54 | PRINCIPLES OF MATHEMATICS 1

✓ 3/18	Withdrawal	30			3,025	
✓ 3/19	Deposit–gift			50	3,075	
✓ 3/31	Interest			1	(3,076)	

3. **Add deposits and subtract payments that haven't been processed yet to the statement's balance.** In order to see if the two balances match, start with the statement's balance and add any deposits you made after the statement date and subtract payments that were not on the statement. (For example, a check you wrote may not have been cashed or processed by the statement date.) You can do this one at a time, or by adding up all the deposits and all the withdrawals and then adding/subtracting once. If you highlighted or put a checkmark next to the transactions that were on the statement, then you can easily find the transactions not yet processed — any transaction left in your checkbook without a checkmark/highlighting wasn't processed by the bank yet as of the statement date.

In this case, notice that check number 173 — the $400 to "My Favorite Charity" was not on the statement (and thus does not have a checkmark next to it in our register). So we need to subtract it from our statement's balance of $3,476.

$$\$3,476 - \$400 = \$3,076$$

4. **Work until reconciled; mark that you've reconciled the statement.** Hopefully, at this point you've verified that, once you add deposits and subtract payments that weren't processed yet by the statement date, the balance in your checkbook matches that in the bank. If it didn't, then you need to recheck your math until you find the error. You may also need to re-compute the balance in your checkbook to see if you made a math error there. Once you have a balance in your checkbook that, with the additions and subtractions not on the statement yet, matches the statement's balance, we would say that you've reconciled that statement. But before you put the checkbook and statement away, be sure to show what balance you reconciled by either a check mark or by highlighting. You might put a checkmark on the statement (or move it into a reconciled folder if it's electronic) and maybe mark the balance in your checkbook that you added to/subtracted from in order to match the bank's. (For example, you could put "rec" next to it to indicate that balance has been *rec*onciled.) Marking it will help you remember what number you ended with previously when you go to balance the next statement.

In this case, since once we subtract the $400 check that hasn't been processed yet our balances tie, we've successfully reconciled the checkbook. We've only to note that somehow on the statement and in our checkbook register.

You could also reconcile your ending balance the opposite way, by starting with your checkbook balance. Only if you do, rather than adding the deposits, you'd have to subtract them (after all, your checkbook balance includes them, but the statement's does not). Likewise, you'd have to add rather than subtract payments that are in your checkbook but not on the statement. For example, here we would start with our checkbook balance of $3,076 and *add* to it the $400 check that has not processed yet. We would then arrive at our statement balance of $3,476.

✓ 3/18	Withdrawal	30			3,025	
✓ 3/19	Deposit–gift			50	3,075	
✓ 3/31	Interest			1	3,076	(rec)

As you might sense, there are a lot of different ways you can balance a checkbook. Why, you might even use software to help you keep track of your finances. The important thing is to know your balance so you don't accidentally write a check for more money than you have in your account! No matter what method you use, you'll want to make sure that the balance you're looking at when you write a check is accurate.

Keeping Perspective

Notice the addition and subtraction involved in balancing the checkbook! Addition and subtraction are "tools" you'll find yourself using throughout life, often without realizing it. Checkbooks are just one example of addition and subtraction serving as a tool we can use to complete the tasks God's given us in real life.

2.5 Different Addition and Subtraction Methods

We saw in 2.3 how our current addition and subtraction method helps us describe place value and keep track of digits so we can add and subtract with less effort. To better understand our method, we also explored how addition and subtraction could be done on an abacus, which uses beads to accomplish the same task.

Since history helps us really understand concepts and view them as tools, let's now take a broader look at some historical addition and subtraction methods.

Adding Bhāskara's Way

This method is attributed to Bhāskara II, a mathematician from India.[10]

Traditional Method for comparison	Bhāskara Method
Tens Ones $\overset{1}{}$ 1 2 + 1 9 ‾‾‾‾ 3 1	Ones $2 + 9 = 1\,1$ Tens $1 + 1 = \underline{2\,0}$ $3\,1$

In this method, the ones column is first added ($9 + 2 = 11$), and then the tens column is added ($1 + 1 = 2$). Lastly, the two columns are added together. Instead of writing the problem vertically, it is written horizontally. Notice in the diagram how a zero was added when adding the tens column ($1 + 1 = \mathbf{20}$) to put the 2 in the tens place.

Subtracting Leonardo's Way

In *Liber Abaci* (circa 1202), a book that was instrumental in bringing the decimal system to Europe, Leonardo Pisano took a slightly different approach to subtracting than we follow today. Here is a simplified version of his approach.[11]

Rather than crossing out a number in a column when he had to rename a quantity, he indicated *with a finger* that he had renamed a quantity (Leonardo had a systematic way of showing different numbers on the left hand; to signify "1," he would bend the left-hand little finger). Then, rather than subtracting that quantity from the tens column of the minuend (the number we are subtracting from — the top number), he *added* it to the subtrahend (the number being subtracted — the bottom number). Adding to the subtrahend does the same thing to the answer as subtracting from the minuend. It is just a different way to think about it. God created us with creativity and the ability to find different mathematical methods that accurately reflect His creation.

Example: Find 21 – 14.

Traditional Method for comparison	**Leonardo's Method**
Tens Ones 1 1 2̸ 1 - 1 4 ——— 7	7 - - - 2 1 1 4

To solve 21 – 14 using Leonardo's method, write the two numbers to be subtracted on top of each other, and then work the ones column. Since 1 is less than 4, mentally move 10 from the tens column to make 11 – 4. Write the answer, 7, above the line. To remember you took from the ten's column, *bend your left-hand little finger* (representing 1 group of 10).

Lastly, work the tens column. Since you had to take ten from the tens column, *add* ten to the number being subtracted (in this case, the 1 in 14). This gives 2 – 2, or 0. Since the answer is zero, you do not have to write anything. (And you can now unbend your left-hand little finger.)

Leonardo Pisano

Leonardo Pisano, nicknamed "Fibonacci," was born in Italy in 1170, right in the middle of the Medieval Ages. Leonardo's father was a "public official"[12] who ended up serving across the Mediterranean Sea in Bugia (a port in northern Africa).

Leonardo wrote this concerning his father and youth: "He [his father] had me in my youth brought to him, looking to find for me a useful and comfortable future; there he wanted me to be in the study of mathematics and to be taught for some days." [13] Leonardo

Leonardo Pisano (i.e., "Fibonacci")

later "travelled considerably afterwards for much study," [14] learning all he could about mathematical methods. Leonardo went on to write a book titled *Liber Abaci* with the goal of sharing the Hindu-Arabic system with others, especially his native Italy. This famous book greatly helped bring the number system we use today not only to Italy, but also to the rest of Europe.

Whether Leonardo acknowledged it or not, God had his hand over his life and used him to accomplish His purposes. Although Leonardo may not have understood why as a youth he had to leave his hometown and go to Africa, God used his father's job to expose the young man to a new system of numbers. He then used Leonardo's work to help bring the Hindu-Arabic number system to Italy, and from there, to Europe! Always remember when there are things in your life you don't understand that God sees the whole picture. You can trust Him.

Edward Stoddard presents a fascinating addition and subtraction approach in his *Speed Mathematics Simplified.* Take a look if you're up for a different way that many find much quicker once they get used to it. Edward Stoddard, *Speed Mathematics Simplified*, Dover ed. (Mineola, NY: Dover Publications, 1994).

Keeping Perspective

Once again, remember that each step-by-step process we follow today is just one way to describe the consistencies around us! Other methods have been — and still are — used. God created man with creativity. Who knows? You might even invent your own method!

Yet, while the method can creatively change, the answer cannot. Methods work only if they accurately describe the consistencies God created and sustains. God ultimately is the One making math possible.

2.6 Problem Solving: An Introduction

Mathematics is like a chest of tools: before studying the tools in detail, a good workman should know the object of each, when it is used, what it is used for. [15] — *Walter W. Sawyer*

Suppose you had a whole garage filled with tools. You knew the name of each one and you could describe each tool in detail, sharing its technical specifications. But when faced with a home-improvement project, those tools will only do you any good if you know how — and when — to use each one.

Likewise, in math, it's important to know how — and when — to use the different concepts (i.e., "tools") at our disposal. It doesn't make sense to learn math without learning how to use it, just as it wouldn't make sense to study a whole bunch of tools without ever actually using any of them to build or repair anything, or to study a whole bunch of kitchen gadgets without ever using them to cook!

58 | PRINCIPLES OF MATHEMATICS 1

You'll also find that the problem-solving skills you learn in math will help you solve non-mathematical problems, too. Solving math problems can help you think through a variety of situations, thereby equipping you to serve God with math wherever He places you.

Today, we're going to take a look at problem-solving steps and apply those steps to some addition and subtraction problems.[16] While the problems right now will be quite simple, focus on learning the steps, as you'll need them for more challenging problems.

Problem-Solving Steps

We solve problems all the time. Perhaps the toast is burnt and we want to know why. Or maybe the garage door won't shut and we need to fix it.

Whatever the case, when we have a problem, we first recognize that it's a problem, figure out what we're going to do about it, do it, and then see if what we did worked.

We follow similar steps within math. Hopefully, these steps become so habitual that we do them automatically. **These steps may seem unnecessary to you, but when you encounter a harder problem, you will be thankful you know them.** Practicing them on easier problems will make harder problems simpler.

1. **Define** — Start by making sure you know what you're trying to find. You won't be able to determine what tools to use if you don't know what you're trying to find out and what information you know. Ask yourself these questions:

    1. What am I trying to find?
    2. What information have I been given?

2. **Plan** — How will you go about finding what you need, given the information you have? Before you start doing any math at all, think through the problem. How will you find the answer? Is there a clear method to use? If you're not sure how to solve a problem, start by trying to find additional information until it becomes obvious what mathematical tool to use. It's often helpful to express the relationship between the information you've been given and the information you're trying to find in words.

3. **Execute** — Execute your plan. Here's where you do the math!

4. **Check** — It's important to always check your work. While a careless mistake in math class might only result in a lower grade, in real life a careless mistake might mean mischarging a customer, making a poor decision, buying the wrong amount of supplies, or something else with more serious ramifications. It is important to train yourself to look at your answers to make sure they make sense. Ask yourself if your answer is reasonable. Checking for reasonableness will help you identify if you added and subtracted the wrong numbers or made a major mistake in your calculations. You can also check your math by reversing it, as we'll see in the examples.

When solving word problems, it is important to **always show your work**. Not only will this make it easier for you to check it for accuracy, but it will also train you to think through problems logically (and to pass college math courses, where you may be asked to show your work).

By showing your work, I mean writing down enough information that you can see how you solved the problem. Sometimes, you will be asked to show each step. While this may seem pointless on simple problems, it will help you develop good problem-solving skills that will serve you well in both math class and, more importantly, in life. Both methods are shown in the example.

Example: You went to the grocery store and bought a can of soup for $3, a package of meat for $6, and a bunch of fruit for $12. How much did you spend in total?

Showing Your Work: $3 + $6 + $12 = $21

Showing Each Step: *These steps are often done mentally, but it's important to be able to write them out when requested.*

1. Define —

Soup Cost (or just *S*) = $3
Meat Cost (or just *M*) = $6
Fruit Cost (or just *F*) = $12
Total Cost (or just *T*) = ?

Notice that **we used letters** in the parentheses to stand for the different food items and for the total. We could have saved our fingers some writing by only using letters or other symbols (whether these or others — the exact abbreviation we choose doesn't matter) and not writing out the words at all, like this:

S = $3
M = $6
F = $12
T = ?

2. Plan —

Soup Cost + Meat Cost + Fruit Cost = Total Cost
Or just $S + M + F = T$
Or just know you'll use addition to find the total cost.

3. Execute — $3 + $6 + $12 = $21

4. Check — $21 − $12 − $6 − $3 = 0

Solving Addition and Subtraction Problems

Are you trying to find a total? Then you need to use addition. Are you trying to find what you would have if we took something away? Then you should use subtraction. And, as we saw back in 2.1, since the two are opposite operations, they're often used in combination or can be thought of either way.

For example, if you spend $5 out of $20 at the store and want to see how much you have left, you can think of this as subtraction ($20 − $5), or as finding a missing

60 | PRINCIPLES OF
MATHEMATICS 1

addend ($5 + __ = $20) . . . which you would find via subtraction. Either way, you'd end up with $15.

Sometimes, solutions to real-life problems might not be obvious. For instance, suppose you had 12 friends who want to ride a horse, but only 7 horses. Assuming only one person could ride each horse, how many people would not have a horse to ride?

Obviously, you don't really want to take away, or minus, the horses from the riders. But in a figurative sense, taking away one horse per person will tell you how many people will not have horses. It still tells us the *difference* between the number of friends and the number of horses.

12 – 7 = 5; 5 friends will not have horses.

Example: Suppose we managed three stores. On a given day, one store made $3,000, one $4,560, and another $3,200. What are our total sales?

1. Define — Here we're trying to find the total sales, and we've been given the daily sales for each store. To show our work, we could write something along these lines:

Sales at First Store (or S_1) = $3,000
Sales at Second Store (or S_2) = $4,560
Sales at Third Store (or S_3) = $3,200
Total Sales (or T) = ?

Again, notice that **we used letters** in the parentheses to stand for the different stores and the total. Once again, we could have saved our fingers some writing by only using abbreviations and not writing out the words at all.

S_1 = $3,000
S_2 = $4,560
S_3 = $3,200
T = ?

2. Plan — Look at the information we've been given and see how it relates. Here, we want to find a total, so we will need to add the sales from each store to find it.

Sales at First Store + Sales at Second Store + Sales at Third Store = Total Sales

Or, if using letters: $S_1 + S_2 + S_3 = T$

Notice how letters save writing!

The main point is just that we know we need to add up the sales at each store to find the total sales.

3. Execute —

$S_1 + S_2 + S_3 = T$
$3,000 + $4,560 + $3,200 = $10,760

4. Check — Does $10,760 sound reasonable? Well, let's simplify the problem for a minute to see. If we ignore everything in the numbers we add except the thousand's place, we'd have 3 + 4 + 3, which equals 10 (or $10,000, since we were adding the thousand's place). The sum $10,760 is very close to that and sounds about right for the sum of all three stores.

(If we had gotten a number such as $2,000 — less than the value of one store — or $29,000 — way different than we would expect — we would want to go back and try again!) We could also double check our math using subtraction, starting with $10,760 and subtracting each of the other amounts to see if we end up at zero. **Because subtraction is the opposite of addition, it can be used to check addition.**

$$
\begin{array}{r}
\$\,1\,0\,,7\,6\,0 \\
-\,\$\quad3\,,2\,0\,0 \\
\hline
\$\quad7\,,5\,6\,0 \\
-\,\$\quad4\,,5\,6\,0 \\
\hline
\$\quad3\,,0\,0\,0 \\
-\,\$\quad3\,,0\,0\,0 \\
\hline
0
\end{array}
$$

Example: Let's say we're on a trip, and we don't have a GPS with us. Our map tells us we have to go 10 miles on a road. We've already gone 4 miles. How much farther do we have remaining?

1. Define — Here we're trying to find how far we have left to go. We know the total distance and the distance we've traveled so far.

$G = 4$ miles
$T = 10$ miles
$R = \,?$

Note: We've used "G" to stand for the distance gone, "T" for the total distance, and "R" for the remaining distance.

2. Plan — To aid in seeing the relationship, let's write it out. We know that the total distance will equal both the distance we've gone and the distance we have left to go.

Total Distance = Distance Gone + Distance Remaining
(or just $T = G + R$)

We know the total (T) and the distance we've gone (G), but we don't know the distance remaining. So we know the sum, and we need to find one of the numbers that will get us to that sum. We can find this by subtracting the distance we have traveled from the total distance. Even though we're not actually taking away quantities, subtraction is the opposite of addition and can help us find the amount we'd have to add to get our total.

3. Execute —

$T = G + R$
10 miles = 4 miles + _____
10 miles – 4 miles = 6 miles
We have 6 miles left to go.

4. Check — Does our answer seem reasonable? Yes, 6 miles is a reasonable answer. If we had come up with 60 miles, we'd want to recheck our steps. Because addition is the opposite of subtraction, we can **check subtraction with addition:** 6 miles + 4 miles = 10 miles. Here we see that if we add the miles left to go to the miles we started with, we end up with the correct total miles (10).

The idea behind the planning step here is just that we figure out what we need to subtract to find the distance we have left to go.

---

62 | PRINCIPLES OF MATHEMATICS 1

Keeping Perspective

Again, while these steps may seem unnecessary on some of the simp[le]
problems you're working with now, familiarizing yourself with them [will]
greatly help you when we get to more advanced concepts. Keep in mi[nd]
throughout this course that our goal is to learn to use each aspect of [math]
as a real-life tool, so expect to have your problem-solving skills stretc[hed]
and enlarged as we proceed, so you'll be equipped to use math to com[plete]
whatever various tasks the Lord sends your way.

2.7 Chapter 1 and 2 Synopsis

While many of the concepts in these first two chapters should have been review for you, we covered a lot of foundational information about math that will continue to apply to the rest of our studies. Let's review the biblical foundation for math, along with how it applies.

Biblical Principles for Math

- **The Bible tells us WHERE math originated.** Since God created everything, He created math, too! This does not mean God created the symbols we write, but He created the consistency those symbols represent. Symbols like $1 + 1 = 2$ record the consistent way God causes objects to operate.

- **The Bible tells us WHY math is possible.** Why does $1 + 1$ consistently equal 2? Because God both created this consistency and keeps it in place! Math's very ability to work is a testimony to God's faithfulness. If this universe were simply a random collection of chemicals, or if we served an inconsistent God, we would have no reason to expect objects to add or subtract consistently.[17]

  In order for math to work, there not only have to be consistencies throughout creation, but we have to be able to recognize those consistencies! Again, the Bible gives us the framework for understanding this. It teaches us that God created man in His image and gave him dominion over the earth (Genesis 1:27–28). Hence, we should expect to be able to develop ways to record the consistencies God placed around us. We should also expect our thinking to, in a very limited way, take after our Creator. Many purely intellectual mathematical theories end up corresponding with reality because *our minds were created by the same Creator who created all things.*

- **The Bible tells us WHAT to expect as we use math.** Although we tend to confine math to a textbook, it actually goes hand in hand with science! Since math records real-life consistencies, we should expect it to help us explore God's creation, showing us both reflections of God's original incredible design and evidence of destruction and death due to the entrance of sin into the world. We

...ould also expect to be able to use math to help us in the various tasks God has given us to do, be they around the house or on the job.

■ **The Bible gives us principles to guide HOW we use math.** Through Jesus, we can again know God and worship Him as we use math! Since God created us, we are accountable to Him for how we use the gift of math He has given us.

Building on the Foundation

As we go forward, we won't always be mentioning that God created and sustains math, but we'll be seeking to let that fundamental understanding affect how we approach the various concepts we learn. Here's a brief overview of some aspects of math and how we'll be approaching them.

■ **Symbols & Terms** — As we learn math's many symbols and terms, we'll be approaching them as a language system to help us describe the quantities and consistencies that God placed around us. To keep them in perspective, we'll sometimes take a look at historical symbols or terms.

■ **Operations** — As we study different operations, we'll see again and again that we live in a consistent universe held together by a consistent God.

■ **Algorithms** — As we learn algorithms, or step-by-step methods, we'll be seeking to understand how the algorithm works so we can see what aspect of God's creation it's describing. We'll often also take a look at the history of algorithms so we can "think outside the box" and really understand the concept and use and view the method as a tool — that is, as one method for describing something God created and sustains.

■ **Problem Solving** — Since we want to learn to use math as a tool and be equipped to use it wherever the Lord might place us, we'll be emphasizing real-life problems and learning problem-solving techniques.

Keeping Perspective

Whether the text mentions it or not, I hope you'll pause often and reflect on the miracle of math's ability to work. Math is shouting out at us that we serve a faithful God we can trust. It's a simple truth, but a life-changing one. The more we reflect on who God is — on His faithfulness, power, wisdom, etc. — the more our pride, fears, and self-reliance will fade into humility, faith, and trust. Let math turn your eyes heavenward to the Creator and sustainer of it all.

[CHAPTER 3]

Mental Math and More Operations

3.1 Mental Math

When I first started working at a bakery, I never dreamed how much math I would use on the job. Giving change to customers, counting bread, inventorying products, writing deposit tickets — the list went on. Not only was math required, but most of the math needed to be done quickly and mentally, as there simply wasn't time to use paper or a calculator.

Being able to perform math calculations in your head is a needed skill, regardless of what the Lord calls you to in life. Because of the consistent way God holds all things together, we can make observations and come up with strategies to help us with mental arithmetic, just as we do with written arithmetic. In this lesson, we'll be looking at a few strategies for mentally adding and subtracting.

Please don't get hung up with the strategies themselves. The important thing is to become proficient in mental arithmetic so you'll have the tools of addition and subtraction at your disposal whenever you need them.

Adding 10

It's very simple in our place-value system to mentally add by 10, or multiples of 10, since our place-value system is based on 10. When adding by 10, 20, etc., we have zeros in the ones place. Thus, we only have to add the tens place (or the hundreds, thousands, etc., if adding larger quantities).

$$
\begin{array}{r}
46 \\
+10 \\
\hline
56
\end{array}
\qquad
\begin{array}{r}
46 \\
+20 \\
\hline
66
\end{array}
\qquad
\begin{array}{r}
489 \\
+100 \\
\hline
589
\end{array}
\qquad
\begin{array}{r}
800 \\
+\ \ 200 \\
\hline
1,000
\end{array}
$$

Making 10

Because our decimal system is based on 10, if we can quickly (and mentally) think about what we need to add or subtract from a number to reach 10, we'll be able to add and subtract easier in our heads.

The number you have to add to a number to reach 10 can be referred to as the **complement** of that number.

> 1 and 9 are complements, as $1 + 9 = 10$.
> 2 and 8 are complements, as $2 + 8 = 10$.
> 3 and 7 are complements, as $3 + 7 = 10$.
> 4 and 6 are complements, as $4 + 6 = 10$.
> 5 and 5 are complements, as $5 + 5 = 10$.

The quicker we can think about what makes 10, the easier it will be to calculate mental math.

The word **complement** also has a more generic meaning than a number that makes 10. It means "a number or quantity of something required to make a group complete."[1] Basically, here we're looking at 10 as the whole group, and then calling whatever number is needed to make 10 the complement.

Turning Subtraction into Addition

Let's say we need to subtract 75 from 83 in our heads. It wouldn't be easy to try to follow our written process mentally.

$$\begin{array}{r} {}^{7}\!\!\not{8}\,{}^{1}3 \\ -\ 7\ 5 \\ \hline 8 \end{array}$$

Instead, we could quickly find the answer by calculating how much we would have to add to 75 to reach 83. The complement of 5 is 5, so we'd have to add 5 to reach 80. Then we'd have to add another 3 to get to 83. Since $5 + 3 = 8$, we have to add a total of 8 to get to 83.

We've just found the difference between 83 and 75 by *adding* from the subtrahend.

Keeping Perspective — God's Capacity and Ours

Hopefully, the strategies we've gone over will help you add and subtract mentally. As with written methods, most of these strategies take advantage of place value. If we were trying to add in a different place value system, such as binary, the tips we studied would need to be modified to work with that system.

Mental strategies, like written methods, are shortcuts to help us describe the consistencies God sustains around us. Picture them as tools that, once mastered, travel with you everywhere.

3.2 Rounding — Approximate Answers

Rounding is a term used to describe approximating a number. We often round in mental arithmetic to find an approximate answer.

For example, suppose you were at the grocery store and wanted to know approximately how much you were spending before you got to the checkout (so you didn't spend more than you had budgeted!). You might not want to actually add $1.99 + $2.35 + $4.98 mentally — but if you round them to the nearest dollar, they would be easy to add mentally.

We'll talk about rounding to the nearest dollar in more depth later when we study decimals. But for now, we can use rounding in mental arithmetic with whole numbers, too. We frequently want to round to the nearest ten, hundred, or thousand.

When rounding a number, **look at the digit to the right of the place you want to round the number to.**

▨ If that digit is **5 or greater**, you **round up**, as the number is closer to the next 10, 100, etc.

▨ If it is **less than 5**, you **round down**, as it is closer to the previous 10, 100, etc.

Note: The 5 is actually midway, so it could round up or down; since we have to pick one, the convention is to round it up.

So . . .

> 47 rounds up to 50.
>
> 455 rounds up to 460 or 500 (depending on if we're rounding to the nearest ten or hundred).
>
> 34 rounds down to 30.

Rounding to Find an Approximate Answer

If all we need is an approximate answer, we can use rounding to greatly simplify mental addition and subtraction.

Example: Find the approximate value of 145 + 890. (Round to the nearest 100.)

> The nearest hundred to 145 is 100 (4 is the digit to the right of the hundreds place, so we round down).
>
> The nearest hundred to 890 is 900 (9 is the digit to the right of the hundreds place, so we round up).

$$145 + 890 \approx 1,000$$

≈ is a way of representing "approximately equals."

Rounding to Find an Exact Answer

Even if we need an exact answer, rounding can still help us find it mentally!

Example: Add 37 and 25.

1. Round one number to the nearest 10, and add the other number to it mentally.

Round the 37 **up 3**:

$$\begin{array}{r} +3 \\ 37 \rightarrow 40 \\ +25 \quad +25 \\ \hline 65 \end{array}$$

or

Round the 25 **up 5**:

$$\begin{array}{r} 37 \quad 37 \\ +25 \rightarrow +30 \\ +5 \quad 67 \end{array}$$

2. Subtract (or add) the amount you added (or subtracted) when you rounded.

Mentally subtract the 3 you added

$65 - 3 = 62$

or

Mentally subtract the 5 you added

$67 - 5 = 62$

Keeping Perspective — God's Capacity and Ours

Rounding is so useful because our brains have to simplify some problems (such as a full cart of groceries) in order to solve them mentally. Our brains can keep track of only so many details at once!

On the other hand, the Bible tells us God keeps track of the number of hairs on every individual's head. "But the very hairs of your head are all numbered" (Matthew 10:30). God's capacity to track details is not limited like ours.

3.3 Multiplication

While I'm sure you already know how to multiply, let's take a quick look at this crucial concept in light of biblical principles. We'll discover that multiplication, like the other operations, loudly proclaims God's praises. Along the way, we'll also begin familiarizing ourselves with some new ways to write multiplication used in upper-level math.

Multiplication in a Nutshell

We often need to add the same numbers over and over again. For instance, if we made $9 dollars an hour and wanted to find out how much we make in 8 hours, we have to add the hourly amount 8 times — $9 + $9 + $9 + $9 + $9 + $9 + $9 + $9.

Imagine that every day you had to do the same tedious task over and over again. You would look for a faster way to get the task done, wouldn't you?

Multiplication is a faster way of looking at repeated additions. Rather than looking at $9 + $9 + $9 + $9 + $9 + $9 + $9 + $9 as addition, we could think of it in terms of taking $9 eight times. It is helpful to think of this as 8 groups of $9.

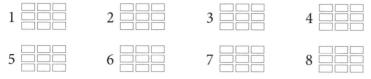

Rather than writing $9 + $9 + $9 + $9 + $9 + $9 + $9 + $9, we could write 8 x $9. This is an agreed-upon shortcut to represent 8 groups of $9, or $9 + $9 + $9 + $9 + $9 + $9 + $9 + $9. When we view repeated additions in terms of how many times we are adding the number (in this case, 8 times), we call it multiplication. As we did with addition and subtraction, we can memorize what 8 x $9 equals and call it a "fact."

Multiplication Facts

1 x 1 = 1	1 x 2 = 2	1 x 3 = 3	1 x 4 = 4	1 x 5 = 5
2 x 1 = 2	2 x 2 = 4	2 x 3 = 6	2 x 4 = 8	2 x 5 = 10
3 x 1 = 3	3 x 2 = 6	3 x 3 = 9	3 x 4 = 12	3 x 5 = 15
4 x 1 = 4	4 x 2 = 8	4 x 3 = 12	4 x 4 = 16	4 x 5 = 20
5 x 1 = 5	5 x 2 = 10	5 x 3 = 15	5 x 4 = 20	5 x 5 = 25
6 x 1 = 6	6 x 2 = 12	6 x 3 = 18	6 x 4 = 24	6 x 5 = 30
7 x 1 = 7	7 x 2 = 14	7 x 3 = 21	7 x 4 = 28	7 x 5 = 35
8 x 1 = 8	8 x 2 = 16	8 x 3 = 24	8 x 4 = 32	8 x 5 = 40
9 x 1 = 9	9 x 2 = 18	9 x 3 = 27	9 x 4 = 36	9 x 5 = 45
10 x 1 = 10	10 x 2 = 20	10 x 3 = 30	10 x 4 = 40	10 x 5 = 50

1 x 6 = 6	1 x 7 = 7	1 x 8 = 8	1 x 9 = 9
2 x 6 = 12	2 x 7 = 14	2 x 8 = 16	2 x 9 = 18
3 x 6 = 18	3 x 7 = 21	3 x 8 = 24	3 x 9 = 27
4 x 6 = 24	4 x 7 = 28	4 x 8 = 32	4 x 9 = 36
5 x 6 = 30	5 x 7 = 35	5 x 8 = 40	5 x 9 = 45
6 x 6 = 36	6 x 7 = 42	6 x 8 = 48	6 x 9 = 54
7 x 6 = 42	7 x 7 = 49	7 x 8 = 56	7 x 9 = 63
8 x 6 = 48	8 x 7 = 56	8 x 8 = 64	8 x 9 = 72
9 x 6 = 54	9 x 7 = 63	9 x 8 = 72	9 x 9 = 81
10 x 6 = 60	10 x 7 = 70	10 x 8 = 80	10 x 9 = 90

The Multiplication Table and Napier's Rods

We can make devices to help us remember and use multiplication facts. One such device is the **multiplication table**, which you may have studied before.

Ivory pieces of two Napier's abacuses, based on models designed by Scottish mathematician John Napier (1550-1617): the so-called "Napier's bones" and a more complex one named Promptuary. (National Archaeological Museum of Spain)

1	2	3	4	5	6	7	8	9
2	4	6	8	10	12	14	16	18
3	6	9	12	15	18	21	24	27
4	8	12	16	20	24	28	32	36
5	10	15	20	25	30	35	40	45
6	12	18	24	30	36	42	48	54
7	14	21	28	35	42	49	56	63
8	16	24	32	40	48	56	64	72
9	18	27	36	45	54	63	72	81

To better understand this device (and see God's power displayed through multiplication), we're going to take a quick look at **Napier's rods** or **Napier's bones**, a device developed by the Scottish Protestant John Napier back in 1617 that is very similar to the multiplication table.

Napier's rods consisted of rods, some of which were made out of ivory or similar animal bones, on which multiplication facts were engraved. For example, different repetitions of 2 would be engraved on one rod. The first box of this rod would record what we would get if we took 2 one time (2 or 1 x 2), the second if we took 2 two times (2 + 2 or 2 x 2), and so forth. Diagonal lines in each box separated the tens place from the ones place. When a user needed to find what 3 x 2 equaled, all he would need to do is look in the third box of the 2 rod. A guide rod made this easy to spot.

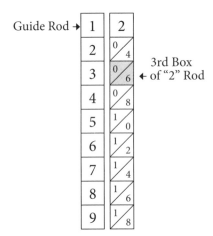

These rods saved people a lot of time, especially before the days of calculators. Using them, men could quickly multiply numbers — even ones with many digits. (We will look at using them to multiply multi-digit numbers in 4.6.)

Notice how the multiplication table is essentially the same thing as Napier's rods, only all the strips have been combined, and the diagonal lines separating the tens places from the ones places have been removed. The multiplication table works the same basic way Napier's rods did — to find 3 x 2, you would look in the third box of the 2 column. The rods are just all connected into a single table, and the one's rod serves as a guide to find the appropriate box.

Napier's Rods (Ones Rod/Guide, Twos Rod)

1	2	3	4	5	6	7	8	9
2	04	06	08	10	12	14	16	18
3	06	09	12	15	18	21	24	27
4	08	12	16	20	24	28	32	36
5	10	15	20	25	30	35	40	45
6	12	18	24	30	36	42	48	54
7	14	21	28	35	42	49	56	63
8	16	24	32	40	48	56	64	72
9	18	27	36	45	54	63	72	81

Multiplication Table (Ones Column/Guide, Twos Column)

1	2	3	4	5	6	7	8	9
2	4	6	8	10	12	14	16	18
3	6	9	12	15	18	21	24	27
4	8	12	16	20	24	28	32	36
5	10	15	20	25	30	35	40	45
6	12	18	24	30	36	42	48	54
7	14	21	28	35	42	49	56	63
8	16	24	32	40	48	56	64	72
9	18	27	36	45	54	63	72	81

Do you catch the glimpse of God's power here? Napier's rods and the multiplication table would be absolutely pointless unless repeated groups of objects add together *so* consistently that we can find the answer once, record it on a rod or in a table, and never have to find the answer again. In a sense, Napier's rods and the multiplication table are a record of God's power in holding all things together!

Different Multiplication Symbols

There are different ways to express multiplication. Although up until now you have probably used the times sign, "x", there are actually other symbols as well. In fact, since an "x" has a different meaning in upper-level math, you'll find that the two ways we are looking at today are more commonly used. One way is to put parentheses around one number that needs to be multiplied and to write the other on the outside of the parentheses. Another way is to use a little dot, "•", instead of a times sign. All three ways listed below are acceptable ways of expressing multiplication and will be used in this course.

8 x $9   or   8($9)   or   8 • $9

Remember, the methods of writing are just agreed-upon conventions to help us communicate about the principle God created and sustains.

> **Term Time**
> Just as with addition and subtraction, we have named the parts of a multiplication problem to make them easy to reference. These names aid in communication.
>
> **Multiplicand** — "The number that is or is to be multiplied by another"[2] — in other words, the quantity in each set or group.
>
> **Multiplier** — "The number by which the multiplicand is multiplied"[3] — in other words, the number of sets, or groups, we have.

3. MENTAL MATH AND MORE OPERATIONS

Customs differ as to whether the first number in a multiplication problem is the multiplier or the multiplicand. Some people would read 9 x 4 as 9 sets of 4 (4 + 4 + 4 + 4 + 4 + 4 + 4 + 4 + 4), while others would read it as 4 sets of 9 (9 + 9 + 9 + 9).

It does not matter which order we use in multiplication, since because of the way God causes objects to add, these products are always equal. The order in which we multiply does not change the answer — it is just a convention/preference as to what we put first — 9 sets of 4 *and* 4 sets of 9 both equal 36. You can easily see this on Napier's rods.

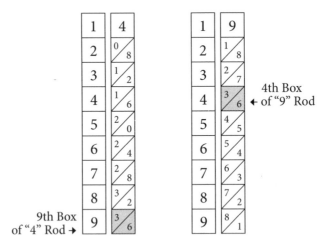

For the purpose of keeping things understandable in this course, we will put the multiplier first, meaning we would write 9 sets of 4 as 9 x 4.

Factor — A name to refer collectively to all the numbers being multiplied.
Product — The result of the multiplication.

In 4 x 3 = 12, 4 and 3 are called factors, and 12 is called the product. Using the convention adopted in this course, 4 is also called the multiplier, and 3 is also called the multiplicand.

Keeping Perspective

As you review multiplication, ponder God's power. Just His Word keeps this universe operating in a predictable enough way that we can write down how repeated sets of objects will add on a table, memorize those facts, and count on them working in real life. Wow.

> *[Referring to Jesus] Who being the brightness of his glory, and the express image of his person, and upholding all things by the word of his power, when he had by himself purged our sins, sat down on the right hand of the Majesty on high (Hebrews 1:3).*

Don't ever forget that nothing is too hard for God — if He can keep this universe in such precise order by His Word, He can definitely give us the grace to handle whatever it is we're facing in a way that honors Him.

> *Behold, I am the Lord, the God of all flesh: is there any thing too hard for me? (Jeremiah 32:27).*

> *There hath no temptation taken you but such as is common to man: but God is faithful, who will not suffer you to be tempted above that ye are able; but will with the temptation also make a way to escape, that ye may be able to bear it (1 Corinthians 10:13).*

3.4 Division

In this lesson, we're going to take a look at another foundational concept you've been using for years: division. As we review the concept, we'll take a look at some history, see again God's power on display, and prepare ourselves to use division practically. As with addition, subtraction, and multiplication, division is an operation you'll need to use over and over again in life — a pocketknife, if you will, that comes in handy for all sorts of tasks.

Reviewing the Concept

Suppose you had 12 paper clips you wanted to share evenly among 3 people. You would end up giving each person 4 paper clips.

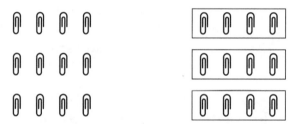

In math, we would call this division. *Division* is a name we use to describe taking a quantity and separating it into "two or more parts, areas, or groups."[4]

Just as multiplication is a shortcut for writing repeated addition, division is really a shortcut for writing repeated subtraction.

In fact, the Egyptians used to solve their division problems by repeatedly subtracting. To divide 12 by 3, they would subtract 3 over and over again, keeping track of the number of subtractions made to completely divide 12 into 3.

$$12 - 3 - 3 - 3 - 3 = 0$$

$$12 \quad - \quad 3 \quad - \quad 3 \quad - \quad 3 \quad - \quad 3 \quad = 0$$

We had to subtract 3 four times in order to completely divide 12. So 12 divided by 3 is 4.

Because God upholds all creation together consistently, we can memorize our division facts just as we did the facts for addition, subtraction, and multiplication!

Division Facts

$1 \div 1 = 1$	$2 \div 2 = 1$	$3 \div 3 = 1$	$4 \div 4 = 1$	$5 \div 5 = 1$
$2 \div 1 = 2$	$4 \div 2 = 2$	$6 \div 3 = 2$	$8 \div 4 = 2$	$10 \div 5 = 2$
$3 \div 1 = 3$	$6 \div 2 = 3$	$9 \div 3 = 3$	$12 \div 4 = 3$	$15 \div 5 = 3$
$4 \div 1 = 4$	$8 \div 2 = 4$	$12 \div 3 = 4$	$16 \div 4 = 4$	$20 \div 5 = 4$
$5 \div 1 = 5$	$10 \div 2 = 5$	$15 \div 3 = 5$	$20 \div 4 = 5$	$25 \div 5 = 5$
$6 \div 1 = 6$	$12 \div 2 = 6$	$18 \div 3 = 6$	$24 \div 4 = 6$	$30 \div 5 = 6$
$7 \div 1 = 7$	$14 \div 2 = 7$	$21 \div 3 = 7$	$28 \div 4 = 7$	$35 \div 5 = 7$
$8 \div 1 = 8$	$16 \div 2 = 8$	$24 \div 3 = 8$	$32 \div 4 = 8$	$40 \div 5 = 8$
$9 \div 1 = 9$	$18 \div 2 = 9$	$27 \div 3 = 9$	$36 \div 4 = 9$	$45 \div 5 = 9$
$10 \div 1 = 10$	$20 \div 2 = 10$	$30 \div 3 = 10$	$40 \div 4 = 10$	$50 \div 5 = 10$

$6 \div 6 = 1$	$7 \div 7 = 1$	$8 \div 8 = 1$	$9 \div 9 = 1$	$10 \div 10 = 1$
$12 \div 6 = 2$	$14 \div 7 = 2$	$16 \div 8 = 2$	$18 \div 9 = 2$	$20 \div 10 = 2$
$18 \div 6 = 3$	$21 \div 7 = 3$	$24 \div 8 = 3$	$27 \div 9 = 3$	$30 \div 10 = 3$
$24 \div 6 = 4$	$28 \div 7 = 4$	$32 \div 8 = 4$	$36 \div 9 = 4$	$40 \div 10 = 4$
$30 \div 6 = 5$	$35 \div 7 = 5$	$40 \div 8 = 5$	$45 \div 9 = 5$	$50 \div 10 = 5$
$36 \div 6 = 6$	$42 \div 7 = 6$	$48 \div 8 = 6$	$54 \div 9 = 6$	$60 \div 10 = 6$
$42 \div 6 = 7$	$49 \div 7 = 7$	$56 \div 8 = 7$	$63 \div 9 = 7$	$70 \div 10 = 7$
$48 \div 6 = 8$	$56 \div 7 = 8$	$64 \div 8 = 8$	$72 \div 9 = 8$	$80 \div 10 = 8$
$54 \div 6 = 9$	$63 \div 7 = 9$	$72 \div 8 = 9$	$81 \div 9 = 9$	$90 \div 10 = 9$
$60 \div 6 = 10$	$70 \div 7 = 10$	$80 \div 8 = 10$	$90 \div 9 = 10$	$100 \div 10 = 10$

Notice that **division is really the opposite of multiplication**. In multiplication, we are adding up sets (3 x 4 means to take 3 sets of 4, or 4 + 4 + 4, which equals 12), while in division, we are starting with the total and dividing it into sets to find the number in each set (12 ÷ 3 means to divide 12 into 3 sets, which gives us 4 in each set).

Thus, as was the case with multiplication, division is shouting out at us that God is still on His throne. Jesus is still upholding all things by the word of His power (Hebrews 1:3).

Term Time

Dividend — The word "dividend" refers to the starting number — the initial quantity we need to divide into smaller parts.

Divisor — The word "divisor" refers to the number of parts we need to divide the initial quantity into.

Quotient — The word "quotient" refers to the ending quantity we will have in each part.

In 12 ÷ 4 = 3, 12 is the dividend, 4 is the divisor, and 3 is the quotient.

These terms allow us to refer to a specific part in a division equation with a single word instead of a lengthy explanation. Without terms, math textbooks would be overwhelmingly lengthy and difficult to read!

Symbols

Throughout the years, men have recorded division problems different ways.[5] Some were quite different from what we use in America today! Even today, we record division different ways, including using our current division sign (÷) and the long-division symbol (⟌).

μορίον	bhâ	12/3	)	÷
•/•	⊂	⌢	:	⌣

Each method of representing division is one useful shorthand for describing the consistent way God causes objects to divide.

Application

So when do we use division?

All the time! Division tells us what happens if we separate out a quantity. This doesn't mean we have to literally separate out the quantity — we can use division anytime we need to look at a quantity in terms of how it separates into another.

For example, suppose we want to find out the price for each tissue box if they're sold in a package of 3 for $6. Here we want to apportion the total cost ($6) among the 3 tissue boxes to see how much each one costs. Division would give us this answer.

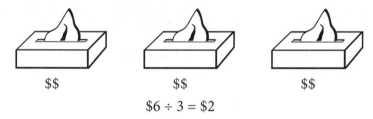

$6 ÷ 3 = $2

It may help to think of division in terms of multiplication. Multiplication helps us do repeated addition; division does the exact opposite. So just as we would use multiplication to find the total for 3 items at $2 each, we would use division to find the unit price for 3 items that cost $6 altogether.

You can think of division as finding a missing factor in multiplication. 3 x ___ = $6 is another way of thinking of $6 ÷ 3 = ___.

Keeping Perspective

Don't let your familiarity with division keep you from being awed by God's handiwork here. The whole reason you can use this useful "pocketknife" is because this universe operates consistently. And the whole reason it operates consistently is because we serve a consistent, all-powerful God. Yet that same God humbled Himself to become a man, claimed our sins as His own, and bore the punishment we deserve.

> *Who his own self bare our sins in his own body on the tree, that we, being dead to sins, should live unto righteousness: by whose stripes ye were healed (1 Peter 2:24).*

3.5 Properties

It's time to take a look at **properties**, or characteristics, that hold true in every situation. As we look, keep in mind that properties are essentially naming consistencies God holds together so we'll be able to easily refer to them and use them to help us solve problems.

Commutative Property of Addition and Multiplication

Addition and multiplication are what we call **commutative** — that is, we can rearrange their order without affecting the meaning.

$$1 + 2 + 3 = 3 + 2 + 1$$

$$2 \times 4 \times 5 = 5 \times 4 \times 2$$

Subtraction and division, on the other hand, are not commutative: 5 – 3 would not give us the same answer as 3 – 5. Likewise, 6 ÷ 3 does not give us the same answer as 3 ÷ 6. The order in which we subtract or divide matters.

You can test the different operations yourself to see if they're commutative by grabbing a handful of paperclips and going through the above problems (or others). You'll find the order doesn't matter for addition and multiplication, but it does for subtraction and division. Thus, we can memorize 4 x 9 and know it will equal the same quantity as 9 x 4, but 4 ÷ 9 and 9 ÷ 4 mean something very different.

We call this property of addition and multiplication the **commutative property**, which basically says we can change the order in addition and multiplication without changing the result.

76 PRINCIPLES OF MATHEMATICS 1

Associative Property of Addition and Multiplication

Not only does the order not matter in addition or multiplication, but the way we group, or **associate**, the numbers when adding or multiplying does not matter either.

In 1 + 4 + 2, it doesn't matter if we add the 1 and the 4 first (which gives us 5, plus the 2, making 7) or the 4 and the 2 first (which would give us 6, plus the 1, making 7). We'll get the same answer either way.

$$(1 + 4) + 2 \qquad or \qquad 1 + (4 + 2)$$
$$5 \;\; + 2 = 7 \qquad\qquad 1 + \;\; 6 \;\; = 7$$

The parentheses show which numbers are added together first. We'll talk more about parentheses in the next lesson.

Likewise, it doesn't matter in 2 x 3 x 4 if we first multiply 2 x 3 to get 6, and then times that 6 by 4 for 24, or if we first multiply 3 x 4 to get 12, and then multiply that by 2 to get 24. We end up with 24 either way.

$$(2 \times 3) \times 4 \qquad or \qquad 2 \times (3 \times 4)$$
$$6 \;\; \times 4 = 24 \qquad\qquad 2 \times \;\; 12 \;\; = 24$$

We call this property of addition and multiplication the **associative property**, which basically says it doesn't matter how we group neighboring numbers in an addition or multiplication problem.

It's important to note that subtraction and division are *not* associative. Consider 8 ÷ 4 ÷ 2. If we divide 8 by 4 first, we'd get 2. That answer, 2, divided by 2 then leaves us with 1.

If, on the other hand, we divided 4 by 2 first, we'd have 2, and then 8 divided by that would leave us 4.

$$(8 \div 4) \div 2$$
$$2 \;\; \div 2 = 1$$

$$8 \div (4 \div 2)$$
$$8 \div \;\; 2 \;\; = 4$$

Notice that the answers are *not* the same. That's because division is *not* associative. The grouping in division matters. Unless there are parentheses changing the grouping, division is solved from left to right.

$$(8 \div 4) \div 2$$
$$2 \;\; \div 2 = 1$$

Subtraction, like division, is *not* associative.

$$(8 - 4) - 2$$
$$4 \;\; - 2 = 2$$

$$8 - (4 - 2)$$
$$8 - \;\; 2 \;\; = 6$$

As with division, unless there are parentheses changing the grouping, solve subtraction from left to right.

$$(8-4)-2$$
$$4 -2 = 2$$

Why left to right instead of right to left? Solving left to right is the convention, or rule, we've adopted. By having a standardized order, everyone knows what is meant by 8 – 4 – 2. Since we are used to reading from left to right, that order makes a lot of sense.

Identity Property of Multiplication and Addition

Addition and multiplication also each have what we call an **identity property**. And what is an identity? Well, think about what the word "identity" means. It's referring to *who* someone is, right? In math, it's referring to two things that may seem different but that are really the same thing (i.e., they have the *same identity*).[6] Let's take a look at what this means in addition and multiplication.

The **identity property of multiplication** describes the fact that any number times 1 equals itself, and that 1 times any number equals that number. So, in other words, 6 x 1 and 1 x 6 both equal 6. Now, "6 x 1" may *seem* different than "6," but they *mean* the same thing.

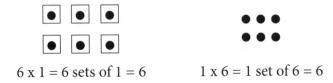

6 x 1 = 6 sets of 1 = 6   1 x 6 = 1 set of 6 = 6

This seems pretty obvious, but you'll be surprised at how often we'll use this property to help us solve problems (especially once we get to fractions).

Since multiplying by "1" doesn't change the value of a number, we call "1" the **multiplicative identity**.

The **identity property of addition** describes the fact that adding 0 to any number doesn't change its value. So, in other words, 0 + 6 and 6 + 0 both equal 6. Again, "0 + 6" may *seem* different than "6," but they *mean* the same thing.

6 + 0 = 6

0 + 6 = 6

Again, while this seems pretty obvious, understanding that adding 0 doesn't change a value is a pretty foundational concept to addition. Since adding "0" doesn't change the value of a number, we call "0" the **additive identity**.

Summarizing the Properties

Commutative property of addition and multiplication

Order doesn't matter

Addition	Multiplication
1 + 2 = 2 + 1	2 x 3 = 3 x 2
3 = 3	6 = 6

Associative property of addition and multiplication

Grouping doesn't matter

Addition	Multiplication
(1+2) + 3 = 1 + (2+3)	(2x3) x 4 = 2 x (3x4)
3 + 3 = 1 + 5	6 x 4 = 2 x 12
6 = 6	24 = 24

The parentheses show which numbers are added or multiplied together first.

Identity property of multiplication

Multiplying by 1 doesn't change the value

2 x 1 = 2

Identity property of addition

Adding 0 doesn't change the value

2 + 0 = 2

You can test these properties yourself with any objects you have around the house. These properties describe what actually happens in real life.

Applying the Commutative and Associative Properties

Let's say you needed to add a whole column of numbers together mentally. Because you know that addition is commutative and associative, you know you can mentally rearrange and group the numbers together to add them up easier. For instance, if asked to add 5 + 8 + 2 + 5, you could mentally rearrange or group them into what makes 10. You know 5 + 5 equals 10, and 8 + 2 equals 10. Add those both together, and you have 20.

On the other hand, if asked to solve 21 − 7 − 3, you would need to subtract each number in order, from left to right, to get the correct answer, as subtraction is *not* commutative or associative.

Notice that, if you instead subtracted 3 from 7 first, you would get 4. Subtract 4 from 21, and you'd have 17...which is *not* the correct answer.

The order in which we subtract matters! Since the agreed-upon order is to solve from left to right, 11 is the correct answer.

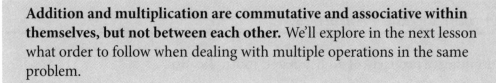

Addition and multiplication are commutative and associative within themselves, but not between each other. We'll explore in the next lesson what order to follow when dealing with multiple operations in the same problem.

Applying the Identity Property of Multiplication and Addition

If multiplying any number by itself equals that number, and if division is the opposite of multiplication, it follows that if we divide any number (0 excluded, as we can't divide by 0 — we'll discuss that more in 4.4) by 1, we'll end up with the number we started with.

$$2 \times 1 = 2 \quad \text{and} \quad 2 \div 1 = 2$$

(Think about it. If you divide 2 bags of candies by 1 person, that person gets both bags . . . and probably a tummy ache!)

Likewise, if we divide any number by itself (0 excluded again), we'll end up with 1.

$$2 \times 1 = 2 \quad \text{and} \quad 2 \div 2 = 1$$

(Think about it. If you divide 2 bags of candies by 2 people, each person gets 1 bag . . . and a little less of a tummy ache.)

Do you see how we can use one property (i.e., characteristic of how God governs an aspect of creation) to make additional observations about math? While these observations about division may not seem important right now, we'll use them extensively once we begin our exploration of fractions.

And as for addition, while you may not hear the term "identity property of addition" often, you often use the knowledge that adding 0 doesn't change the value! For example, when you add 20 + 5 and know your answer is 25, you've just applied the identity property of addition (0 + 5 = 5, so you know that the 0 in 20 plus 5 leaves you with 5 ones). You'll use the knowledge that 0 added to a number doesn't change the number even more as you get into upper-level math and work with unknown numbers.

Worldview Matters — Be on Guard

A property is a description of one of the ordinances by which God sustains the universe. Unfortunately, however, properties are often presented as some sort of independent fact instead. Here's an example of a property presentation:

> "A property describes the way something is. We can't change properties. We are stuck with properties because they are what they are."[7] — Saxon's Algebra 2

Notice this quote states a truth: We cannot change properties. Yet "they are what they are" gives the impression that math properties sustain themselves, leaving us viewing a property as something *independent* from anything or anyone else.

In contrast, in a biblical worldview, far from being independent from God, a property is entirely dependent on God! God's whole message is one of dependence — the Bible teaches us that we are dependent upon Him for our every breath. Satan, on the other hand, urges us that we can be independent from Him. Back in the Garden of Eden, part of his temptation to Eve was to ". . .be as gods, knowing good and evil" (Genesis 3:5). Wherever you are — including in a math class — be on guard against subtle *independent* thinking.

Keeping Perspective

As you review the **associative**, **commutative,** and **identity properties** of addition and multiplication, keep in mind that properties are truths about the ordinances God holds in place.

3.6 Conventions — Order of Operations

It's time now to consider a very important mathematical convention: the order of operations. **Conventions**, unlike properties, are agreed-upon protocols or rules that aid us in communication. Much as we could use different grammar rules (and do, in other languages), we could use different conventions in math. Math rules, however, are a little more unforgiving than grammar rules. While people can often still guess their way through a poorly punctuated sentence, using the wrong convention in math will result in a totally different answer!

The Order of Operations

What happens when we need to perform more than one operation on a group of numbers? How do we know which operation to perform first?

Consider $4 + 12 \div 2$. What does this equal? Do we need to divide 12 by 2, and then add 4, in which case we would get 10, or do we need to add the 4 and 12 and then divide the sum (16) by 2, in which case we would get 8?

To avoid this sort of vagueness, we follow an agreed-upon order. This convention is known as the **order of operations** and is very important to memorize.

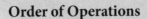

> **Order of Operations**
> 1. Solve anything within parentheses first, using the following order (which is the same order we'll follow outside of parentheses):
>   a. Solve exponents and roots, from left to right.*
>   b. Multiply or divide, from left to right.
>   c. Add or subtract, from left to right.
> 2. Next, solve exponents and roots, from left to right.*
> 3. Now multiply or divide, from left to right.
> 4. Lastly, add or subtract, from left to right.
>
> *There will not be any exponents or roots in your problems until they are presented in 16.1 and 16.2.

Thus, in 4 + 12 ÷ 2, we would do the division first, and then the addition, giving us 10. Below are a few examples of how this convention works.

Example: Solve 36 ÷ 3 + 3 x 3 − 10 ÷ 5

There are no parentheses, so we first work the division or multiplication in the order it appears, from left to right.

$$\begin{aligned}
\mathbf{36 \div 3} + 3 \times 3 - 10 \div 5 \\
\mathbf{12} + 3 \times 3 - 10 \div 5 \\
\\
12 + \mathbf{3 \times 3} - 10 \div 5 \\
12 + \mathbf{9} - 10 \div 5 \\
\\
12 + 9 - \mathbf{10 \div 5} \\
12 + 9 - \mathbf{2}
\end{aligned}$$

Now we do the addition and subtraction in order, from left to right.

$$\begin{aligned}
\mathbf{12 + 9} - 2 \\
21 - 2 = 19
\end{aligned}$$

Example: Solve 36 ÷ (3 + 3 x 3) − 10 ÷ 5

Notice that this is the *same* problem, only this time we have parentheses! Our first step is to solve what's inside the parentheses, applying the same order we would if the entire problem was just what was inside the parentheses. So here, we'll multiply before we add.

$$\begin{aligned}
36 \div (3 + \mathbf{3 \times 3}) - 10 \div 5 \\
36 \div (3 + \mathbf{9}) - 10 \div 5 \\
36 \div (\mathbf{12}) - 10 \div 5
\end{aligned}$$

Now that we've solved inside the parentheses, we'll solve the division or multiplication in order, from left to right.

$$\begin{aligned}
\mathbf{36 \div 12} - \mathbf{10 \div 5} \\
3 - 2
\end{aligned}$$

Now it's time for the addition or subtraction — in this case, all we have left is subtraction.

$$3 - 2 = 1$$

Combining the Conventions:
Using Parentheses for Multiplication

Remember from 3.3 that "•" and putting a number next to the outside of a parenthesis are both ways of representing multiplication. 2 x 3 can be written 2 • 3 and 2(3).

Can you see now why putting a number next to a parenthesis is helpful? It comes in handy when we need to use parentheses to show which numbers need to be added first. It saves a step to write 2(4 + 6 − 8) instead of writing 2 x (4 + 6 − 8), although they both mean the same thing.

$$2(4 + 6 − 8) = 2(10 − 8) = 2(2) = 4$$

Keeping Perspective — God's Capacity and Ours

As you review the order of operations and use of parentheses, keep in mind that conventions are agreed-upon ways of expression — part of the language used in mathematics to communicate clearly.

3.7 Chapter Synopsis

We've covered/reviewed a lot of territory this week. Here are a few important takeaways:

- Different strategies — such as making 10 (using the **complement** of the number), turning subtraction into addition, and rounding — can help us **add and subtract mentally.**

- **Multiplying** is a shortcut for adding the same number over and over again; **division** is a shortcut for subtracting the same number over and over again. They both work because of God's faithfulness in holding all things together. We can represent both operations different ways, including ÷ and $\overline{)}$ for division, and these different notations for multiplication:

    2 x 3
    2 • 3
    2(3)

- We use terms to refer to different parts of multiplication and division problems: **multiplicand, multiplier, factor, product, dividend, divisor,** and **quotient.**

- **Properties** are truths about the ordinances God holds in place. Addition and multiplication, unlike subtraction and division, are **commutative** (order doesn't matter) and **associative** (grouping doesn't matter). In multiplication, multiplying by 1 (the **multiplicative identity**) doesn't change the value (we call this the **identity property of multiplication**). In addition, adding by 0 (the **additive identity**) doesn't change the value (we call this the **identity property of addition**).

■ **Conventions** are agreed-upon ways of **expression. Parentheses** are a convention used to **show portions** of an equation that need to be dealt with first. We follow an agreed-upon **order of operations** so we all know **what to solve first.**

Order of Operations

1. Solve anything within parentheses first, using the following order (which is the same order we'll follow outside of parentheses):

   a. Solve exponents and roots, from left to right.*
   b. Multiply or divide, from left to right.
   c. Add or subtract, from left to right.

2. Next, solve exponents and roots, from left to right.*

3. Now multiply or divide, from left to right.

4. Lastly, add or subtract, from left to right.

*There will not be any exponents or roots in your problems until they are presented in 16.1 and 16.2.

Now for the biggest takeaway. . . .

You've now learned the four basic operations: addition, subtraction, multiplication, and division. We've seen how we can memorize facts and develop methods/tools because creation operates consistently. We've discussed the power and faithfulness of God that these operations reveal. As you review today, think about whether you're remembering who God is today. Are you living like you believe He is the Faithful, Almighty Creator and Sustainer of all?

[CHAPTER 4]

Multi-digit Multiplication and Division

4.1 Multi-digit Multiplication

In this chapter, we're going to take a look at the **algorithms**, or step-by-step methods, we use to multiply and divide. While I'm sure you're already familiar with these methods, we're going to take a closer look at them and seek to understand the mechanics behind them. We'll also take a look at some different methods and continue to build problem-solving skills. Our journey will reinforce once again that math methods are not independent entities of their own, but useful tools for describing the consistencies God created and sustains.

The Concept behind the Methods

Rather than jumping to the method we typically use to multiply, let's first explore the concept of what we're doing. Suppose we needed to multiply 9 x 12 (which is another way of saying to take 12 nine times) but had only memorized our multiplication facts through 10.

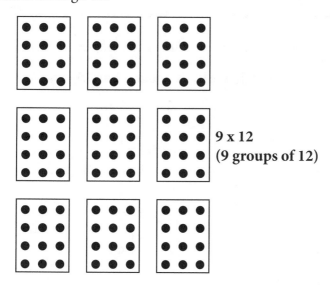

9 x 12
(9 groups of 12)

Since 12 equals 10 + 2, we could think of each group of 12 as a group of 10 and a group of 2.

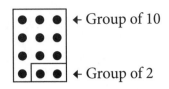

We could find the product of 9 x 12 by multiplying 9 x 10 and 9 x 2 and adding their products together.

9 x 12 (9 groups of 12) =
9 x 10 (9 groups of 10) + 9 x 2 (9 groups of 2)
90 + 18 = 108

Notice how we just solved this multi-digit problem by treating each digit as a separate problem. This, in essence, is all multiplication methods do — except they use place value to automate the process more for us.

Keeping this in mind, let's take a look now at two different multiplication methods and see their different approaches.

Gelosia Method

Here are some basic steps to find 9 x 12 using the Gelosia method:

1. Draw two squares and divide those squares into triangles.

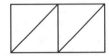

2. Write the two numbers being multiplied above and to the right of the squares.

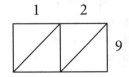

3. Write the answer to 9 x 2 in the first square.

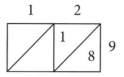

4. Write the answer to 9 x 1 in the second square.

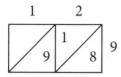

5. Add the numbers in the little triangles diagonally from right to left. In the first diagonal, we have 8, so we write 8 underneath the first triangle.

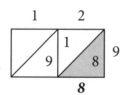

While you'll likely never use the Gelosia method after this lesson, it's helpful in understanding what we're really doing when we use our traditional method . . . and how both are merely techniques for describing a consistency God is holding together.

Note: The triangles are shaded in the pictures to illustrate which numbers were added together.

The triangles with the 1 and the 9 are both within the same diagonal, so we add them together and get 10. We record the 10 as pictured below, using the diagonals to keep track of the place value for us.

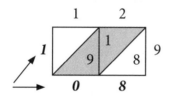

6. Reading the number from left to right we get our answer of 108.

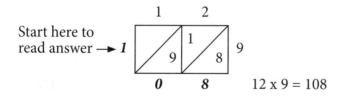

Now let us take a closer look at what happened here. In the Gelosia method, the squares and triangles separated our ones, tens, and hundreds. When we multiplied 9 x 2, we wrote the 1 part of 18 in the upper triangle because that 1 really represented 1 group of 10. Likewise, when we multiplied 9 x 1, we wrote the 9 in the tens section of the second square because the 9 really represented 9 tens (the 1 in 12 stands for 1 ten, so when we multiplied 9 x 1, we were really multiplying 9 x 1 ten). When we added up the triangles diagonally, we were really adding up our ones and our tens to find our total answer.

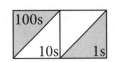

Notice how the Gelosia method is a simple way to keep track of place value while we multiply. Rather than thinking "9 x 2" and "9 x 10," we thought of "9 x 2" and "9 x 1." The method made sure we wrote the answer to 9 x 1 in the tens place, since that 1 really stood for 1 ten.

To solve problems involving more digits in the Gelosia method, we would just draw more squares. In the figure below, since we have a three-digit number, we draw an extra row of squares. Notice that whenever we got an answer of 10 or greater when adding the diagonals, we mentally added the tens digit to the next diagonal, thus letting our diagonals keep track of place value for us.

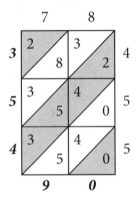

78 x 455 = 35,490

Traditional Method

A very similar thing happens in the traditional method with which most of us are familiar. Rather than jumping right to the final way we multiply, let us *develop the method step by step* by breaking apart 9 x 12 into smaller problems (9 x 2 and 9 x 10) and solving.

$$\begin{array}{r} 1\,2 \\ \times\ \ 9 \\ \hline 1\,8 \\ +9\,0 \\ \hline 1\,0\,8 \end{array}$$
 (9 x 2) Ones column
 (9 x 10) Tens column

Now, when we have a task to do over and over again, it is always good to find ways to save steps! Instead of writing 18 under the line, we could write only the 8 under the line, and put the 1 up on top of the 1 in 12. This reminds us we have 1 ten we will need to add to our tens column.

$$\begin{array}{r} {\scriptstyle 1} \\ 1\,2 \\ \times\ \ 9 \\ \hline 8 \end{array}$$

Now we can work the tens column — 9 times 1 ten equals 9 tens, or 90, plus the 1 ten we need to add gets us to 10 tens, or 100. If we add 100 and 8, we will get the answer, 108.

$$
\begin{array}{r}
^{1}1\,2 \\
\times\ \ 9 \\
\hline
8 \\
+\ \mathbf{1\,0\,0} \\
\hline
\mathbf{1\,0\,8}
\end{array}
$$

Now there is still a way to save steps. Instead of writing the 100 underneath the 8 and adding it, we could have written 10 next to the 8. Since the 10 is in the tens column, it stands for 10 groups of 10, or 100.

$$
\begin{array}{r}
^{1}1\,2 \\
\times\ \ 9 \\
\hline
\mathbf{1\,0\,8}
\end{array}
$$

To multiply a two-digit number by a two-digit number, we could follow a similar procedure, except we would have to break it down a little further still. Let's say we want to multiply 29 x 12.

We can think of this as 9 x 12 and 20 x 12 (remember, the 2 in 29 represents 2 sets of 10, or 20). Now we can break down both of these problems.

Let's start by multiplying 9 x 12 just as we did before.

$$
\begin{array}{r}
^{1}1\,2 \\
\times\ 2\,9 \\
\hline
\mathbf{1\,0\,8}
\end{array}
$$

Now let's multiply 20 by 12. We can do this the same way we did with the 9, except we need to first write a 0 underneath to make sure we begin writing our answer in the tens column. We should also scratch out the 1 we carried last time so as not to accidentally add it to our tens column again.

$$
\begin{array}{r}
^{\cancel{1}} \\
1\,2 \\
\times\ 2\,9 \\
\hline
1\,0\,8 \\
\mathbf{0}
\end{array}
$$

Now we can start multiplying. We know 2 x 2 equals 4, so we write a 4 next to the zero. Notice how we're writing the 4 automatically in the tens column — the method is keeping track of place value for us, so even though we're thinking 2 x 2, we're really finding the answer for 20 x 2, which is 40.

$$
\begin{array}{r}
^{\cancel{1}} \\
1\,2 \\
\times\ 2\,9 \\
\hline
1\,0\,8 \\
\mathbf{4\,0}
\end{array}
$$

Next we multiply the 2 by the 1, and put a 2 next to the 40 below. Notice again how we're writing the answer in the hundreds column. We can think "2 x 1" instead of "20 x 10" (which is what we're really finding) because the method is automatically putting our answer in the correct place.

$$
\begin{array}{r}
1\ 2 \\
\times\ 2\ 9 \\
\hline
1\ 0\ 8 \\
2\ 4\ 0
\end{array}
$$

Now we add up our products to find our total product.

$$
\begin{array}{r}
1\ 2 \\
\times\ 2\ 9 \\
\hline
1\ 0\ 8 \\
+\ 2\ 4\ 0 \\
\hline
3\ 4\ 8
\end{array}
$$

We can use this algorithm, or method, to multiply numbers, regardless of the number of digits.

■ **Write the numbers vertically, making sure the digits line up in the correct places** (the ones columns, tens columns, etc.). While technically using the convention we've been following, the multiplicand (number being multiplied) goes on top and the multiplier (amount of times we're taking the other number) goes on bottom, it doesn't really matter due to the commutative property of multiplication. You will find it easiest to put the number with the least number of digits on the bottom.

■ Multiply the **bottom number** by the **top number** digit by digit, from right to left. Multiply the ones digit in the bottom number by the entire top number, then the tens digit in the bottom number by the entire top number, etc. This is essentially breaking down the multiplication into a series of smaller multiplications. To keep track of place value, though, add a zero when you multiply the tens place of your bottom number so as to write your result in the tens place, two zeros when you multiply the hundreds place so as to write your result in the hundreds place, etc.

For example, if you have a three-digit number on the bottom, record the first digit's multiplication in the ones column, but add one zero before multiplying the second digit (so your answer will start in the tens place) and two zeros before starting the third digit (so your answer will start in the hundreds place). The easiest way to make sure you are starting with the correct number of zeros is to remember that you should start with the **same number of zeros as there are digits to the right of the digit in the bottom number you are currently multiplying.** (If you are multiplying the 5 in 2,561, two zeros would get your answer in the correct place; for the 2 in 2,561, you would need three zeros to get in the correct place.) This ensures your answer is being recorded in the same place as the digit you are multiplying.

90 | PRINCIPLES OF MATHEMATICS 1

```
      1 2 3
  x 2,5 6 1        When multiplying ones,
          3        start here.


      1 2 3
  x 2,5 6 1        When multiplying tens,
      1 2 3
  7 3 8 0          put the 0 as a place marker and start where bolded.
                   Notice how you start underneath the digit in the
                   bottom number you are multiplying.
      1 2 3
  x 2,5 6 1        When multiplying hundreds,
      1 2 3
  7 3 8 0
6 1 5 0 0          put the 0s as a place marker and start where bolded.
                   Notice how you start underneath the digit in the
                   bottom number you are multiplying.
      1 2 3
  x 2,5 6 1        When multiplying thousands,
      1 2 3
  7 3 8 0
6 1 5 0 0
2 4 6 0 0 0        put the 0s as a place marker and start where bolded.
                   Notice how you start underneath the digit in the
                   bottom number you are multiplying.
```

■ **Add the results you obtained from each digit to find the total product.**

```
        1 2 3
    x 2,5 6 1
        1 2 3
      7 3 8 0
    6 1 5 0 0
  +2 4 6 0 0 0
    3 1 5,0 0 3
```

Keeping Perspective

The rules we follow when we multiply keep track of place value, thereby allowing us to break multi-digit multiplication problems we do not have memorized into a series of smaller problems we do have memorized. A multiplication method will work only if it accurately describes the way God causes objects to multiply. If God were not faithfully holding all things together, reducing multiplication to a method would be impossible!

4.2 The Distributive Property — Understanding Why

Today, it's time to explore and name a property that we unknowingly use when solving multi-digit multiplication problems using our traditional method. Before we see how the distributive property applies to multi-digit multiplication, though, let's take a look at it with a simple problem: 2(3 + 4).

The Distributive Property of Multiplication

Up until now, you've solved problems such as 2(3 + 4) by adding the numbers inside the parentheses, and then multiplying. If we do this, we get 2(7), which equals 14.

Add 3 + 4 first

2 sets of 3 + 4

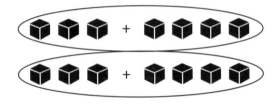

2(3 + 4)
2(7) = 14
14 total

Notice that we get the same answer, though, if, instead of adding the numbers inside the parentheses first, we multiply the 2 by each number within the parentheses and then add them together — **in other words, if we multiply by each addend and then add those results together.**

Multiplying the 2 by *each* number

2 sets of 3 and 2 sets of 4

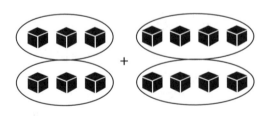

2 (3 + 4)
2(3) + 2(4)
6 + 8 = 14
14 total

Notice that when we added the parentheses first and then multiplied, we got the *same answer* (14) as when we multiplied by each of the addends in the parentheses and then added those results together. *Both approaches gave us the same answer.*

It turns out that, no matter what the specific problem, multiplication can be broken out between addends this way without affecting the result. We call multiplication's ability to be distributed between addends that make up a sum the **distributive property of multiplication**, commonly abbreviated to the distributive property. Remember, a property is a characteristic — something that holds true in every situation. It's essentially a way of naming an aspect of how God holds something together. Calling this property the distributive property makes sense since the word "distribute" means "to parcel out."[1] When we distribute, we're "parceling out" the multiplication over several addends.

The Distributive Property of Multiplication

If you distribute multiplication among addends and then add those results together, you'll get the same results as if you first did the addition and then multiplied by the sum.

$$2 (3 + 4) = 2 \bullet 3 + 2 \bullet 4$$

Note: It doesn't matter which side of the parentheses the multiplier is on. The following would also hold true:

$$(3 + 4)2 = 2 \bullet 3 + 2 \bullet 4$$

Using the Distributive Property of Multiplication

Later on in math, you'll learn many different ways the distributive property helps us solve and simplify problems. In the meantime, though, let's look at two simple ways this property helps us use multiplication.

In Multi-digit Multiplication

The principle of distributing multiplication to addends that make up a sum applies whether represented with parentheses or not. For example, would it surprise you to learn that you've been distributing every time you multiply multi-digit numbers? As we've already seen, because of the consistent way God causes objects to add repeatedly, we can break up multi-digit multiplications into a series of smaller multiplications.

$$\begin{array}{r} {}^{2}\;\; \\ 2\,9 \\ \times \;\; 3 \\ \hline 7 \end{array} \rightarrow 3 \bullet 9 = 27$$

$$\begin{array}{r} {}^{2}\;\; \\ 2\,9 \\ \times \;\; 3 \\ \hline 8\,7 \end{array} \rightarrow 3 \bullet 2 = 6 \text{, plus the 2 we are carrying}$$

equals 8. We write the 8 in the tens column, as we are dealing with 8 sets of 10.

When we multiply 29 by 3, we're really multiplying 3 times 9 and then 3 times 2 sets of 10, or 20. In other words, we're distributing the 3 to both the ones and the tens digit in 29.

Just as we can distribute multiplication between the ones and tens digits, we can distribute multiplication between the different components inside parentheses.

$$3(20 + 9)$$
$$3(20) + 3(9)$$
$$60 + 27 = 87$$

← The parentheses shows 29 broken into 20 (2 sets of 10, or 2 • 10) and 9 ones.

In Mental Math

You can also use the distributive property to aid in mental math. Say you needed to mentally multiply 4 by 48.

$$4(48)$$

One way to find this would be to mentally break it down into a series of addends, each one of which would be easier to mentally multiply.

$$4(40 + 8)$$
$$4(40) + 4(8)$$
$$160 + 32 = 192$$

Notice that this gives us the same answer as if we multiplied 4 directly by 48.

So, while you may not think about the distributive property of multiplication often, it applies in many, many places — including in our traditional multiplication method itself and in mental math!

> ## Keeping Perspective
>
> The **distributive property of multiplication** is a way of describing a characteristic of multiplication that holds true because of the consistencies God created and sustains and the notations we use to describe them. Because of this attribute of multiplication, we have the option of breaking down multiplication problems into smaller problems, whether on paper or mentally. While you have probably used this property for years without realizing it, it's important to understand, as you'll see it again later and learn how to use it further in describing the quantities and consistencies God has placed around it.

4.3 Multi-digit Division

As with a lot of what we've looked at so far, I imagine you have already studied how to divide multi-digit numbers. But if we don't understand the why behind the methods, it's all too easy to end up viewing that method itself as division or some sort of man-made truth rather than as one way we, using the creativity God gave us, keep track of place value. Let's look at *why* our division method works and *how* it leads us to the correct answer.

Dividing Manually

Let's start by first taking a look at a different way to solve the problem 48 ÷ 2 using circles as an illustration. I know — this is such a simple problem, you could probably solve it in your head. I purposefully chose an easy problem because we want to break it apart and see if we can find a simple method to solve it that can transfer to paper and help us work with harder problems.

Rather than trying to divide 48 directly by 2, look at the 48 circles as 4 groups of 10 and 8 individual circles.

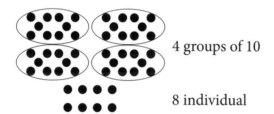

Next, divide the 4 groups of 10 by 2. Notice that just as 4 divided by 2 equals 2, 4 groups of 10 divided by 2 equals 2 groups of 10.

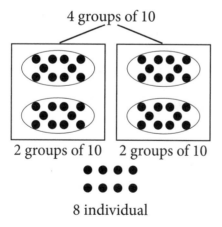

Now divide the 8 individual circles into two piles, getting 4 in each pile, and add these piles to the ones obtained from dividing 4 groups of 10, forming 2 piles of 24. 48 divided by 2 equals 24.

$$48 \div 2 = 24$$

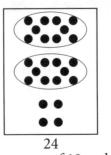
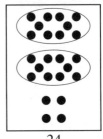

24
2 groups of 10 and
4 individual

24
2 groups of 10 and
4 individual

Dividing with the Traditional Division Method

The traditional division method completes this same basic process on paper. To solve 48 ÷ 2 using it, start by writing the 48 and the 2 separated by a partial box that looks a bit like a stair. This "stair" format is a popular way to keep the numbers separated. As with the division sign itself, it's just one convention for representing division.

$$2 \overline{)48}$$

Now we're going to start by dividing our sets of 10. What we're really trying to find out is *how many sets of 10* we can make if we divide 4 sets of 10 by 2. We can make 2 sets of 10, or 20. To represent this, put a 2 over the 4 in 48. Notice how we're writing the 2 in the tens place *because it represents sets of 10*.

$$2 \overline{)48}^{\,2}$$

Now multiply 2 x 2 (sets of 10) to find out the quantity divided so far. The purpose of this multiplication is to see what of our tens have been divided: 2 sets of 10 times 2 equals 4 sets of 10, or 40. Write 4 underneath the 48 *in the tens column* to represent the 4 sets of 10 that have been divided.

$$\begin{array}{r} 2 \\ 2\overline{)48} \\ 4 \end{array}$$

Now subtract the 4 we've divided from the total we're dividing to see how many tens we have left to divide (0).

$$\begin{array}{r} 2 \\ 2\overline{)48} \\ -\,4 \\ \hline 0 \end{array}$$

We can now deal with the ones place, or the individual units. Write the ones we have to divide (8) next to any tens we still need to divide (in this case, 0).

$$\begin{array}{r} 2 \\ 2\overline{)4\,8} \\ -\,4 \\ \hline 0\,8 \end{array}$$

We now divide 8 by 2 and write the answer (4) in our answer next to the 2.

$$\begin{array}{r} 2\,4 \\ 2\overline{)4\,8} \\ -\,4 \\ \hline 0\,8 \end{array}$$

Once again, use multiplication to see what has been divided. Multiply 2 x 4 and write an 8 underneath to represent the number used to make 2 groups of 4. Subtract it out of the 8 to see that 0 remain undivided. The answer to 48 ÷ 2 is 24.

$$\begin{array}{r} 2\,4 \\ 2\overline{)4\,8} \\ -\,4 \\ \hline 0\,8 \\ -\,8 \\ \hline 0 \end{array}$$

The division method and rules to divide numbers on paper break a division problem into little steps, letting us divide the groups of 10 first and then the single items. Because this method keeps track of place value for us, we can divide groups of 10 (or 100, 1,000, etc.) as easily as we would individual items.

Let's walk through one more division problem and make sure we thoroughly understand how the method works.

$$5\overline{)3\,6\,5}$$

In this problem, our divisor (5) is larger than the first digit (3) of our dividend (365). What do we do?

It may be tempting to think we can divide the 3 by the 5, since we know the 3 represents 3 sets of 100, or 300. However, in order to use our division method and not have to think through each problem on our own, we need to find the number of *hundreds* we can make, and we cannot divide 300 into 5 sets *of a hundred*.

So we move on to the next digit and ask what 36 divided by 5 equals (note that the 36 represents 36 sets of 10, or 360, but because the method is keeping track of place value, we don't have to think about its place). We know 5 x 7 = 35, so we put a 7 in the *tens* column of the quotient (which is saying that we can make *7 sets of ten*) and do the subtraction below to see that we have 1 ten remaining to divide. (Note: We don't necessarily have to think about writing the 7 in the tens place. If we just write it over the top of the last digit we're currently dividing, the method will automatically keep track of the place value for us.)

$$\begin{array}{r} 7 \\ 5\overline{)3\,6\,5} \\ -\,3\,5 \\ \hline 1 \end{array}$$

Term Connection
Remember, we use terms to refer to different numbers in a division problem. The dividend is the number being divided, the divisor is the number we're dividing by, and the quotient is the answer.

$$\text{divisor}\overline{)\,\text{dividend}}^{\,\text{quotient}}$$

dividend ÷ divisor = quotient

We now have 1 set of 10 still to deal with. We can't make any more sets of 10 with it (10 divided by 5 doesn't make any sets of 10), so we're going to deal with it *at the same time we deal with our ones column*, looking at how many sets of 1 instead of sets of 10 it can be broken into. We bring down the 5 to show we have 5 more ones. The new quantity, 15, is the quantity we now need to divide.

$$
\begin{array}{r}
7 \\
5\overline{)3\,6\,5} \\
-\ 3\,5 \\
\hline
1\,5
\end{array}
$$

If we divide 15 by 5, we get 3. So we write that in the ones place up in the quotient and do the math below to see that we have divided our entire dividend.

$$
\begin{array}{r}
7\,3 \\
5\overline{)3\,6\,5} \\
-\ 3\,5 \\
\hline
1\,5 \\
-\ 1\,5 \\
\hline
0
\end{array}
$$

As with multiplication, there are other division methods we could use. Our current method is just one way to keep track of place value.

The Rule

Below is the "algorithm," or step-by-step method, written out as a rule.

1. Starting at the left, divide the first digit of our dividend (number being divided — the one underneath the stair step) by our divisor (the number we're dividing by). If we can't divide it, we couple it with the next digit to the right until we find a quantity we can divide. Once divided, we write the answer in the quotient over the top of the right-most digit in the dividend we used when dividing.

2. Multiply the digit in the quotient we just wrote down by the divisor, writing the answer underneath the dividend. This shows the amount we have just divided.

3. Subtract the amount we just divided from the amount we had to divide in the first place. This remainder gets coupled with the next digit in the dividend and divided to find the next digit in our quotient.

4. We continue this process until we have divided the entire dividend.

Although this process should already be so habitual that you can do it without thinking, and while you may sometimes do the steps in your head, it's important to know what it's really doing — how we're really just breaking down our dividend and using place value to automate the process.

98 | PRINCIPLES OF MATHEMATICS 1

Keeping Perspective

Division methods are good examples of how we can use the creativity and reason God gave us to observe and record His creation. Although the word "work" has a negative connotation in today's society, God created man to work! Even before the Fall, God gave Adam the task of ruling over His creation and subduing the earth.

> *And God blessed them, and God said unto them, Be fruitful, and multiply, and replenish the earth, and subdue it: and have dominion over the fish of the sea, and over the fowl of the air, and over every living thing that moveth upon the earth (Genesis 1:28).*

> *And the LORD God took the man, and put him into the garden of Eden to dress it and to keep it (Genesis 2:15).*

I find it encouraging that God didn't just give mankind tasks to do — He also gave us the ability to do them. He equipped us with the capacity to develop techniques such as division methods to help us in the various tasks we'd encounter. There's a common, but very true, saying that where God calls, He equips.

4.4 More Multi-digit Division, Remainders, and Operations with Zero

Yesterday, we explored our traditional long-division method and saw how it keeps track of place value as we divide. Today, we're going to look a little more at this method and at some of the different situations we might encounter when we use division as a tool.

Multi-digit by Multi-digit Division

Companies use division and other math concepts to help them determine how much to charge for their products. For instance, suppose you were a carpenter, and you spent $1,189 to produce 41 dollhouses. You want to find out how much each dollhouse costs you. What would you do?

You would use division! Even though you don't want to *physically divide* $1,189 by 41, you want to *divide the cost* between all the *dollhouses* to see how much each one really cost. Let's do this division together.

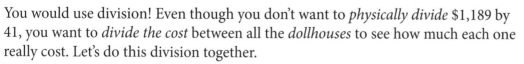

Working from left to right, we see we can't divide our thousands or hundreds, but we can divide our tens. We have 118 sets of 10. How many times can they be divided by 41? Since this isn't a division fact we have memorized, we will have to figure out the answer another way.

It's often easiest to think of the problem in terms of multiplication, rounding as we go. We need to figure out how many times 41 is contained in 118. If we round 41 to 40, we can more easily estimate in our heads — 40 x 2 equals 80, and 40 x 3 equals 120. Since 120 (the result of multiplying 40 by 3) is greater than the 118 we're trying to divide, let's assume we're only going to be able to divide by 2 (our estimate showed we'd need 120 to divide by 3). We'll write a 2 above the tens place to represent that we were able to form 2 sets of 10, perform the multiplication, and do the subtraction below to show how much we still have remaining to divide.

$$\begin{array}{r} 2 \\ 41\overline{)\$1,189} \\ -82 \\ \hline 36 \end{array}$$

If we had ended up with 41 or greater, we would know 118 could divide into 41 more than 2 times. But in this case, 2 was the correct number, as we were left with a number less than 41.

We now bring down our ones place (the 9) and ask ourselves how many times 41 fits into 369. Again, we may have to give an estimated guess, checking ourselves with multiplication until we find the answer — 9.

$$\begin{array}{r} \$29 \\ 41\overline{)\$1,189} \\ -82 \\ \hline 369 \\ -369 \\ \hline 0 \end{array}$$

It costs us $29 per dollhouse.

Remainders

What happens if we can't divide a number evenly? Let's say, for instance, that we have 11 cookies we want to divide by 4 people.

$$\begin{array}{r} 2 \\ 4\overline{)11} \\ -8 \\ \hline 3 \end{array}$$

Oops! We have a problem, don't we? We *cannot* divide 3 evenly into 4. To show that we had a quantity left over that did not divide evenly, we will write "r3" next to the 2 — this stands for "remainder of 3" and means that each person would get 2 cookies, and we'd have 3 left over.

$$\begin{array}{r} 2\,r3 \\ 4\overline{)11} \\ -8 \\ \hline 3 \end{array}$$

In a few chapters, we'll cover how to use fractions and decimals to record remainders, but use remainders for now.

Working with Zero

Before we move on from division, we need to talk about division and zero. First, though, let's look at the other operations and what happens when working with zero.

Addition and Subtraction: What happens when we add or subtract by zero? Absolutely nothing! If we add or subtract nothing to a number, nothing happens. *Remember, we already saw this with the identity property of addition!*

> 2 + 0 = 2
> 2 − 0 = 2

Multiplication: What about if we multiply by 0? Here we need to sit back and think about what multiplication is representing.

> 2 x 0 means 2 sets of 0 . . . which would obviously be 0.
> 0 x 2 means 0 sets of 2 . . . which would also be 0.

Zero times any number equals zero.

So if there were an item that cost $5, and we chose not to buy any of them, we would spend nothing.

> $5 x 0 = 0

Division: How about division? Let's think through what we're asking in a division problem.

> 0 ÷ 2 means 0 (nothing) divided into 2 . . . which would be 0. We can check this using multiplication: 0 x 2 = 0.

Zero divided by any number equals zero.

What if we try to divide *by* zero? In this case, we run into a problem. Remember, division is the opposite of multiplication — 12 ÷ 4 equals 3, and 4 x 3 (our divisor times our quotient) equals 12 (our dividend). One expresses repeated addition; the other expresses repeated subtraction. Keeping this in mind, think about what 2 ÷ 0 would equal. You might think it should equal 2, but if 2 ÷ 0 equaled 2, then 0 x 2 (our divisor times our quotient) should equal 2, and it doesn't. It equals 0.

For another example, take 0 ÷ 0. What would this equal? Think about the opposite multiplication. What times 0 equals 0? Well, 1 x 0 = 0, 2 x 0 = 0, 3 x 0 = 0 . . . so 0 ÷ 0 could equal 1 . . . or 2 . . . or 3 . . . or — the list goes on and on. So how could we define 0 ÷ 0? The answer could be *any* number.

Dividing a number by 0 is considered an invalid operation. If you ever come across a problem that makes you divide by 0 (including 0 ÷ 0), write "Unsolvable, as cannot divide by 0." Knowing that you can't divide by zero will prove important in upper-level math.

Keeping Perspective — Combining Tools

As we continue to explore math, we're going to continue to build on the information we've learned so far, combining it with new tools to help us better record different situations we encounter when using math to record real life. When multiplying multi-digit numbers, the method used addition (adding the results of each digit). When dividing, we use multiplication (multiplying to see how many tens, etc., we've divided so far) and subtraction (seeing if we have any tens, etc., left to divide).

Math builds on itself. The more complex concepts in math can all be broken down into simple pieces. And all of it rests on the consistency God created and sustains.

4.5 More Problem-Solving Practice/ Enter the Calculator

In real life, situations don't come neatly located under an "addition" or "multiplication" heading — they often require a variety of math concepts to solve . . . and a thorough knowledge of which concept, i.e., mathematical "tool," to use when. The solution — and the tools necessary to find it — becomes more obvious when you break apart problems using the problem-solving steps we went over in 2.6. To help solidify these steps, we're going to take today to review them. Along the way, we'll look at checking our work and introduce the calculator, which you'll be allowed to use occasionally as we move into more in-depth problems.

Take the time now to develop good problem-solving skills — you'll be thankful you did when you need to solve a real-life problem!

Example: Three businesses, one worth $104,000, another $42,000, and the third $210,000, get bought out by two men and merged into one business, with each man owning an equal share of the business. What is each person's share?

1. Define — What do we know? What are we trying to find out?

We know the starting values of the businesses. We know they got merged into one business.

> Business 1 (or B_1) — $104,000
> Business 2 (or B_2) — $ 42,000
> Business 3 (or B_3) — $210,000

We want to find out the share of each of two partners. We know each partner's share will be equal.

2. Plan — We need to break this problem down into steps and look at how the information we have relates.

In order to find out each of the partner's shares, we need to find out what the total value of the final merged business is. We can find the total by adding up the total from all the businesses that merged.

Business 1 + Business 2 + Business 3 = New Business
or
$B_1 + B_2 + B_3 = B_4$

Then, since we know each share is equal, we just need to divide the total by the number of partners to find each share.

New Business ÷ Number of Partners = Each Partner's Share
or
$B_4 ÷ P = S$

> It's not necessary to write out each step like this — the point is just to think through that we need to use addition to find the new business total and then division to figure out each partner's share.

3. Execute —

Step 1: Find the value of the merged business.

$B_1 + B_2 + B_3 = B_4$
$104,000 + $42,000 + $210,000 = $356,000

Step 2: Find each partner's share of the merged business.

$B_4 ÷ P = S$
$356,000 ÷ 2 = $178,000

4. Check — Does $178,000 seem reasonable, given the individual values of the businesses? Yes . . . if we had come up with an answer of a million dollars or $10,000, we would have wanted to check our math again. We can also check our division by multiplying 2 x $178,000 to make sure we didn't make a careless mistake.

Example: If there are about 10 words to a line and 20 lines to a page, about how many pages will be occupied by a 100,000 word novel?[2]

1. Define — What do we know? We know each page has 20 lines, and each line has 10 words.

What are we trying to find? We want to find out how many pages we'll have in a 100,000 word novel.

Lines on 1 page = 20 lines
Words on each line = 10 words
Total words = 100,000
Total pages = ?

2. Plan — Again, let's look at how the information we have relates. The relationships in this problem are not as obvious — we have to figure out how to find what we need using information that doesn't instantly connect.

We know our total words, but we don't know how many words fit on a page. However, we *do* know how many words fit in a line, so let's start by finding the total number of lines needed for our novel.

Total words ÷ words on each line = total number of lines

Again, the point with the planning step is to just think through what operation(s) we need to use to find the desired information — and it might take more than one!

Now, we also know the number of words that fit on a line. We can use that information to find the total pages needed!

Total number of lines ÷ lines on 1 page = total pages

Notice that **we worked from the relationships we did know to find the one we didn't.**

3. Execute —

Step 1: Find the number of lines needed.

Total words ÷ words on each line = total number of lines
100,000 ÷ 10 = 10,000

Step 2: Find the number of pages we would need.

Total number of lines ÷ lines on 1 page = total pages
10,000 ÷ 20 = 500

4. Check — Does 500 pages sound reasonable for 100,000 words? To answer that, let's think through the problem backward, using multiplication to check our work. We can see that each page would have 10 (number of words per line) x 20 (number of lines per page) or 200 words. 200 x 500 equals 100,000.

A Note on Checking Your Work

We can use multiplication to check division, and division to check multiplication. If we take our product and divide it by one of the factors, it should equal the other factor. And if we multiply a divisor and quotient, it should equal the dividend.

$$4 \times 3 = 12$$
$$3 \times 4 = 12$$
$$12 \div 3 = 4$$
$$12 \div 4 = 3$$

You do not always have to check your answer by using the opposite operation, but you should know how, and you should **always check to see if the answer seems reasonable.**

Meet the Calculator

Because we will be working with extremely large numbers in some problems, you can use a calculator for the problems in your *Student Workbook* marked with a 🖩. (This holds true for the rest of this course — anytime you see a 🖩, you are permitted to use a calculator.) To perform a calculation on most basic calculators, just enter the first number, press the appropriate operation symbol (+, -, x, or ÷), and enter the next number. Continue until you've entered all the numbers, and then press "=" to see the answer.

Example: Solve 4 + 6 − 3 on the calculator.

> Press 4, then the + sign, then 6, then the − sign, then 3, then the = sign.
> The answer is 7.

Note: If you need more help with using a calculator, ask your parent/teacher or look at your calculator's manual. There are also lots of resources online.

Don't get too attached to your calculator! There will be many times in your life when you will need to solve a problem and won't have a calculator on hand (or time to pull it out), so it's important to be comfortable doing math without calculators. But to avoid busywork, calculators will be allowed on some problems.

Keeping Perspective

I hope by now you're beginning to appreciate math's usefulness. It aids us when making a variety of real-life decisions we encounter, such as planning an event or selecting an apartment. (Your *Student Workbook* will give you practice.)

As you practice using math to make decisions, remember to let the Lord, not just the bottom-line dollar, guide your decisions. Mary Magdalene was blessed for pouring out expensive perfume on Jesus' feet, a decision that didn't make financial sense at all. Math can help us understand and evaluate situations, but we want the Lord to lead each step, not our own understanding.

> *Trust in the LORD with all thine heart; and lean not unto thine own understanding. In all thy ways acknowledge him, and he shall direct thy paths (Proverbs 3:5–6).*

4.6 Chapter Synopsis

Congratulations! You've now reviewed the four basic operations of math: addition, subtraction, multiplication, and division. As we continue to explore other areas of math, we'll use these tools over and over again. Because they describe the core consistencies of how objects combine, these four operations aid us in countless real-life settings.

Here's a quick review of what we learned this chapter:

■ There are many different ways to **multiply** — each one helps us keep track of place value.

■ Multi-digit multiplication uses a property of multiplication known as the **distributive property**. This property expresses the characteristic of multiplication that we can "parcel out" the multiplication over several addends.

■ Our **division algorithm**, or method, like our multiplication one, automates the process of division by keeping track of place value for us.

■ Many real-life problems require multiple steps and multiple tools to solve. It helps to **define** the problem and **plan** the solution first, and then **execute** and **check** our work.

As we get ready to move on to other aspects of mathematics, always remember why all of this is even possible — because of who God is. You could find yourself in the darkest dungeon, and 1 + 1 would still equal 2; 4 – 1 would still equal 3; 2 x 3 would still equal 6; and 10 ÷ 2 would still equal 5 because God would still be on His throne and in charge. Whatever happens, nothing can change God.

Napier's Rods and Multi-digit Multiplication

Last chapter, we looked at Napier's rods and how they recorded multiplication facts by writing the result of 1 x 2 on the first row of the 2 strip, 2 x 2 on the second, etc. The rods were a way of simplifying the Gelosia method of multiplying that we learned earlier in this chapter. Let's look at how they could be used to multiply multi-digit numbers using the multiplication facts already recorded on the rods.

Example: Solve 5 x 782 using Napier's rods.

Arrange the appropriate rods next to the guides. Read the 5 row, letting the diagonal lines keep track of place value. (You would do this step mentally, but it is shown below for clarity.)

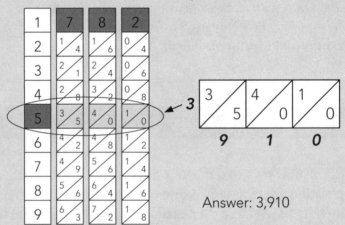

Answer: 3,910

What about multiplying 2 multi-digit numbers, such as 25 x 782? We can follow the same process, except we'll also need to find the answers from the 2 row, adding a zero to the end since the 2 represents 2 tens.

Example: Solve 25 x 782 using Napier's rods.

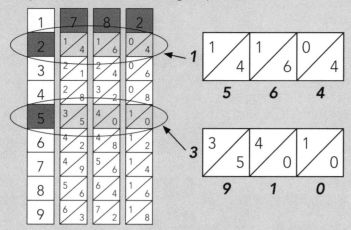

1 5 , 6 4 0   the result of 20 x 782 (We found 2 x 782 and added a zero to the result to find 20 x 782 instead.)
+ 3 , 9 1 0   the result of 5 x 782
1 9 , 5 5 0
Answer: 19,550

Napier's rods are an example of a counting machine: a device to simplify (or totally perform) math operations. (We learned how to use another counting machine this chapter: the calculator.)

John Napier, the inventor of these rods, was an inventor both in and out of mathematics. His rods proved enormously popular during the period when Europe was still adapting to the decimal system. Although John Napier had some mixed-up theology, he acknowledged that his skills came from God. He wrote in reference to some of his defense inventions: "Designed by the Grace of God, and the work of expert craftsmen."[3]

[CHAPTER 5]

Fractions and Factoring

5.1 Understanding Fractions

Now that we've reviewed the basic operations (addition, subtraction, multiplication, and division), it's time to explore **fractions**. In this lesson, we'll be taking a closer look at what fractions are and at how they apply. Along the way, we'll take a quick look at some terms we use when dealing with fractions, along with a tad of history.

As we view fractions from a biblical worldview, we're going to see that, like the concepts we've covered so far, fractions are a way of describing God's creation and a real-life tool. Ready? Let's begin.

Defining Fractions

Fractions Are Partial Quantities

The word "fraction" is used many different ways in our English language. You're probably used to associating the word with a *partial quantity*. And a fraction is that.

For example, say we cut a pie into six same-size slices. We can look at those six slices individually as 6. But what if we want to think of each slice as a *part* or *portion* of the whole pie?

Each pie would then represent a *partial quantity* — a portion of the whole pie. We'd say that each piece is a *fraction* — or part — of the whole. The word *fraction* has its root in the Latin word *frangere*, meaning "to break."[1] So in the case of our pie, we're breaking up the whole into 6 pieces. A fraction can thus be used to mean a partial quantity, no matter how we actually represent it. In this sense of the word, both $\frac{1}{6}$ and 0.17 (another way of representing $\frac{1}{6}$) would be considered fractions, as they both represent a partial quantity.

> Now that we have introduced fractions, we'll refer to nonfractional quantities 1 and higher as **whole numbers.** Note: The definition of whole numbers varies; some definitions include 0 as a whole number.

Fractions Are a Notation

You're also probably familiar with associating the word "fraction" with a particular notation. Using this **notation** (i.e., "series or system of written symbols used to represent numbers, amounts, or elements . . .")², we would represent 1 slice out of 6 as $\frac{1}{6}$ (read "one sixth"). Notice that rather than just saying we have 1 slice, we have expressed the portion of the whole pie that slice represents. Fractions in this sense help us see how **a quantity compares to the whole.** The top number — what we call the **numerator** — tells us how many parts of the whole we're considering, while the bottom number — called the **denominator** — tells us how many parts are in a whole. So in $\frac{1}{6}$, there are 6 parts to a whole, and we're looking at 1 of those parts. In $\frac{1}{3}$, there are 3 parts to a whole, and we're looking at 1 part. In $\frac{4}{3}$, there are 3 parts to a whole, and we're looking at 4 parts.

 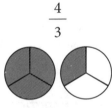

1 part; 6 parts to a whole   1 part; 3 parts to a whole   4 parts; 3 parts to a whole

Fractions give us an easy way to record quantities as parts, or portions, of others. For example, saying you have read 100 pages of a book does not really give a clear idea of how close to the end you are (the book could have 103 pages or 500 pages). But saying you have read 100 pages out of 360 pages, or $\frac{100 \; pages}{360 \; pages}$, tells us what part of the whole book you've finished. Likewise, $\frac{1}{4}$ cup or $\frac{1}{4}$ yard gives a clear understanding of how the quantity relates to the whole cup or the whole yard.

Fractions Are Division

As surprising as it may seem, another way of looking at a fraction is as a division problem.

Let's return to our pie example. When we divide a pie evenly among 6 people, we're really breaking, or dividing, the pie by 6.

1 pie ÷ 6 pieces = ?

We have a bit of a problem, don't we? How do we represent the result of 1 divided by 6? After all, 1 divided into 6 doesn't give us a whole quantity!

One way would be to use a fraction:

$$\frac{1}{6}$$

We've already seen how to look at this number as 1 piece out of 6 pieces (the quantity of pie each person got), but we can also look at it as 1 pie *divided by* 6. **The fraction line can be thought of as another symbol for division!**

In fact, fractions can be used to show *any* division problem. Say we wanted to divide $10 among 5 people. So far, we've been representing this as $10 ÷ 5. Another way, though, would be to represent it as a fraction: $\frac{\$10}{5}$.

$$\frac{\$10}{5} = \$10 \div 5$$

Notice that our answer (2) could be found by dividing or by thinking of the whole as 5 and taking 10 of those parts. Fractions can be thought of as parts of a whole as well as division problems because we'll get the same result either way.

Viewing $\frac{10}{5}$ as division:

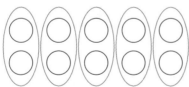

10 divided into 5 piles give us 2 in each pile.

Viewing $\frac{10}{5}$ as a part of a whole:

Each whole has 5 pieces.

10 pieces gives us 2 wholes.

In a way, the *numerator* of a fraction can be thought of as a *dividend*, and the *denominator* as a *divisor*!

$$\frac{20}{4} \qquad 4\overline{)20}$$

20 = dividend
4 = divisor

5. FRACTIONS AND FACTORING | 111

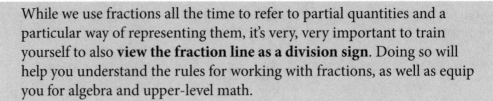

Pronouncing Fractions

There are several ways (surprise!) to pronounce fractions.

1. Pronounce the numerator as you normally would pronounce the number, but use an **ordinal number** (i.e., "a number defining a thing's position in a series, such as 'first,' 'second,' or 'third'"[3]) to pronounce the denominator.

   $\frac{1}{3}$ = "one-third"

   $\frac{2}{3}$ = "two-thirds"

   Exceptions include pronouncing $\frac{1}{2}$ as "one-half," $\frac{1}{4}$ as "one-quarter" (although "one-fourth" is also acceptable), and using "a" instead of "one" when you have a numerator of 1, as in "a third."

2. Pronounce the numerator and the denominator as normal with the word "over" in between.

   $\frac{1}{3}$ = "one over three"

   $\frac{2}{3}$ = "two over three"

3. Pronounce the numerator and the denominator as normal with the words "divided by" in between.

   $\frac{1}{3}$ = "one divided by three"

   $\frac{2}{3}$ = "two divided by three"

Thinking of Whole Numbers as Fractions

If you think of fractions as division, you'll have no trouble representing any quantity as a fraction, including whole numbers!

For example, since dividing by 1 doesn't change the value (see 3.5), we can think of 6 as 6 ÷ 1 . . . which can be written $\frac{6}{1}$!

Any whole number can be written as a fraction with the whole number in the numerator and a denominator of 1.

Example: Express 35 as a fraction.

$\frac{35}{1}$ means 35 ÷ 1, *which equals 35.*

> You should be able to find lists of ordinal numbers and more examples of pronouncing fractions online.

Categorizing Fractions

We have names to refer to different types of fractions. We call fractions that have an equal or greater numerator (top number) than denominator (bottom number) **improper fractions**, and ones with a greater denominator than numerator **proper fractions**. In other words, proper fractions are fractions that represent a partial quantity, while all other fractions are improper.

$$\frac{1}{4} = \text{proper fraction}$$

$$\frac{5}{4} = \text{improper fraction}$$

$$\frac{4}{4} = \text{improper fraction}$$

Fractions in History

Throughout history, men have expressed partial quantities in various ways. Seeing these different methods reminds us that our current method is just a tool — one standardized method for recording partial quantities on paper.

Egyptian — The Egyptians had more than one way of expressing $\frac{1}{4}$. In fact, according to Florian Cajori, "there are three forms of Egyptian numerals: the hieroglyphic, hieratic, and demotic."[4] The symbols for $\frac{1}{4}$ in these three forms are shown below. The first one on the left is the hieroglyphic form, the middle one the hieratic, and the one on the right the demotic.[5]

Babylonian — The Babylonian number system was based on 60. They wrote $\frac{1}{4}$ by writing 15, which is $\frac{1}{4}$ of 60. This could get confusing, however, since $\frac{1}{4}$ and 15 both were written the same way.

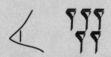

Greek — According to Florian Cajori, "Greek writers often express fractional values in words. . . . When expressed in symbols, fractions were often denoted by first writing the numerator marked with an accent, then the denominator marked with two accents and written twice."[6] The picture below shows $\frac{1}{4}$ in Greek symbols.

$$\alpha'\ \delta''\ \delta''$$

Roman — The Romans avoided fractions by using smaller and smaller units of measure. For example, rather than saying they had $\frac{1}{4}$ of an *as* (a specific Roman coin), they would say they had a *quadrans*. Below are some Roman symbols for a *quadrans*, or $\frac{1}{4}$.[7]

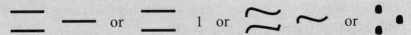

Seeing the different ways other cultures recorded partial quantities helps guard against thinking of our modern method of math as math itself. It reminds us that our current fractional notation is just one system we can use to help describe God's complex universe!

Keeping Perspective

While often thought of as partial quantities or parts of a whole, fractions are also another a way of representing division. They're an incredibly versatile tool!

5.2 Mixed Numbers

Since one of the ways to think of fractions is as representing division, we can use fractions to represent remainders to division problems. When we represent a remainder as a fraction, we're basically just completing part of the division, but leaving the part that's not a whole number as a fraction.

For example, if we divide 5 muffins among 4 people, which we could write as $\frac{5}{4}$, we'd end up giving each person $1\frac{1}{4}$ muffin.

$$\frac{5}{4} = 4\overline{)5}^{\,1} \;\; 1r1 \quad or \quad 1\frac{1}{4}$$
$$\phantom{\frac{5}{4} =}\underline{-\;4}$$
$$\phantom{\frac{5}{4} =}\;\;1$$

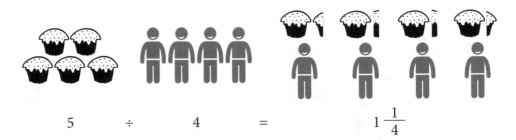

$$5 \;\; \div \;\; 4 \;\; = \;\; 1\frac{1}{4}$$

We would call $1\frac{1}{4}$ (read "one and a quarter" or "one and one quarter") a **mixed number**. A mixed number is a whole number and a fraction combined. Note that

$\frac{5}{4}$ and $1\frac{1}{4}$ represent the same quantity. They both represent the results of 5 divided by 4. Only in $1\frac{1}{4}$, we've completed some of the division.

If we were to think of $\frac{5}{4}$ as 5 parts with 4 to a whole, we would again have $1\frac{1}{4}$.

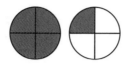

5 parts with 4 parts to a whole: $\frac{5}{4} = 1\frac{1}{4}$

Converting from Improper Fractions to Mixed Numbers

It's easy to convert between improper fractions and mixed numbers. Since a fraction represents division, all we need to do is complete the division, leaving the remainder as a fraction!

Example: Rewrite $\frac{21}{4}$ as a mixed number.

$\frac{21}{4}$ means "21 divided by 4." When we divide 21 by 4, we end up with 5 wholes, and a remainder of 1. In other words, we have 1 we still need to divide by the 4. So we have $5\frac{1}{4}$.

$$\frac{21}{4} \quad = \quad 4\overline{)21}\ r1 \quad = \quad 5\frac{1}{4}$$

If we think of $\frac{21}{4}$ in terms of parts of a whole, we would get the same answer — 21 parts with 4 parts to a whole gives us 5 wholes and $\frac{1}{4}$.

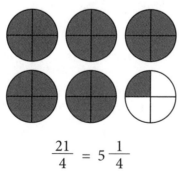

$$\frac{21}{4} = 5\frac{1}{4}$$

(21 parts, 4 parts to a whole)

Converting from a Mixed Number to an Improper Fraction

How do we express a mixed number as an improper fraction? Again, if we remember what a mixed number represents, it's not hard. Let's take the mixed number $5\frac{1}{4}$ again as an example.

We want to re-express 5 wholes as division by 4. Since 5 x 4 = 20, $\frac{20}{4}$ is another way of writing 5 wholes.

$$5\frac{1}{4}$$

5 x 4 = 20
Tells us what numerator would express 5 wholes as a division by 4.

But wait just a minute — we can't forget about the $\frac{1}{4}$ part of our mixed number! We need a numerator that will give us 5 wholes *plus* 1 left over. In other words, we have to add 1 to the numerator we'd need to give us 5 wholes, as we want to have 1 left over. Our numerator needs to be 20 + 1, or 21.

20 + 1 = 21
Finds the numerator that will give us 5 wholes plus 1 left over.

$$5\frac{1}{4} = \frac{21}{4}$$

We could also think of this in terms of parts of a whole. By multiplying the whole number by the denominator, we're finding out how many parts we have in that number of wholes, and adding to it the parts we already had in the numerator to give us the total parts.

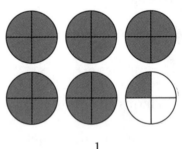

$$5\frac{1}{4}$$

5 x 4 = 20
Tells us how many parts are in 5 wholes.

20 + 1 = 21
Adds the parts in 5 wholes to the 1 additional part.

$$5\frac{1}{4} = \frac{21}{4}$$

To summarize, to convert from a mixed number to an improper fraction, just **multiply the whole number by the denominator, and add the product to the numerator.** Know that the reason this "rule" works is because it's really just re-expressing the whole number as a division.

Keeping Perspective — Why So Many Ways?

Improper fractions, proper fractions, mixed numbers, division sign (÷), division stairs ($\overline{)}$) — all are different ways of representing division. You might be wondering by now why we have to have *so* many different ways to record and work with quantities. The answer is because each one proves helpful in different areas of real life.

Take improper fractions and mixed numbers, for example. Sometimes we want to look at a quantity as a mixed number (it's easier to talk about getting $1\frac{1}{2}$ yards of fabric than $\frac{3}{2}$ yards), but other times we need to look at it as an improper fraction (it's often helpful when adding, subtracting, multiplying, or dividing fractions).

Bottom line: God created a complex universe, and it takes us a lot of different tools in our toolbox to work with and describe it. If you start to feel overwhelmed by math's complexity, don't fret; simply lift your eyes upward and think about how much greater is the God who created all the things we can hardly even comprehend or describe.

> *O the depth of the riches both of the wisdom and knowledge of God! how unsearchable are his judgments, and his ways past finding out! (Romans 11:33).*

5.3 Equivalent Fractions and Simplifying Fractions

As you know, the same quantity can be represented many different ways. Notice how if we were to complete the division in $\frac{6}{3}$, $\frac{8}{4}$, or $\frac{12}{6}$, we'd end up with the same quantity: 2.

$$\frac{6}{3} = 6 \div 3 = 2 \qquad \frac{8}{4} = 8 \div 4 = 2 \qquad \frac{12}{6} = 12 \div 6 = 2$$

Likewise, the shaded area of these rectangles and pies can be looked at as $\frac{2}{3}$, $\frac{4}{6}$, or $\frac{8}{12}$. All these fractions have the same value.

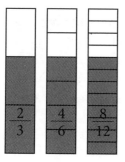

$$\frac{1}{3} = \frac{2}{6} = \frac{4}{12}$$

We call fractions that represent the same quantity **equivalent**.

When working with fractions, we often need to re-express a fraction as an equivalent fraction. But how do we do this? We certainly don't want to have to draw a bar or pie each time! Instead, we want to find a way to make it simpler.

> *Mathematics is being lazy. Mathematics is letting the principles do the work for you so that you do not have to do the work for yourself.*
> — *George Pólya*[8]

While we don't really want to be lazy, we do want to find a way to simplify tasks so we can accomplish them more efficiently. While you probably already know an efficient way to find equivalent fractions, let's see if we can understand how our simplification process is really just observing something about how God governs this universe and seeing how that fits with the mathematical conventions we've adopted.

Finding Equivalent Fractions

Back in 3.5, we mentioned the identity property of multiplication — that multiplying any number by 1 doesn't change its value.

$$6 \times 1 = 6$$

We also mentioned that this would also mean that dividing any number by 1 would not change its value, and that dividing a number (other than 0 — as we saw in 4.4, we can't divide by 0) by itself would equal 1.

$$6 \div 1 = 6$$
$$6 \div 6 = 1$$

Well, since fractions are another way of expressing division, notice how we could rewrite $6 \div 6$ as a fraction worth 1.

$$6 \div 6 = \frac{6}{6} = 1$$

We can use this knowledge to help us find equivalent fractions. Let's take a look.

Since any number (except 0) divided by itself equals 1, fractions with the same numerator and denominator (such as $\frac{4}{4}$ or $\frac{2}{2}$) all equal 1. So any fraction (except $\frac{0}{0}$) with the same numerator and denominator is going to equal 1!

$$\frac{1}{1}, \frac{2}{2}, \frac{3}{3}, etc. = 1$$

If we couple this with the knowledge that multiplying or dividing by 1 doesn't change the value of a number, we now know that we can multiply or divide a fraction by a fraction with the same numerator and denominator without changing its value. In other words, to find an equivalent fraction, we've only to **multiply or divide a fraction by a fraction with the same numerator and denominator (i.e., a fraction worth 1)!**

For example, we can multiply $\frac{4}{5}$ by $\frac{2}{2}$ to find an equivalent fraction: $\frac{8}{10}$. Note that when we multiply fractions, we **multiply both the numerator and the denominator.** (We'll learn more about multiplying fractions in the next chapter.)

$$\frac{4}{5} \times \frac{2}{2} = \frac{8}{10}$$

Or we could have gone the other way, starting with $\frac{8}{10}$ and dividing to find an equivalent fraction. Again, we'll divide both the numerator and the denominator.

$$\frac{8}{10} = \frac{8}{10} \div \frac{2}{2} = \frac{4}{5}$$

When we multiply or divide by a fraction worth 1, we're not changing the end result of the division. We're just expressing it differently. We could do the same thing if the division were written using a different method.

$$5\overline{)10} = 2$$

Multiplying the dividend (10) and the divisor by 5:
$$25\overline{)50} = 2$$

Dividing the dividend (10) and the divisor by 5:
$$1\overline{)2} = 2$$

Notice the quotient (2) is the same! Multiplying or dividing both our dividend and our divisor by the *same number* doesn't change the results.

> Notice how we just took a characteristic God set in place (the identity property of multiplication) and used it to find a way with fractional notation to express the same quantity in a different way.

Simplifying Fractions

In the next chapter, we'll use equivalent fractions to help us add and subtract fractions. For now, though, let's take a look at using them to express fractions as simply as possible.

If I've read 13 pages out of 26, it would be easier to quickly assess my progress if I wrote $\frac{1}{2}$ than if I wrote $\frac{13}{26}$. Equivalent fractions make it simpler to take in a fraction at a glance.

This simplicity gets magnified when working with larger numbers. For instance, if our book had 332 pages and we'd read 166, it would be much easier to say we're halfway finished with the book ($\frac{166}{332} \div \frac{166}{166} = \frac{1}{2}$) than to say we've read $\frac{166}{332}$ of the book.

To simplify a fraction where the numerator and denominator are whole numbers, keep dividing by fractions worth 1 until you reach a point where the numerator and denominator will not divide completely anymore by the same whole number (except, of course, for 1, as any number can be divided by 1). We would then say that the fraction is in its **lowest terms.**

$$\frac{350}{600} \div \frac{10}{10} = \frac{35}{60} = \frac{35}{60} \div \frac{5}{5} = \frac{7}{12}$$

7 and 12 can't be divided any further by the same number, so this fraction is now in lowest terms. $\frac{7}{12}$ represents the same value as $\frac{350}{600}$, but in lowest terms. Of course, we could have divided by 50 right from the start if we had realized both numbers were divisible (i.e., could be divided) by 50.

$$\frac{350}{600} \div \frac{50}{50} = \frac{7}{12}$$

To make your answers easier to read, from now on in this course **you should always convert improper fractional answers to mixed or whole numbers and simplify fractional answers to their lowest terms.**

Keeping Perspective

You may be wondering why we're spending so much time with fractions. After all, do we really use them that often in everyday life?

Yes, we use fractions in everyday life. Just take a look at the fractions on a measuring cup, go to the sewing store and buy $\frac{1}{4}$ yard of fabric, build a bookcase and measure fractions of an inch, watch the 4 *quarters* of a football game, play a *quarter note* on an instrument, or measure out $\frac{1}{4}$ inch on that next greeting card you make.

But the main reason we're studying fractions in depth is because fractions are incredibly useful in representing division. It's easier to work with division when it's expressed as a fraction. For instance, today we saw how fractions make it easy to simplify the division. While this may not seem that useful at first glance, it becomes indispensible in algebra.

5.4 Understanding Factoring

When simplifying fractions, it can get challenging to figure out what to even divide by to reduce the fraction to its simplest terms. In this lesson, we're going to look at a useful technique to help called **factoring**.

Beyond its application within fractions and algebra, factoring helps us better understand numbers and get comfortable with mental multiplication. Factoring is also used in fields such as cryptography ("the art of writing or solving codes"[9]). A common cryptography application is keeping information secure online — secure websites encrypt your information so hopefully only the intended receiver gets it.

Defining Factoring

Back when we studied multiplication, we talked about how the numbers in a multiplication problem can be referred to as **factors**.

factor x *factor* = *product*

Factoring is the process of looking at a number as a product and seeing what non-fractional numbers could be multiplied to make that product (i.e., what are the number's *factors*). This would in turn tell us what nonfractional numbers we can divide the number by.

For instance, consider the number 12. Ignoring 12 x 1, 12 is the product of 3 x 4 and 2 x 6. 12, 6, 4, 3, 2, and 1 are all *factors* of 12 — they can all be multiplied by a whole number to make 12.

Prime Factors

When looking at the factors that could be used to make a number, we're often interested in what we call its **prime factors**. A prime factor is a factor that's a **prime number**. And just what is a prime number? A prime number is what we call a whole number greater than 1 that cannot be evenly divided by any whole number except by itself and 1.

So 3 and 2 are the *prime factors* of 12. Both 3 and 2 can't be divided by any whole number but themselves and 1.

Sometimes it's obvious whether a factor is prime, but other times it's not. While there's a way to figure it out mathematically, you won't be asked to do that. Instead, you can use the given list of all the prime numbers under 100. After working with prime numbers, you'll start to have a good feel for what is prime and what is not.

Prime Numbers Under 100
2, 3, 5, 7, 11, 13, 17, 19, 23, 29, 31, 37, 41, 43, 47, 53, 59, 61, 67, 71, 73, 79, 83, 89, 97

Using Prime Factors

Notice that all the other factors of 12 (with the exception of 1, which is a factor of every number), could have been found by multiplying our prime factors.

Factors of 12: 2, 3, 4, 6, 12

> 2 and 3 are the prime factors; the other factors are all products of 2 and 3.
>
> $$2 \times 3 = 6 \qquad 2 \times 2 = 4 \qquad 2 \times 2 \times 3 = 12$$

Because of this, factoring a number down to its prime factors can help us compare and work with numbers, as we'll see in the next lesson.

When we factor, we **keep factoring until we have broken the number down into its prime factors.** At that point, we can't factor any further, as all our factors are only divisible by 1 and themselves!

> **Here's a summary of how to find the factors of a number:**
> 1. Start with any two factors (other than 1 — we're not interested in knowing that 1 is a factor). If you don't know what factors make up the number, try seeing if it will divide by 2 or 3.
> 2. Find factors of any non-prime factors.
> 3. Continue finding factors until each factor is a prime number.

Factoring with Factor Trees

Factors are often easier to find using what we call **factor trees**. Notice how we've shown how we could have found the prime factors of 12 using a tree.

$$
\begin{array}{ccc}
12 & \qquad & 12 \\
\wedge & & \wedge \\
3 \times 4 & & 2 \times 6 \\
\quad \wedge & & \quad \wedge \\
\quad 2 \times 2 & & \quad 3 \times 2
\end{array}
$$

Notice that it didn't matter what factors we started with — we eventually got down to the same **prime factors** (2 and 3) — 12 can be thought of as 2 x 2 x 3.

Let's take a look at a larger number: 188.

1. Since we have not memorized any factors that equal 188, let's begin by dividing it by 2.

$$
\begin{array}{r}
9\,4 \\
2\overline{)1\,8\,8} \\
-\ 1\,8 \\
\hline
0\,8 \\
8 \\
\hline
0
\end{array}
$$

188 divides evenly by 2, so 2 and 94 will be our first factors.

122 | PRINCIPLES OF MATHEMATICS 1

$$188$$
$$\wedge$$
$$2 \ \times \ 94$$

2. 2 is already a prime number, but we need to keep factoring 94. Let's divide it by 2.

$$188$$
$$\wedge$$
$$\mathbf{2} \ \times \ 94$$
$$\wedge$$
$$\mathbf{47} \times \ \mathbf{2}$$

3. Since 47 is a prime number (see the list of prime numbers), we've finished factoring 188.

188 can be thought of as 2 x 2 x 47.

Factoring without Trees

While drawing a factor tree can sometimes help factor a number, you don't have to draw one. It's a tool, not a requirement. Many times factoring can be done mentally.

In fact, we'll see in the next lesson that we can replace numerators and denominators with its factors, and then more easily simplify the fraction.

$$\frac{\$12}{\$6} = \frac{2 \ \times \ 3 \ \times \ 2}{3 \ \times \ 2}$$

The point is simply to realize that **a number can be thought of as the product of its factors.**

Keeping Perspective — Abstract Mathematics

Factoring helps us see what numbers could be multiplied together to obtain a certain number . . . which in turn tells us what that number could be divided by (i.e., what it is **divisible** by).

We're beginning to cross the line between math concepts that are very concrete (it's very easy to see what aspect of creation they describe) and those that are more **abstract** (apparently separate from real life). After all, while we add, subtract, multiply, and divide quantities around the house, we don't usually think about factoring them.

It's easy to get lost in abstract math and view it as some sort of intellectual exercise. But the beauty of mathematics is that because creation is so orderly and consistent and because God made us in His image, we can take a concrete concept, work with it abstractly, and then bring back a concrete application.

You'll see how factoring helps us reduce real-life fractions in the next lesson — for now, just get familiar with looking at a number as a product of prime factors.

5.5 Simplifying Fractions, Common Factors, and the Greatest Common Factor

Okay, let's put what we learned about factoring and prime factors yesterday to some practical use. Let's say that you've read 56 out of 360 pages. You could express this as a fraction: $\frac{56}{360}$.

Now, it will be easier to see how far along you are in the book if we simplify this fraction. What fraction worth one ($\frac{2}{2}$, $\frac{3}{3}$, etc.) should we divide it by to get it into its lowest terms?

It's sometimes obvious what to divide by, but when it's not, factoring can help us think it through. To see what we can divide both the numerator (56) *and* the denominator (360) by, let's see what prime factors make up the numbers.

$$
\begin{array}{ccccc}
56 & & & 360 & \\
\wedge & & & \wedge & \\
2 \;\; x \;\; 28 & & 36 & x & 10 \\
\wedge & & \wedge & & \wedge \\
14 \; x \; 2 & & 6 \;\; x \;\; 6 & & 5 \; x \; 2 \\
\wedge & & \wedge \quad \wedge & & \\
7 \; x \; 2 & & 3 \, x \, 2 \quad 3 \, x \, 2 & &
\end{array}
$$

Now let's replace the numbers in our fraction with these prime factors. To make it easier to compare the factors, we'll list them in descending order.

$$\frac{56}{360} = \frac{7 \;\; x \;\; 2 \;\; x \;\; 2 \;\; x \;\; 2}{5 \;\; x \;\; 3 \;\; x \;\; 3 \;\; x \;\; 2 \;\; x \;\; 2 \;\; x \;\; 2}$$

Notice how both our numerator *and* our denominator have a 2. In this case, 2 is what we call a **common factor** — a factor that two or more numbers share in common. This tells us that both our numerator and our denominator can be divided by 2.

$$\frac{56 \text{ pages}}{360 \text{ pages}} \div \frac{2}{2} = \frac{28}{180}$$

> Another term for *common factors* is *common divisors*, and another term for *greatest common factor* is *greatest common divisor*. Terms sometimes vary, but the idea is the same. A divisor is a number we can divide by — a factor is a number we can multiply by. Since we can divide a number by any of the factors that make it up, they are both different ways of describing the same thing.

But just dividing by $\frac{2}{2}$ doesn't reduce the fraction to its lowest terms. To do that, we have to divide by what's called the **greatest common factor** — the *greatest* factor that two or more numbers share in common.

So how do we find the greatest common factor? We look at how many times the common factor(s) are repeated.

$$\frac{7 \;\; x \;\; 2 \;\; x \;\; 2 \;\; x \;\; 2}{5 \;\; x \;\; 3 \;\; x \;\; 3 \;\; x \;\; 2 \;\; x \;\; 2 \;\; x \;\; 2}$$

Notice that in this case, both our numerator and our denominator have "2 x 2 x 2" as part of their makeup. This means that both our numerator *and* our denominator are divisible (i.e., can be divided) by 2 x 2 x 2, or 8. Thus, 8 is our greatest common factor — it's the greatest factor both the numerator and the denominator can be divided by.

Dividing both our numerator and our denominator by it will reduce our fraction to its lowest terms.

$$\frac{56 \text{ pages}}{360 \text{ pages}} \div \frac{8}{8} = \frac{7}{45}$$

Keeping Perspective

When we need to see what we could divide two or more numbers evenly by (such as when simplifying fractions), we can look for **common factors** or the **greatest common factor** (i.e., the *greatest* factor two or more numbers have in *common*).

Of course, you do not always need to consciously factor in order to simplify fractions. But it's important to understand that, because of the consistent way objects multiply, **any number *can* be thought of as the product of its factors.** This knowledge about numbers will help with mental multiplication and other math operations, along with one day finding the value for quantities you don't know.

No matter how abstract a concept may seem, it has concrete applications. You will learn more about factoring's applications as you continue your math studies.

> . . . all of the mathematics and science topics taught in school have workplace applications. — From the book Connecting Mathematics and Science and Workplace Contexts[10]

5.6 Adding and Subtracting Fractions

Say you are planning a party and spend $9 on 12 plates, $5 on 10 cups, $2 on 50 napkins, and $8 on 45 sets of silverware. How much are you spending altogether per place setting (1 plate, 1 cup, 1 napkin, and 1 set of silverware)? An easy way to find out would be to express each division as a fraction, and then add those fractions together.

$$\frac{\$9}{12} + \frac{\$5}{10} + \frac{\$2}{50} + \frac{\$8}{45} = \ ?$$

Or say you go to the hardware store and find a piece of wood on sale that's $\frac{11}{12}$ feet long. You only need $\frac{1}{2}$ foot for your project. How much will you have left? Would you be saving money if you bought it instead of buying a different piece that's exactly the length you need?

Or if you have ever made yeast bread, you have probably added flour to a mixing bowl a little at a time. Say you added $\frac{1}{2}$ cup, then a $\frac{1}{4}$ cup. How would you figure out the total quantity of flour used?

Clearly, we need to add and subtract fractions . . . but how? Consider adding $\frac{4}{2} + \frac{8}{4}$. In this case, we could easily complete the divisions, and see that this is the same thing as 2 + 2, which equals 4.

$$\frac{4}{2} = 4 \div 2 = 2$$

$$\frac{8}{4} = 8 \div 2 = 2$$

$$\frac{4}{2} + \frac{8}{4} = 2 + 2 = 4$$

But let's take a look and see if there's a way we could have found this *without* completing the division — a way that would allow us to add and subtract fractions in fractional form. While we could add and subtract most everyday fractions by converting them to decimals (which we'll cover in a couple of chapters), it's essential to know how to add and subtract fractions as fractions, as it's a concept you'll use extensively in algebra to explore the relationships and consistencies in God's creation at an even deeper level. And while you may already know the mechanics, hopefully our exploration will help you see what's really happening when we add and subtract fractions, and how it is ultimately possible because of God's faithfulness.

Thinking It Through

We want to find a way to add fractions. Let's take $\frac{1}{2}$ and $\frac{1}{4}$ for examples — $\frac{1}{2}$ represents 1 *divided by* 2, which would result in a quantity equivalent to 1 part out of 2 parts (in other words, half of a whole). Likewise, $\frac{1}{4}$ represents 1 *divided by* 4, which would result in a quantity equivalent to 1 part out of 4 parts (in other words, a quarter of a whole).

126 | PRINCIPLES OF MATHEMATICS 1

We *can't* just add the numerators or denominators together, as that would be like adding apples and oranges together. Instead, we have to first **look at these fractions as divisions by the same quantity/parts of the same whole!** Once we do, we can simply add the numerators together to find the sum . . . or subtract the numerators to find the difference.

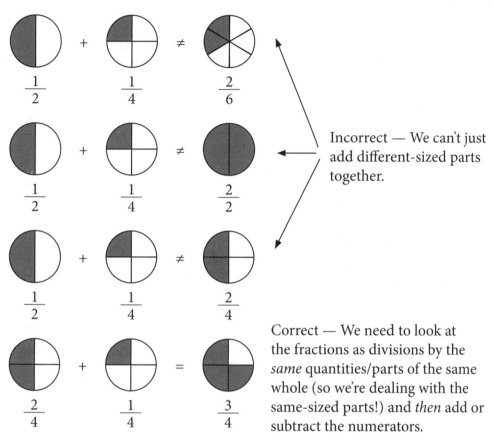

Incorrect — We can't just add different-sized parts together.

Correct — We need to look at the fractions as divisions by the *same* quantities/parts of the same whole (so we're dealing with the same-sized parts!) and *then* add or subtract the numerators.

Subtracting fractions follows the same principle as adding them. To subtract $\frac{1}{2}$ yard from $\frac{2}{3}$ yard, we have to first express them as divisions of the same quantity/parts of the same whole. Then we can subtract the numerators to find the answer.

Original problem

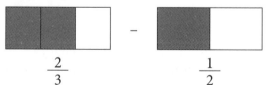

Rewritten with same denominators:

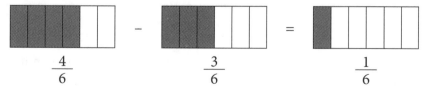

We would have $\frac{1}{6}$ of a yard left.

5. FRACTIONS AND FACTORING 127

> In short, to add or subtract fractions, **we rename the fractions by multiplying or dividing by a fraction worth one to get all our denominators the same, and then add or subtract the numerators.**

Picking a Denominator

How do we know what denominator to use when adding or subtracting fractions? Any number that all the denominators we're trying to add are a factor of will work. When in doubt, just multiply the denominators you're trying to add together and use that number.

Example: Add $\frac{7}{8}$ pound, $\frac{4}{5}$ pound, and $\frac{1}{2}$ pound.

We need to find a number that contains 8, 5 and 2. We know 8 x 5 is 40, and 40 is also divisible by 2. So we'll use 40 as our new denominator. (We could have also used 80, as 8 x 5 x 2 = 80. It doesn't matter which denominator you use, so long as all the fractions can be rewritten as an equivalent fraction of it.)

Now we just have to rewrite each fraction as a fraction of 40. Let's start with $\frac{7}{8}$. How do we know what to multiply by? To figure it out, stop and think through what we're trying to find. We want to find the factor that, multiplied by 8 (our current denominator), gets us 40 (our desired denominator). We can mentally divide 40 by 8, which gives us 5. If we multiply $\frac{7}{8}$ by $\frac{5}{5}$ (a fraction worth 1), we'll end up with 40 as our denominator.

$$\frac{7}{8} \times \frac{5}{5} = \frac{35}{40}$$

We can do the same thing for the other fractions.

$$\frac{4}{5} \times \frac{8}{8} = \frac{32}{40}$$
$$\frac{1}{2} \times \frac{20}{20} = \frac{20}{40}$$

Now we can add.

$\frac{35}{40} + \frac{32}{40} + \frac{20}{40} = \frac{87}{40}$ pounds. Simplified, that equals $2\frac{7}{40}$ pounds.

Enter Whole Numbers

What happens when we want to add a fraction to a whole number? For instance, say we find 2 yards of fabric on one bolt, and $\frac{1}{4}$ on another bolt. How much fabric is there altogether?

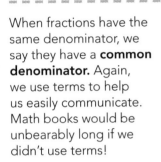

When fractions have the same denominator, we say they have a **common denominator**. Again, we use terms to help us easily communicate. Math books would be unbearably long if we didn't use terms!

128 | PRINCIPLES OF MATHEMATICS 1

Adding fractions to whole numbers is pretty intuitive — just put the two numbers together. In our fabric scenario, we'd have $2\frac{1}{4}$ yards of fabric.

$$\frac{1}{4} + 2 = 2\frac{1}{4}$$

Subtracting fractions from whole numbers works much the same way, except we often have to rename the whole number as a fraction. Say we found a bolt of fabric with 2 yards on it, and we want $\frac{2}{3}$ yards. To find how much fabric would be left, we first rewrite the 2 yards as a fraction, and then subtract as normal.

Example: Solve $2 - \frac{2}{3}$

Rewrite the whole number as a fraction of the same denominator:

$$\frac{6}{3} - \frac{2}{3}$$

Solve and simplify: $\frac{6}{3} - \frac{2}{3} = \frac{4}{3} = 1\frac{1}{3}$

Note: In this case, we probably could have found the answer mentally — just know that you can rewrite a whole number as a fraction when needed.

Keeping Perspective

When adding and subtracting fractions, we have to first make sure those fractions are written as portions of the same quantity / divisions by the same quantity (i.e., that they have a **common denominator**). We can then add or subtract the numerators to find the answer.

Adding and subtracting fractions illustrates how helpful it can be to look at quantities different ways. For instance, we may need to view $\frac{1}{2}$ as $\frac{6}{12}$ or $\frac{10}{20}$ in order to add or subtract it.

As you review adding and subtracting fractions, remember that just as with whole numbers, our ability to add and subtract real-life quantities on paper assumes a consistent universe . . . which we can safely rely on because God established and sustains the "ordinances" around us.

> *Thy faithfulness is unto all generations: thou hast established the earth, and it abideth. They continue this day according to thine ordinances: for all are thy servants (Psalm 119:90–91).*

Ponder this today as you do your worksheet: we can only rely on ordinances like addition and subtraction because *all* things are God's servants. He is in control of every atom and molecule. He is in control of our lives and the things that happen to us — both the large and the small. And He bids us cast our every care on Him.

Casting all your care upon him; for he careth for you (1 Peter 5:7).

5.7 Least Common Denominator/Multiple

When adding and subtracting fractions so far, it has been fairly easy to find a common denominator. Sometimes, though, it may not be intuitively obvious what denominator fractions have in common. We know we could always multiply every denominator together to find a common denominator, but that's sometimes pretty tedious.

For example, suppose we need to rewrite these fractions to add them:

$$\frac{1}{156} \qquad \frac{1}{24}$$

While we *could* just multiply the denominators together to find a common denominator, that would result in a rather large denominator (3,744). Is there a way to find a smaller common denominator?

Yes, there is a way! Let's start by looking at the prime factors of the denominators.

156
12 x 13
6 x 2
3 x 2

24
12 2
6 x 2
3 x 2

Now let's replace our denominators with their prime factors.

$$\frac{1}{156}$$
2 x 2 x 3 x 13

$$\frac{1}{24}$$
2 x 2 x 2 x 3

To find the least number that both of these numbers are factors of, we need to multiply **each prime factor the same number of times it is included in any *one* of the numbers.**

So if we start at the left and circle each prime factor for our first number, we could then ignore any factors in the next number(s) that we've already circled . . . unless it appears more times in the next number(s) than it did in the first (such as "2" in 24 — it appears *three* times in 24, and we only accounted for it *twice* in the factors we circled for 156).

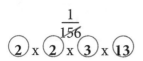 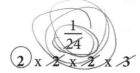

We already had two 2s and a 3 in 156.

Now if we multiply together all the numbers we've circled, we'll have found the lowest number we can use as a denominator.

$$2 \times 2 \times 3 \times 13 \times 2 = 312$$

We'd refer to 312 as the lowest or **least common multiple (LCM)** of 24 and 156. It's the lowest number of which both numbers are factors. A multiple is "a number that can be divided by another number without a remainder."[11] The least, or lowest, one is just the smallest one. Since both 24 and 156 are denominators, we could also call 312 our **least common denominator (LCD)** — least common denominator means the same thing as the least common multiple, only it lets us know we're specifically dealing with denominators.

> ## Keeping Perspective
>
> The **least common multiple** (also called the **least common denominator** when referring specifically to a denominator) is found by finding all the prime factors of each number, and then multiplying **each prime factor the same number of times it is included in any *one* of the numbers.** This will find the least number of which two or more numbers are factors.
>
> Once again, we're using our God-given abilities to observe aspects of numbers (and the consistent multiplications that can be used to arrive at those numbers) and using that to simplify a task (in this case, finding a common denominator).

5.8 Chapter Synopsis and Expanding Our Worldview

Well, we've explored a lot about both fractions and factoring this chapter. It's time for a few key takeaways:

- While often thought of as a partial quantity or part of a whole, **fractions are also a way of describing division.**

- **Numerator and denominator** are terms used to refer to different parts of a fraction; **proper fractions, improper fractions, and mixed numbers** refer to different types of fractions.

- **Equivalent** fractions are fractions that have the same value; we can rewrite fractions by multiplying or dividing both the numerator and the denominator by the same quantity — a fraction worth 1.

- **Factoring** helps us see numbers as products of other numbers. **Prime factors** are factors that are prime numbers — whole numbers greater than 1 that cannot be evenly divided except by themselves and 1. **Common factors** or **common divisors** are factors two or more numbers share in common; the **greatest common factor** or **greatest common divisor** of two or more numbers is the greatest factor they share in common. The **least common multiple** or **denominator** is the lowest number of which two or more numbers or denominators are factors.

- To **simplify a fraction** so it is in its **lowest terms**, divide both the numerator and the denominator by the same common factor (using the greatest common factor will simplify the fraction as much as possible).

- To add and subtract fractions, we have to first rewrite them so they have **common denominators**. We can then add or subtract the numerators.

In the next chapter, we'll continue to build on what we learned in this chapter as we dig even deeper into fractions. As we do, you'll get a chance to apply fractions to more settings. Keep in mind, though, that fractions prove most helpful as a way of representing division. The more comfortable you can become with using them to represent division, the easier it will be when you build on that knowledge in future math courses.

Expanding Our Worldview — Why There is Suffering

We have discussed over and over that it's God's faithfulness in holding all things together that makes math possible. I hope you've let that thought inspire you — we do indeed serve an all-powerful God!

You, or someone you know, might wonder why, since God *is* so powerful and in charge of everything, there's so much death and suffering in the world. Isn't He a God of love? Let's take a look at what the Bible has to say about this.

The Bible makes it clear that in God's original creation, there wasn't *any* death or suffering. It was all "very good" (Genesis 1:31). Death was the punishment for sin, which is rebellion against God (Romans 6:23; Genesis 2:17). It is *man's* fault that we have suffering and death in this world, not God's.

God keeps His Word. He warned Adam, the first man, that the punishment for eating the fruit of the tree He'd forbidden was death, yet Adam still ate. The moment he did, he died. His soul, which had been in perfect fellowship with his Creator, was now dead spiritually. His body — and all of creation — began to physically die. One day, Adam would die — and, unless God intervened, his soul would be eternally dead and separated from his Creator.

But man's sin did not surprise God. Before the beginning of the world, God already had a plan. He planned to send His own Son — Jesus Christ — to bear the sin of the entire world upon His shoulders. Jesus lived as a perfect man (He was fully God, yet fully man), endured the wrath of God and death, and rose from the dead, conquering death's power forever so that man could have eternal life.

Not only did God Himself become a man to die in our place to redeem us from the suffering and death we deserve, but He also patiently waits to return and end this falling-apart world so those who do not know Him may yet respond. He is "not willing that any should perish" (2 Peter 3:9).

We get to experience life in a fallen world in which He has allowed us to taste the effects of sin . . . yet still extends the mercy of His grace. For those who know Jesus Christ, God promises to use the suffering we experience in this fallen world for good.

> *And we know that all things work together for good to them that love God, to them who are the called according to his purpose (Romans 8:28).*

In short, God is in charge of everything, but that doesn't make suffering and death His fault. They came as the result of man's sin. God in His love and mercy provided a way of salvation — and lovingly uses even the brokenness of this universe that we earned to draw us unto Himself and His eternal life.

[CHAPTER 6]

More with Fractions

6.1 Multiplying Fractions

It's time to take a look at multiplying fractions. While you probably already know how to multiply fractions, hopefully our exploration will help you understand better what we're really doing when we multiply fractions and why the method actually works.

What We're Doing When We Multiply

As we've already explored, multiplication is a way of representing repeated additions — of taking one quantity a certain number of times. Thus, $3 \times \frac{1}{2}$ means $\frac{1}{2}$ taken 3 times.

$\frac{1}{2} + \frac{1}{2} + \frac{1}{2} = \frac{3}{2}$, which equals $1\frac{1}{2}$

$3 \times \frac{1}{2}$

a shortcut for writing the same thing

This can be a little trickier to see when our multiplier is a fraction, such as if we wanted to take $\frac{1}{2} \times \frac{1}{2}$. When we have a fraction as our multiplier, we are taking the quantity a partial number of times. What would that mean?

When you think about it, all multiplication can be thought of in terms of *of*. When we multiply 6×5, we're taking 6 sets *of* 5. When we multiply $3 \times \frac{1}{2}$, we're taking 3 sets *of* $\frac{1}{2}$. Viewing multiplication this way helps make sense out of multiplication when our multiplier is a fraction.

$\frac{1}{2}$ of $\frac{1}{2}$ = $\frac{1}{2}$ x $\frac{1}{2}$ = $\frac{1}{4}$

If we take half *of* a half, we end up with a quarter of a whole.

If we take $\frac{1}{2}$ *of* 3, we end up with $1\frac{1}{2}$ wholes.

 $\frac{5}{6}$ of $\frac{1}{2}$ = $\frac{5}{6}$ x $\frac{1}{2}$ = $\frac{5}{12}$

If we take $\frac{5}{6}$ *of* a half, we end up with $\frac{5}{12}$ of a whole.

Notice that when a partial quantity is our multiplier, our product is a smaller quantity instead of a larger. This is because we're taking the quantity a partial number of times.

Remember that multiplication is commutative — $\frac{1}{2}$ x 3 (i.e., $\frac{1}{2}$ *of* 3) results in the same quantity as 3 x $\frac{1}{2}$ (i.e., 3 sets *of* $\frac{1}{2}$).

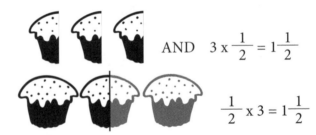

Getting to a Rule

The "rule" for multiplying fractions is to multiply the numerators and the denominators.

$$\frac{5}{6} \times \frac{1}{2} = \frac{5 \times 1}{6 \times 2} = \frac{5}{12}$$

When multiplying a whole number by a fraction, think of the whole number as having 1 as its denominator — after all, dividing by 1 doesn't change the value, and $\frac{3}{1}$ means $3 \div 1$.

$$3 \times \frac{1}{2} = \frac{3}{1} \times \frac{1}{2} = \frac{3 \times 1}{1 \times 2} = \frac{3}{2} = 1\frac{1}{2}$$

We don't necessarily have to bother to rewrite the whole number as a fraction — just remember to view the whole number as a numerator.

While the mathematical proof of why we can multiply fractions by multiplying the numerators and the denominators is beyond the scope of this course, we can see in general that it makes sense. It's obvious to see why it gives us the correct answer in simple cases, such as $3 \times \frac{1}{2}$. And 3×1 is doing the same thing as adding up the numerators in $\frac{1}{2} + \frac{1}{2} + \frac{1}{2}$ would. The obvious answer is $\frac{3}{2}$, which simplifies to $1\frac{1}{2}$. But what about $\frac{5}{6}$ of $\frac{1}{2}$? Why does multiplying by the numerators and denominators here give us the correct answer?

Remember, while we often use fractions to represent partial quantities, they're also a way of writing division. Thus, $\frac{5}{6}$ means $5 \div 6$. To take $\frac{5}{6}$ of a quantity (i.e., find out what we'd have if we took $\frac{5}{6}$ of it), we need to divide the quantity by 6 (which we can do by multiplying the denominator by 6), and then take 5 of those parts (which we can do by multiplying the numerator by 5).

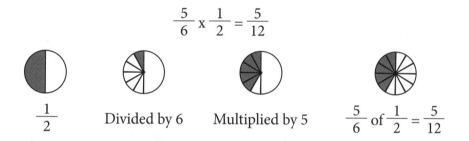

> One of the neat things about fractions is that we can view them both as an operation (division) and a quantity (the result of that division). While we pictured fractions as partial quantities in order to understand multiplication better, know that **each of the fractions we looked at was still representing division.**

Fractions in Music

Did you realize that music utilizes fractions? Well, it does! Hum any tune. Notice that we linger on some notes longer than others. When we write music for songs, we need a way of specifying how long to hold each note.

Musicians use different notes to specify different lengths of time. The notes are based on a consistent rhythm called a *beat*. When you clap your hands to a song, you are clapping the beat of the song. The symbol called a whole note is most commonly used to specify 4 beats.

In music, the time signature determines how many beats a note will have. If you know how many beats one note has, you can use the fractional values of notes to figure out how many beats the other notes have!

A variety of other notes represent a fraction of the whole note. When a whole note is worth 4 beats, a half note is worth 2 beats ($\frac{1}{2}$ of 4 is 2), a quarter note is worth 1 beat ($\frac{1}{4}$ of 4 is 1), and an eighth note is worth $\frac{1}{2}$ of a beat ($\frac{1}{8}$ of 4 is $\frac{1}{2}$).

On the other hand, if the whole note is worth 8 beats, then the half note is worth 4 beats ($\frac{1}{2}$ of 8 is 4), and so on for the other notes.

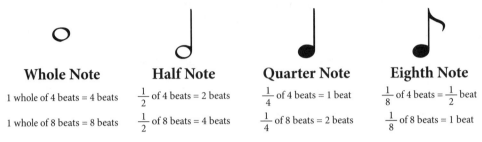

Whole Note	Half Note	Quarter Note	Eighth Note
1 whole of 4 beats = 4 beats	$\frac{1}{2}$ of 4 beats = 2 beats	$\frac{1}{4}$ of 4 beats = 1 beat	$\frac{1}{8}$ of 4 beats = $\frac{1}{2}$ beat
1 whole of 8 beats = 8 beats	$\frac{1}{2}$ of 8 beats = 4 beats	$\frac{1}{4}$ of 8 beats = 2 beats	$\frac{1}{8}$ of 8 beats = 1 beat

Notice all the times we used the word *of*. You could substitute multiplication for each one.

The application of fractions in music is just one example of how we can use math to help us praise the Lord, serve and encourage others, and simply refresh ourselves.

Keeping Perspective

Today we looked at multiplication again — at the consistent way even partial quantities multiply. Let's pause for a moment and consider the One holding all this together.

Day in and day out, Jesus is holding everything together so consistently that we can develop methods based on that consistency to work with fractions. Is anything too hard for Him? Of course not! God is perfectly capable of doing everything He says He will do in His Word.

*Behold, I am the L*ORD*, the God of all flesh: is there any thing too hard for me? (Jeremiah 32:27)*

6.2 Working with Mixed Numbers

It's time to apply what you've learned about adding, subtracting, and multiplying fractions to mixed numbers. As we do, you'll see that the same principles apply!

Addition

Say we have $1\frac{30}{36}$ yards on one bolt, and $2\frac{9}{36}$ on another. Can you guess how to add these numbers together?

If you said to add the whole numbers together and the fractions together, you were correct; that is one way to solve the problem.

Example: Solve $1\frac{30}{36} + 2\frac{9}{36}$ by adding the whole numbers and fractions together.

Add whole numbers: $1 + 2 = 3$

Add fractions: $\frac{30}{36} + \frac{9}{36} = \frac{39}{36} = 1\frac{3}{36}$

Add results together: $3 + 1\frac{3}{36} = 4\frac{3}{36}$ (which reduces to $4\frac{1}{12}$)

Instead, we could have converted both mixed numbers into improper fractions and then added them together; $1\frac{30}{36}$ would convert to the improper fraction $\frac{66}{36}$ and $2\frac{9}{36}$ would convert to $\frac{81}{36}$. We could then add them as fractions.

Example: Solve $1\frac{30}{36} + 2\frac{9}{36}$ by converting to improper fractions first.

Convert to improper fractions:

$$1\frac{30}{36} = \frac{66}{36} \qquad 2\frac{9}{36} = \frac{81}{36}$$

Solve: $\frac{66}{36} + \frac{81}{36} = \frac{147}{36}$

Simplify: $\frac{147}{36} = 4\frac{3}{36} = 4\frac{1}{12}$

Subtraction

When subtracting mixed numbers, it's helpful to **rename mixed numbers as improper fractions in order to solve the problem**. This will avoid confusion when the fractional quantity we're subtracting is greater than the fractional quantity we're subtracting it from.

Example: Solve $1\frac{1}{3} - \frac{2}{3}$

Notice that $\frac{2}{3}$ is greater than $\frac{1}{3}$ ... but not greater than $1\frac{1}{3}$!

Convert the mixed number to an improper fraction:

$$1\frac{1}{3} - \frac{2}{3} = \frac{4}{3} - \frac{2}{3}$$

Solve: $\frac{4}{3} - \frac{2}{3} = \frac{2}{3}$

Example: Solve $2\frac{1}{3} - 1\frac{5}{8}$

Convert both mixed numbers to improper fractions:

$$2\frac{1}{3} - 1\frac{5}{8} = \frac{7}{3} - \frac{13}{8}$$

Rewrite so the denominators are the same: $\frac{7}{3} - \frac{13}{8} = \frac{56}{24} - \frac{39}{24}$

Solve: $\frac{56}{24} - \frac{39}{24} = \frac{17}{24}$

Multiplication

Let's say we didn't quite want to double a cookie recipe. Instead, we wanted to make $1\frac{1}{2}$ batches. If the recipe calls for $\frac{1}{4}$ cup of sugar, we would need $1\frac{1}{2}$ times that quantity. We could write this as $1\frac{1}{2} \times \frac{1}{4}$.

What does this equal? While you may already know how to find the answer mathematically, let's draw the problem out using pies as a visual to see what's really happening. We want to take $\frac{1}{4}$ one-and-a-half times. This results in $\frac{3}{8}$ of a whole.

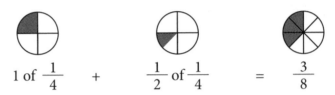

Or say we wanted to quarter a recipe that used $1\frac{1}{2}$ cups of sugar. In this case, we want to find $\frac{1}{4}$ of $1\frac{1}{2}$. What would that equal?

Again, let's use pies as a visual to find the answer. Notice that this time the pies have been cut into eighths, and there are 12 pieces in $1\frac{1}{2}$ pies (the shaded section).

If we take $\frac{1}{4}$ of those 12 pieces, we would have 3 pieces, or $\frac{3}{8}$ of a whole pie.

$$\frac{1}{4} \text{ of } 1\frac{1}{2} = \frac{3}{8}$$

Notice that $\frac{1}{4} \times 1\frac{1}{2}$ equals the same quantity ($\frac{3}{8}$) as $1\frac{1}{2} \times \frac{1}{4}$ equaled. It doesn't matter which quantity we use as the multiplier; the answer will be the same, since multiplication is commutative.

Now, it's not practical to draw a visual every time we want to solve a problem. Mathematically, we can easily multiply mixed numbers by **converting mixed numbers to improper fractions and multiplying as usual.**

Example: Solve $\dfrac{1}{4}$ x $1\dfrac{1}{2}$

Convert the mixed number to an improper fraction: $\dfrac{1}{4}$ x $\dfrac{3}{2}$

Multiply: $\dfrac{1}{4}$ x $\dfrac{3}{2}$ = $\dfrac{1 \times 3}{4 \times 2}$ = $\dfrac{3}{8}$

Example: Solve $1\dfrac{1}{2}$ x $\dfrac{1}{4}$

Convert the mixed number to an improper fraction: $\dfrac{3}{2}$ x $\dfrac{1}{4}$

Multiply: $\dfrac{3}{2}$ x $\dfrac{1}{4}$ = $\dfrac{3}{8}$

Another Approach

Since multiplication is distributive (see 4.2), we could have solved $1\dfrac{1}{2}$ x $\dfrac{1}{4}$ using distribution, distributing the multiplication to the whole number portion and the fractional portion.

$$1\dfrac{1}{2} \times \dfrac{1}{4} = (1 \times \dfrac{1}{4}) + (\dfrac{1}{2} \times \dfrac{1}{4})$$

$$\dfrac{1}{4} \quad + \quad \dfrac{1}{8}$$

$$\dfrac{2}{8} \quad + \quad \dfrac{1}{8} \quad = \quad \dfrac{3}{8}$$

Now, if you're multiplying two mixed numbers, it can get a little more complicated to make sure you've distributed all the multiplication. So in general, stick with converting your mixed numbers to improper fractions and multiplying as usual.

Just know that there's often more than one way to go about finding the answer. Don't settle for just memorizing rules as to how to work with fractions — make sure you understand the principle.

Keeping Perspective

Much of math is focused on applying the same basic principles to different notations (first to fractions and now to mixed numbers). Keep remembering that we have these different notations because God created an amazing world, and the different notations help us describe different situations we encounter.

6.3 Simplifying While Multiplying

One of the goals of mathematics is to make every operation as simple as we possibly can. Much as an inventor tries to invent machines to simplify tasks, we try to think of methods to simplify the process of recording quantities.

Notice how when we multiply fractions, we often get a fraction that needs to be simplified in order to make it more understandable. For example, let's say that you learned from one news source that $\frac{3}{5}$ of the people who voted in a specific election were male, and from another that $\frac{5}{14}$ of the male voters voted for your candidate. You want to know what fraction of all the voters — both male and female — were male voters who voted for your candidate. You can find this by finding $\frac{5}{14}$ of $\frac{3}{5}$.

Example: Solve $\frac{5}{14}$ x $\frac{3}{5}$

$$\frac{5}{14} \times \frac{3}{5} = \frac{5 \times 3}{14 \times 5} = \frac{15}{70}$$

Now we need to simplify: $\frac{15}{70} \div \frac{5}{5} = \frac{15 \div 5}{70 \div 5} = \frac{3}{14}$

Imagine how nice it would be if we could multiply in such a way as to get a simplified answer right off the bat. The question is, can we, and if so, how?

Again, while you may already know how to do this, thinking through why this shortcut works will bring better understanding to what is really happening. Let's take a look.

Coming to the Shortcut

Take a closer look at the multiplication of $\frac{5}{14}$ and $\frac{3}{5}$. When we multiply the numerators and the denominators, we get this:

$$\frac{5 \times 3}{14 \times 5}$$

Notice how we have a 5 in the numerator and a 5 in the denominator. So *before* we complete the multiplication, we could divide the whole equation by $\frac{5}{5}$.

$$\frac{5 \times 3}{14 \times 5} \div \frac{5}{5} = \frac{1 \times 3}{14 \times 1} = \frac{3}{14}$$

Since 5 divided by 5 equals 1, dividing by 5 replaced the 5 in both the numerator and the denominator with 1 . . . which we can ignore, since any number times 1 equals itself (we didn't really even need to write out that step). We now have

142 | PRINCIPLES OF MATHEMATICS 1

our answer ($\frac{3}{14}$), and we never had to bother to complete any unneeded multiplication or perform anything but very simple division.

But it gets even easier. We could have done all our simplification without rewriting the problem at all.

For example, we could show that we're dividing by 5 by simply crossing a line through each 5, replacing it with a 1 (the result of 5 ÷ 5).

$$\frac{\overset{1}{\cancel{5}} \times 3}{14 \times \underset{1}{\cancel{5}}} = \frac{3}{14}$$

And now we have a simplified answer all in one step!

Making It a Little More Complex

In $\frac{5 \times 3}{14 \times 5}$, it was easy to see there was a 5 in both the numerator and the denominator. Sometimes simplifying fractions isn't quite so obvious — but what you've learned about factoring can help.

Let's change the numbers in our voting problem. Suppose this time that you learned that $\frac{2}{15}$ of all the voters in the election were Independents, and that $\frac{5}{14}$ of the Independents voted for the Republican candidate. You want to find what fraction of all the voters were Independent voters who voted for the Republican candidate. You need to find $\frac{5}{14}$ of $\frac{2}{15}$.

$$\frac{5}{14} \times \frac{2}{15}$$

When we multiply the numerators and the denominators, we get this:

$$\frac{5}{14} \times \frac{2}{15} = \frac{5 \times 2}{14 \times 15}$$

At first glance, none of the numerators and the denominators are the same. However, let's think about the factors that make up 14: 2 x 7. Notice the 2. So we have a 2 in the numerator and a number that can divide by 2 (the 14) in the denominator. So before we finish multiplying, we can divide by $\frac{2}{2}$.

$$\frac{5 \times 2}{14 \times 15} \div \frac{2}{2} = \frac{5 \times 1}{7 \times 15}$$

Now, we did not have to do all that writing. We could have shown the division in the original problem ($\frac{5}{14} \times \frac{2}{15}$) like this:

$$\frac{5}{\underset{7}{\cancel{14}}} \times \frac{\overset{1}{\cancel{2}}}{15}$$

6. MORE WITH FRACTIONS | 143

Notice how we divided the denominator by dividing the 14 by 2, which equals 7. So instead of a 1, we put down a 7, as that's what we get when we divide 14 by 2.

Now notice that we have a 5 in the numerator and a number that can be divided by 5 in the denominator. Let's divide the whole equation by $\frac{5}{5}$.

$$\frac{5}{7 \times 15} \div \frac{5}{5} = \frac{1}{7 \times 3}$$

or, more simply, like this:

$$\frac{\cancel{5}^{1}}{\cancel{14}_{7}} \times \frac{\cancel{2}^{1}}{\cancel{15}_{3}}$$

Again, notice how we divided the denominator by 5 by dividing 15 by 5. After all, 15 can be thought of as 5 x 3.

Now we can easily solve to get the simplified answer.

$$\frac{1}{7 \times 3} = \frac{1}{21}$$

or

$$\frac{\cancel{5}^{1}}{\cancel{14}_{7}} \times \frac{\cancel{2}^{1}}{\cancel{15}_{3}} = \frac{1}{21}$$

Make the Connection

We just used the concept of factoring to find a simplified answer to $\frac{5}{14} \times \frac{2}{15}$. We saw the 14 in the first denominator as 2 x 7, and the 15 in the second denominator as 5 x 3, and used that knowledge to simplify prior to multiplying. It's important to understand that you can **think of numbers in terms of the factors that compose them**.

Notice that $\frac{1}{21}$ is the *same answer* we would have arrived at if we completed the multiplication first, and then divided by $\frac{10}{10}$ to simplify.

$$\frac{5}{14} \times \frac{2}{15} = \frac{10}{210}$$

$$\frac{10 \div 10}{210 \div 10} = \frac{1}{21}$$

Why is this? When we simplified as we went, we really divided by $\frac{2}{2}$, and then by $\frac{5}{5}$. Since $\frac{2}{2} \times \frac{5}{5} = \frac{10}{10}$, we essentially divided the fraction by $\frac{10}{10} \ldots$ only we did it as we went instead of at the end.

Understanding Check

Be sure you understand the principle behind simplifying fractions: since dividing by 1 doesn't change the value (a fact about how God causes objects to divide we learned back in 3.5), we can divide any fraction by a fraction worth 1 without changing its value. It doesn't matter if we do this at the end of multiplying fractions, or while performing the multiplication.

But be careful. We would *not* be able to simplify the division in this fraction:

$$\frac{5 \times 2}{(2 \times 7) + (3 \times 5)}$$

Why? The fraction line means to divide. But here we want to divide 5 x 2 by (2 x 7) *plus* (3 x 5). We want to divide the *sum* of the two products.

When all we have is multiplication within a fraction, we can divide before we complete the multiplication, as a number can be thought of as the product of its factors. But don't misapply this shortcut when there's another operation involved!

$$\frac{5 \times 2}{(2 \times 7) + (3 \times 5)} = \frac{1}{10} \qquad\qquad \frac{5 \times 2}{(2 \times 7) + (3 \times 5)} = \frac{10}{14 + 15} = \frac{10}{29}$$

Incorrect   Correct

Keeping Perspective

In this chapter, we've been looking at a lot of mechanics — i.e., how to solve different problems on paper. It's easy to lose sight of what's going on in the rules and the terms, so let's pause for a moment.

Simplifying fractions as we multiply them together is an example of how men used their God-given observational skills and ability to reason to find a time-saving method. It not only makes it easier to multiply fractions, but it also saves us from having to do as much simplification on the answer!

Shortcuts like this ultimately work because they describe a real-life consistency God holds in place year after year, decade after decade. And that same God offers His strength and power to His people. Now there's a thought to ponder.

> *O God, thou art terrible out of thy holy places: the God of Israel is he that giveth strength and power unto his people. Blessed be God (Psalm 68:35).*

6.4 Reciprocal/Multiplicative Inverse and More

We've now seen how to add, subtract, and multiply using the fractional notation. Before we look at our final operation — division — there are a couple of aspects of fractions we need to explore in order to understand why we divide fractions the way we do. You may at first wonder why you need to know these things, but we'll see in the next lesson that they apply to one degree or another every time we divide a fraction using the common shortcut. Math builds on itself, and mathematicians are always looking for ways to shorten a process by making observations about how a notation works with the consistencies God created and sustains.

Reciprocal / Multiplicative Inverse

Notice that the second fraction in each pair below is the exact opposite of the first one — the numerator has become the denominator, and vice versa.

$$\frac{4}{3} \qquad \frac{3}{4}$$

$$\frac{1}{2} \qquad \frac{2}{1}$$

$$\frac{2}{3} \qquad \frac{3}{2}$$

Notice that we could write whole numbers this way too. After all, since dividing by 1 doesn't change the value, we can think of any whole number as a fraction divided by 1. We could then flip that fraction just as we did the others.

$$2 \quad \text{or} \quad \frac{2}{1} \qquad \frac{1}{2}$$

$$3 \quad \text{or} \quad \frac{3}{1} \qquad \frac{1}{3}$$

etc.

What would happen if we multiply a fraction by its flipped version? We'd essentially be multiplying the fraction by its opposite, which effectively turns it into a number divided by itself, giving us 1.

$$\frac{2}{3} \times \frac{3}{2} = \frac{6}{6} = 1$$

Notice how if we simplify as we solve, both the numerator and the denominator always cancel out, leaving us with 1.

$$\frac{\overset{1}{\cancel{2}}}{\underset{1}{\cancel{3}}} \times \frac{\overset{1}{\cancel{3}}}{\underset{1}{\cancel{2}}}$$

146 | PRINCIPLES OF MATHEMATICS 1

Although it may not seem so at first glance, knowing what to multiply a number by to make 1 is a very helpful thing. It's so helpful, in fact, that we have a name to describe the **number that, when multiplied by another number, equals 1**. Well, more than one name, actually. Some people call it the **reciprocal** of a number, while others refer to it as the **multiplicative inverse**, or, so long as the context is clear, the **inverse**.

In a fraction, the inverse will always be a flipped version of the fraction (where the numerator and denominator have been switched), while for a whole number, it will be 1 over the number (since a whole number can be written as a fraction divided by 1).

In the next lesson, you'll take a look at using the multiplicative inverse to help divide fractions. Because it can essentially reverse the division in a fraction, leaving us with 1, we'll also end up using the multiplicative inverse of fractions as we work with fractions down the road.

Fractions Represent Division — Expanding Our Thinking

Since fractions themselves are a way of representing division, we could represent dividing by fractions as a fraction.

$$4 \div 2 = \frac{4}{2} \qquad 1 \div 2 = \frac{1}{2}$$

$$\frac{1}{2} \div 2 = \frac{\frac{1}{2}}{2} \qquad \frac{1}{2} \div \frac{1}{4} = \frac{\frac{1}{2}}{\frac{1}{4}}$$

Again, you'll need to know this when we explore how to solve fractional division problems in the next lesson.

Keeping Perspective

Notice how we keep building our understanding of math by thinking through how the notation and consistencies we know apply to different situations. Sometimes you might wonder why you have to learn something, but, taken in conjunction with other aspects of math, it might be just the observation you'll need in the next concept to accurately reflect the real-life consistencies God created. In the next lesson, we'll see how both inverses and using a fraction line to represent division of a fraction by a fraction combine to help us arrive at a shortcut to easily divide fractions. While you probably already know the shortcut, understanding why it works will serve you well, both in seeing how math builds and in understanding fractions to a greater level.

6.5 Dividing Fractions

Ready for some more fraction fun? Well, ready or not, today's topic is dividing fractions. You may have already learned to "invert and multiply," but did you realize there are other methods we could use to divide fractions? As we did with multiplying fractions, we're going to start by taking a look at what we're really doing when we divide fractions. As we do, we're going to discover once again that the "rule" we follow is just a way of automating the process of describing a consistency God created and sustains.

Understanding What We're Doing When We Divide

As we saw in 3.4, division means separating into "two or more parts, areas, or groups."[1] We also saw that because of God's faithfulness in holding all things together, division is consistent enough for us to memorize answers and divide using algorithms, or step-by-step methods.

Before we jump to *how* to divide fractions, though, let's explore the concept of dividing fractional quantities. Although fractional division can sometimes be a little more abstract, it too ultimately works because of God's faithfulness.

While we typically think of division as dividing a quantity into smaller parts, it is also "the process of finding out how many times one number is contained in another."[2]

When we divide 10 by 2, our answer, 5, tells us both what would happen if we took 10 and divided it into 2 piles, and also how many times 2 is contained inside 10 (5 times).

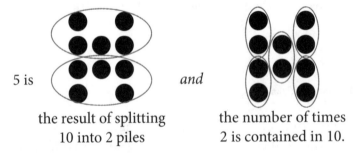

5 is the result of splitting 10 into 2 piles   and   the number of times 2 is contained in 10.

Likewise, when we divide 6 by 2, the answer, 3, not only tells us how many we will have if we divide 6 into 2, but also how many times 2 is contained in 6.

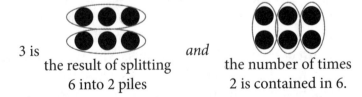

3 is the result of splitting 6 into 2 piles   and   the number of times 2 is contained in 6.

The same holds true when dividing a fractional amount by a whole amount — we can look at the division as what would happen if we divided *and* as how many times one number is contained in the other.

$\frac{1}{2} \div 2 = \frac{1}{4}$

$\frac{1}{4}$ is the result of splitting $\frac{1}{2}$ into 2 portions

and the number of times 2 is contained in $\frac{1}{2}$. Only 1 out of 4 parts of 2 (that is, $\frac{1}{4}$) is contained in $\frac{1}{2}$.

$1\frac{1}{2} \div 2 = \frac{3}{4}$

$\frac{3}{4}$ is the result of splitting $1\frac{1}{2}$ into 2 portions

and the number of times 2 is contained in $1\frac{1}{2}$. 3 out of 4 parts (that is, $\frac{3}{4}$) of 2 is contained in $1\frac{1}{2}$.

It proves extremely helpful for understanding dividing by a fraction if you think of dividing fractions as figuring out how many times one number is contained in another.

$2 \div \frac{1}{2} = 4$  $\frac{1}{2}$ is contained in 2 a total of 4 times.

$\frac{1}{2} \div \frac{1}{4} = 2$ $\frac{1}{4}$ is contained in $\frac{1}{2}$ 2 times.

Getting to a Rule

So how do we reduce division by fractions to something we can do mechanically without having to draw diagrams? You've probably been taught to **invert and multiply** — invert (i.e., use the **inverse** of — the flipped version we learned about in the previous lesson) the second fraction and then multiply.

Example: Solve $\frac{1}{2} \div \frac{1}{4}$

$$\frac{1}{2} \div \frac{1}{4} = \frac{1}{2} \times \frac{4}{1} = \frac{4}{2} = 2$$

But why does this rule work? Let's step back for a minute and examine dividing fractions more carefully.

6. MORE WITH FRACTIONS | 149

Division Means Division

When we multiply, we multiply both our numerators and our denominators to get our answer. We can do the same thing in division if we have a problem where the numerators and denominators divide evenly.

$$\frac{8}{6} \div \frac{2}{3} = \frac{8 \div 2}{6 \div 3} = \frac{4}{2} = 2$$

There's just one problem: sometimes we run into fractions that don't divide so easily, like our original example.

$$\frac{1}{2} \div \frac{1}{4}$$

Following the same process as before, we would end up with a fraction in the denominator.

$$\frac{1}{2} \div \frac{1}{4} = \frac{1 \div 1}{2 \div 4} = \frac{1}{\frac{2}{4}}$$

We still haven't found a very useful answer. But remember how we learned that we can multiply any fraction by a fraction worth 1 without changing its value? Well, even though $\frac{1}{\frac{2}{4}}$ has an additional fraction in the denominator, the whole thing is still a fraction, so we could multiply it by a fraction worth 1 without changing its value. Notice that if we multiply the fraction by $\frac{\frac{4}{2}}{\frac{4}{2}}$ (the inverse of the denominator), the denominator will cancel out.

Multiplying by a fraction worth 1:

$$\frac{1}{\frac{\cancel{2}}{\cancel{4}}} \times \frac{\frac{4}{2}}{\frac{\cancel{4}}{\cancel{2}}} = \frac{\frac{4}{2}}{1}$$

Since any number divided by 1 equals itself, this leaves us with $\frac{4}{2}$, or 2. The answer to $\frac{1}{2} \div \frac{1}{4}$ is 2. Notice we found the same answer we did with the cupcake illustration and with the "invert and multiply" rule, only it took us more steps.

Understanding the Rule

Don't worry if you didn't follow the whole process we went through above. Just notice how when we multiplied by a fraction worth 1, the fraction in the denominator cancelled out, essentially leaving us with **1 x $\frac{4}{2}$**.

$$\frac{1}{\frac{\cancel{2}}{\cancel{4}}} \times \frac{\frac{4}{2}}{\frac{\cancel{4}}{\cancel{2}}} = \frac{\frac{4}{2}}{1}$$

While it may appear different, this is the same basic multiplication we get when we invert and multiply.

Result we just obtained	Result of inverting and multiplying
$1 \times \dfrac{4}{2} = \dfrac{1 \times 4}{2} = 2$	$\dfrac{1}{2} \div \dfrac{1}{4} = \dfrac{1}{2} \times \dfrac{4}{1} = \dfrac{1 \times 4}{2 \times 1} = 2$

The "rule" for dividing by fractions by inverting the divisor and multiplying is just a shortcut to jump to the end result and save ourselves a lot of work! It makes it possible for us to quickly find an answer without rethinking through as much math each time.

Another Way to Divide Fractions

The "invert and multiply" method for dividing fractions is just one way to figure out on paper what happens in real life when we divide fractions. Let's take a quick look at another method of solving fractional division.

Example: Solve $\dfrac{1}{2} \div \dfrac{1}{4}$

We could divide by first renaming the fractions until our denominators are equal, as we did for addition and subtraction. If our denominators are equal, then when we divide the denominator, it will equal 1, as any number divided by itself equals 1. (See 5.3.)

Rename so there are equal denominators:

$$\frac{1}{2} \div \frac{1}{4} = \frac{2}{4} \div \frac{1}{4}$$

Divide the numerators and the denominators: $\dfrac{2 \div 1}{4 \div 4}$

Simplify: $\dfrac{2 \div 1}{1} = 2 \div 1 = \dfrac{2}{1} = 2$

In other words, to divide fractions, we could rewrite the fractions so they both have the same denominators, and then divide the numerators. While the "invert and multiply" method is faster, remember that it's just one way to help us work with the consistency around us — a consistency God created and sustains.

Dividing Mixed Numbers or Fractions with Whole Numbers

Any guesses on how to divide whole or mixed numbers? Convert them to improper fractions and then follow one of the methods discussed above.

Keeping Perspective

Today, we explored the "**invert and multiply**" division method, which uses the identity property of multiplication (that multiplying by 1 doesn't change the value), coupled with what we know about multiplying fractions, inverses, and more.

Each "rule" in math is a tool — a way of building on what we know to make it easy next time we have to solve a similar problem. And remember, making rules like this is only possible because of the consistency with which God governs all things.

And as a reminder, if you're wondering why you need to be familiar with fractions, keep in mind that you're going to need a thorough understanding of fractions in algebra, where you'll learn to use fractions to explore God's universe at an even deeper level.

6.6 Chapter Synopsis

Congratulations! You've made it to the end of our fractions study — for now. We'll be using fractions throughout the rest of the course in various capacities. We've only touched on fractions' uses here, as many of them require other concepts we haven't studied yet. Fractions are a tool you'll pull out many times!

Don't worry if you find fractions a little harder to get your mind around than some other concepts. They are more abstract. Just remember that all the many rules for working with them are ways of describing the consistent way God governs all things.

Here are the "rules" we learned:

- **Multiplying fractions:** Multiply the numerators and the denominators. (Rewrite any whole or mixed numbers as fractions first.)

- **Reciprocal / Multiplicative Inverse:** A reciprocal or multiplicative inverse is a name to describe a number that, when multiplied by another number, equals 1. In a fraction, the inverse will always be a flipped version of the fraction (where the numerator and denominator have been switched), while for a whole number, it will be 1 over the number.

- **Dividing fractions:** Invert the second fraction and multiply.

- **Mixed Numbers:** We can use the same processes for mixed numbers as we do for fractions if we convert them to improper fractions first.

The most important thing to remember is that these rules ultimately work because of the power of God that's holding all things together. In the *Truth Project*, Del Tackett asks a very profound question: Do we really believe that what we believe is really real? It's one thing to say God is in charge and holding all things together — it's a very different thing to let that belief permeate our lives and govern all our actions. As you complete the worksheet in your *Student Workbook*, ask God to show you one specific area where you need to remember His might and power.

[CHAPTER 7]

Decimals

7.1 Introducing Decimals

Throughout this course, we've been using the decimal system (i.e., our base-10 place-value system) to represent whole numbers. It's time now to look at extending this system to also represent partial quantities. (The word *decimal* actually comes from the Latin root *decimus*, which means "tenth."[1])

Most of us were exposed to representing partial amounts in the decimal system since we were little, mainly because we use them to write parts of a dollar. But let's take a closer look at this notation and at how to use it as an effective "tool" in our mathematical toolbox.

Decimals — Extending the Notation

To better understand how to use our decimal system to describe partial quantities, picture a grocery store without decimals. How would you describe the price of items less than $1? You could use fractions.

$\$\frac{4}{10}$ $\$\frac{2}{5}$ $\$\frac{1}{2}$

Most likely, you would write all your fractions with the same denominator to make them easier to compare. Since there are 100 cents in $1, it would make sense to use 100 as the denominator.

$\$\frac{40}{100}$ $\$\frac{40}{100}$ $\$\frac{50}{100}$

Now it is easier to compare the cost of each item. We could further simplify expressing these costs by writing these costs in our base-10 decimal system, letting the place, or location, of the number show its denominator.

$ 0.40 $ 0.40 $ 0.50

Let's take a look at what we just did. We basically added what we call a **decimal point** to the right of the ones digits and extended place value to represent partial quantities.

Each place to the left in our decimal system is worth 10 times the previous place, and each place to the right is worth $\frac{1}{10}$ of the previous place. The decimal point separates the whole numbers from the partial ones.

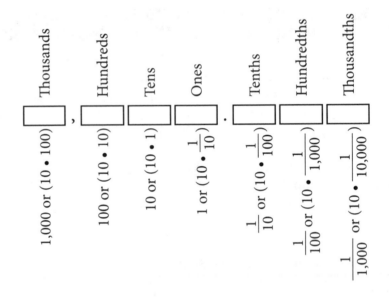

> We're going to begin using a dot (•) rather than the times sign (x) to represent multiplication in our presentations. Since an "x" has a different meaning in algebra, it's important to become familiar with other ways to show multiplication. Just remember, 4 x 2, 4 • 2, and 4(2) all mean four *times* two!

Whether representing a partial quantity or a whole one in the decimal system, the place, or location, of each digit determines its value. Notice how the place of the digit "6" gives it radically different meanings!

154 | PRINCIPLES OF MATHEMATICS 1

$$0.006 = 6 \text{ thousandths or } \frac{6}{1,000}$$

$$0.06 = 6 \text{ hundredths or } \frac{6}{100}$$

$$0.6 = 6 \text{ tenths or } \frac{6}{10}$$

$$6 = 6 \text{ ones or six}$$

$$60 = 6 \text{ tens or sixty}$$

$$600 = 6 \text{ hundreds or six hundred}$$

From now on, we'll refer to partial quantities written in the decimal system as **decimals.**

Switching Between Fractions and Decimals

Example: Express 0.6 as a fraction.

What fraction does 0.6 represent? Well, the 6 is in the tenths place ($\frac{1}{10}$), so the 6 represents 6 tenths, or $\frac{6}{10}$.

Example: Express 0.61 as a fraction.

Again, the 6 is in the tenths place, so we have 6 tenths ($\frac{6}{10}$). However, we also have a 1 in the hundredths place, giving us $\frac{1}{100}$. So we have $\frac{6}{10} + \frac{1}{100}$. Rewriting with the same denominator and adding gives us $\frac{60}{100} + \frac{1}{100}$, or $\frac{61}{100}$.

Now, in the last example, we didn't really need to do the addition. Because of how place value works, we could have looked at 0.61 as $\frac{61}{100}$ to begin with. All we really needed to do was remove the decimal point and add the denominator from the right-most place value represented.

You can always look at the entire decimal part of a number as a fraction of the right-most decimal place represented.

Example: Express 0.612 as a fraction.

We can view this as a fraction of the right-most decimal place represented. So we can think of it as $\frac{612}{1,000}$.

We could also have found that via addition:

$$\frac{6}{10} + \frac{1}{100} + \frac{2}{1,000} = \frac{600}{1,000} + \frac{10}{1,000} + \frac{2}{1,000} = \frac{612}{1,000}$$

7. DECIMALS | 155

You may already know how to convert $\frac{4}{5}$ to a decimal by dividing. We'll get to that later in this chapter, but for now, practice the way shown in the example.

Note: We could then simplify the fraction if we need a simple answer.

$$\frac{612}{1,000} \div \frac{4}{4} = \frac{153}{250}$$

Example: Express $\frac{4}{5}$ as a decimal.

Remember, our decimal place-value system expresses partial quantities in terms of tenths or multiples of tenths. So to write this quantity as a decimal, we first need to make its denominator 10 or a multiple of 10.

$$\frac{4}{5} \cdot \frac{2}{2} = \frac{8}{10}$$

Now we can rewrite it in the decimal system.

0.8

Remember, the first place to the right of the decimal point represents fractions of 10, so 0.8 is another way of saying 8 tenths, or $\frac{8}{10}$.

Applying It: Writing Checks

When writing a check, we write the amount using our decimal system, and then rewrite it in words. Instead of using words, though, the cents are typically written as a fraction (notice that the fraction takes up less space than the words "twenty-seven hundredths" would).

Adding Zeros

You can always add zeros to the right of a decimal point.

$$0.6 = 0.60 = 0.600 = 0.6000 = \frac{6}{10} = \frac{60}{100} = \frac{600}{1,000} = \frac{6,000}{10,000}$$

Adding (or removing) a zero to the right of the last number of a decimal does not change its value; adding (or removing) a 0 actually multiplies (or divides) the fraction by $\frac{10}{10}$, which is worth 1.

$$0.6 = \frac{6}{10}$$

$$\frac{6}{10} \cdot \frac{10}{10} = \frac{60}{100} = 0.60$$

For example, when working with money in America, we represent amounts in terms of dollars and cents. Since there are **100 cents in a dollar**, we always represent the partial amount of a dollar in terms of hundredths.

For example, rather than writing $0.1, we would add a zero, making this $0.10 (which is essentially just rewriting $\frac{1}{10}$ as $\frac{10}{100}$). Now it's easy to see that this represents 10 cents.

Reading Partial Quantities in the Decimal System

0.1 could be read "one tenth" just as you would read $\frac{1}{10}$; 3.24 could be read "three and twenty-four hundredths" (notice the "and" used to break up the whole and partial portion of the number).

However, you'll sometimes hear 0.1 read as "point one" and 3.24 as "three point two four" or "three point twenty-four" instead.

Or, if we were dealing with money, we'd read 0.1 as "10 cents" and 3.24 "three dollars and twenty-four cents."

A Look at History

While we're quite used to seeing quantities written with a decimal point today, that hasn't always been the case. Notice some of the different ways "2.16" has been expressed — one mathematician even used our current equal sign as a way of separating whole numbers from partial amounts![2]

2⓪1①6②	216②	2.1.6.	2,16	2:16
2ǀ16	2=16	2▲16	2¹⁶	2'16

Over the years, mathematicians have tried to standardize notation. After all, it is a lot easier not to have to learn a whole new set of symbols for every math book you read! In America, we use a decimal point. In some other countries, however, a comma is used instead of a decimal point. Our current notation is only *one* way to describe quantities on paper. God gave man the creativity to develop different ways to express quantities.

7. DECIMALS | 157

Keeping Perspective

Decimals (i.e., partial quantities written in our base-10 decimal system) are another way of describing and working with quantities. We are only able to explore creation this way *because God gave us this ability*. Mice do not do math, but man can because God created man in His image. Isn't it wonderful that God designed us differently than animals and gave us the ability to fellowship with Him?

The account of King Nebuchadnezzar (Daniel 4) really brings this point home. King Nebuchadnezzar boasted to himself about Babylon, taking credit for building such a great empire. God humbled him and made him like the beasts of the earth. God was showing him *he could not even think or function apart from God's enabling*. We are utterly and completely dependent on God for *everything*, including the ability to think and name quantities!

7.2 Adding and Subtracting Decimals

One of the biggest advantages to expressing partial quantities using the decimal system is that we can then use the same basic methods to work with them as we do with whole numbers! Again, although you probably already have used these methods with decimal numbers, we're going to take a deeper look at *why* we're able to apply these methods to decimals and *what* we're really doing when we do.

Addition and Subtraction

Notice that *each* digit — including those to the right of the decimal point — in the decimal system is 10 times the previous one.

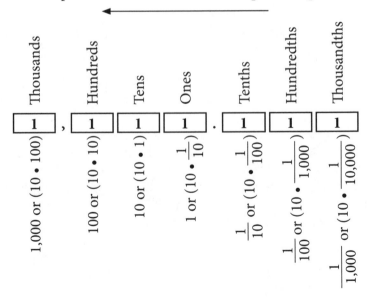

Each place to the left is 10 times the previous place.

Keep this in mind as we look at some simple additions.

Say we need to add 0.9 + 0.1. Notice that if we rewrite this as fractions, we see that it equals 1.

$$0.9 + 0.1$$

$$\frac{9}{10} + \frac{1}{10} = \frac{10}{10} = 1$$

Now say we want to add 0.09 + 0.01. Again, let's do it with fractions.

$$0.09 + 0.01$$

$$\frac{9}{100} + \frac{1}{100} = \frac{10}{100}$$

We could simplify the answer down to $\frac{1}{10}$, as $\frac{10}{100} \div \frac{10}{10} = \frac{1}{10}$.

Notice how we could have found both of these answers using the standard addition algorithm we use for whole numbers. **We can rename partial quantities just as we can whole quantities** because *each* place is still 10 times the previous one. Thus, we can add and subtract decimals (i.e., partial quantities written in the decimal system) using the *same basic processes* we do for whole numbers.

Rename $\frac{10}{10}$ as 1:

$$\begin{array}{r} \overset{1}{}0.9 \\ +0.1 \\ \hline 1.0 \end{array}$$

Rename $\frac{10}{100}$ as $\frac{1}{10}$:

$$\begin{array}{r} \overset{1}{}0.09 \\ +0.01 \\ \hline 0.10 \end{array}$$

Now let's picture adding 0.26 + 0.07 on an abacus for a moment. Notice how we labeled the bottom three rows of our abacus as partial quantities, and that we exchanged 10 from one row for 1 from the row above. This is *exactly* what we do on paper when we add partial amounts using our traditional addition algorithm.

Example: Find 0.26 + 0.07 on an abacus.

Step 1: *Form the starting quantity of 0.26.*

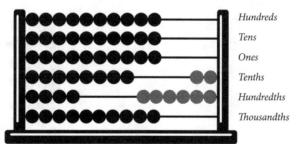

Step 2: *Add 0.07.*

Notice how we run out of beads after adding 4 out of the 7 hundredths.

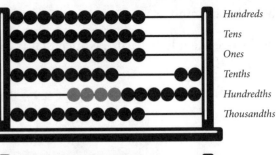

We have to rename 10 hundredths as 1 tenth.

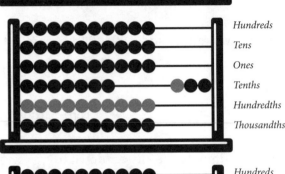

And then we can add the remaining 3 hundredths.

Answer: *0.33*

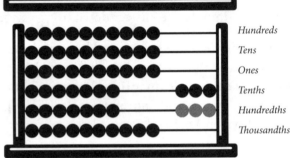

Example: Find 0.26 + 0.07 on paper.

Rename $\frac{10}{100}$ as $\frac{1}{10}$:

$$\begin{array}{r} \overset{1}{}0.26 \\ +\,0.07 \\ \hline 0.33 \end{array}$$

Likewise, the same process works for subtracting partial amounts written as decimals that we use for whole numbers.

Example: Find 0.26 − 0.07 on an abacus.

Step 1: *Form the starting quantity of 0.26.*

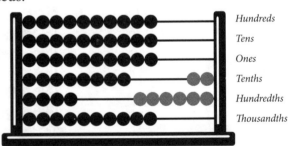

Step 2: *Subtract 0.07.*

We only have 6 hundredths and we need to subtract 7. So we have to rename 1 tenth as 10 hundredths and mentally subtract 7 from that, leaving 3 additional hundredths.

Another way of thinking about it is that renaming the 1 tenth to 10 hundredths gives us a total of 16 hundredths, which minus 7 equals 9.

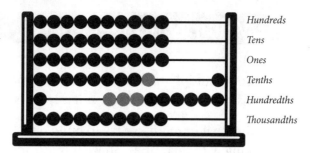

Answer: 0.19

Example: Find 0.26 − 0.07 on paper:

Rename $\frac{1}{10}$ as $\frac{10}{100}$:

$$\begin{array}{r} \overset{1\ \ 1}{0.\cancel{2}6} \\ -\ 0.07 \\ \hline 0.19 \end{array}$$

Expanding It Further

It doesn't matter if the numbers we're adding and subtracting contain whole numbers, partial quantities, or a combination — if they're written using the decimal system, we can use our standard algorithm to keep track of place value.

$$\begin{array}{r} \overset{1\ \ \ 1\ \ 1}{4.157} \\ +\ 5.878 \\ \hline 10.035 \end{array} \qquad \begin{array}{r} \overset{0\ \ 11\ \ 12\ 13\ 1}{\cancel{1}\cancel{2}.\cancel{3}\cancel{4}2} \\ -\ 8.978 \\ \hline 3.364 \end{array}$$

Adding 0s

Note that it's important to keep our digits lined up so we subtract tenths from tenths, hundredths from hundredths, etc. Otherwise, we could end up with a totally incorrect answer! Adding zeros to the right of a number so it has the same number of total digits as the other numbers we're working with can help guard against accidentally lining up digits incorrectly.

$$\begin{array}{r} 48.23 \\ -\ \ \ 4.1 \\ \hline \end{array} \qquad \begin{array}{r} 48.23 \\ -\ \ 4.10 \\ \hline \end{array}$$

Incorrect   Correct

Keeping Perspective

Place value allows us to describe partial quantities in a way that makes it possible for us to use the *same algorithms*, or methods, to easily add or subtract them on paper. As a result, we'll find decimals an invaluable tool in describing the quantities God has placed all around us and in serving Him wherever He calls us.

7. DECIMALS | 161

7.3 Multiplying Decimals

What about multiplying numbers with decimals? While you probably already know this rule, let's try to arrive at it step-by-step as if you didn't.

Remember that in math, we try to build on what we know to find simple methods for working with quantities. So we want to adapt our traditional multiplication method to work with decimals. If we could find a simple way to remove the decimal point temporarily and then add it back again when we were finished, we would be able to apply the method we already use to numbers with decimals.

Let's take a look at a problem:

$$7.5 \bullet 5$$

Since our place-value system is based on 10, multiplying by 10, 100, 1,000, etc., is just a matter of moving the decimal point to the right the appropriate number of times, which increases the value of each digit by 10, 100, 1,000, etc. So if we multiply 7.5 by 10, this would remove the decimal point, leaving us 75.

$$7.5 \bullet 10 = 75. \quad \longrightarrow \text{Decimal point moved to the right.}$$

We can now find the answer to 75 • 5 using the method we have already learned. After we have finished, we can divide by 10 again to put the decimal point back in the correct place. Again, since our decimal system is based on 10, to divide by 10, we just move the decimal over one place to the left.

Since we are both multiplying and dividing by the same number (in this case, 10), the multiplication and division will cancel each other out and not affect the final result.

	Multiply to remove the decimal.	Solve.	Divide to add decimal back.
7.5 x 5	7.5 x 5	$\overset{2}{7}5$ x 5 375	$\overset{2}{7}5$ x 5 37.5
	(7.5 x 10 = 75)		(375 ÷ 10 = 37.5)

What if we have more than one digit to the right of the decimal? We would follow the same guideline, just multiplying by 100, 1,000, etc., as needed. Multiplying by 100 would move the decimal two places to the right, by 1,000 three places, etc.

	Multiply to remove the decimal.	Solve.	Divide to add decimal back.

$$\begin{array}{r} 7.50 \\ \times\ \ \underline{\quad 5} \\ \end{array}$$

$$\begin{array}{r} 7.50 \\ \times\ \underline{\ 100} \\ 750 \end{array}$$

$$\begin{array}{r} \overset{2}{750} \\ \times\ \ \underline{\quad 5} \\ 3750 \end{array}$$

$$\begin{array}{r} 3750 \\ \div\ \underline{\ \ 100} \\ 37.50 \end{array}$$

$$\begin{array}{r} 7.500 \\ \times\ \ \underline{\quad 5} \\ \end{array}$$

$$\begin{array}{r} 7.500 \\ \times\ \underline{\ 1000} \\ 7500 \end{array}$$

$$\begin{array}{r} \overset{2}{7500} \\ \times\ \ \underline{\quad 5} \\ 37500 \end{array}$$

$$\begin{array}{r} 37500 \\ \div\ \underline{\ 1000} \\ 37.500 \end{array}$$

$$\begin{array}{r} 7.5000 \\ \times\ \ \underline{\quad 5} \\ \end{array}$$

$$\begin{array}{r} 7.5000 \\ \times\ \underline{\ 10000} \\ 75000 \end{array}$$

$$\begin{array}{r} \overset{2}{75000} \\ \times\ \ \underline{\quad 5} \\ 375000 \end{array}$$

$$\begin{array}{r} 375000 \\ \div\ \underline{\ 10000} \\ 37.5000 \end{array}$$

Now that we've found a way to multiply decimals, we want to find a way to simplify the process. Rather than actually writing out the multiplication and the division, we can just ignore the decimal point to start with and multiply as normal. After multiplying, we could then count the number of digits to the right of the decimal points in the numbers being multiplied, and add a decimal point in the answer so as to keep the same number of total digits to the right of the decimal point. This would reduce multiplying and dividing by 10, 100, 1,000, etc., to a mechanical process we do not even have to think about.

Example: Solve 7.51 • 5

$$\begin{array}{r} 7.51 \\ \times\ \ \underline{\quad 5} \\ 37.55 \end{array}$$

7.51 ←——— Total of two digits to the right of the decimal.

37.55 ←——— Total of two digits to the right of the decimal.

Example: Solve 7.51 • 6.45

$$\begin{array}{r} 7.51 \\ \times 6.45 \\ \hline 3755 \\ 3040 \\ \underline{45060\ } \\ 48.4395 \end{array}$$

7.51 ←——— Total of four digits to

×6.45 ←——— the right of the decimal.

Total of four digits to

48.4395 ←——— the right of the decimal.

Keeping Perspective

Once again, we're building on what we know about place value to find an easy way of working with quantities. Don't forget, though, that we can only do this because of the amazingly consistent way God governs all things. If 10 times a number didn't always equal the same thing, we couldn't multiply decimals like this! Multiplication ultimately rests on God's faithfulness.

7.4 Dividing and Rounding with Decimals

It's time now to explore division yet again. This time, we're going to combine what we know about decimals and rounding to find yet another way to express remainders.

Division — From Remainder to Decimal

As we've seen before, when we divide two numbers, we don't always end up with a whole number — sometimes we have a remainder.

Let's say you spent $46 for a package containing 5 DVDs. If you wanted to find out how much each DVD cost, you would need to divide $46 by 5.

$$
\begin{array}{r}
9 \\
5\overline{)4\,6} \\
-\,4\,5 \\
\hline
1
\end{array}
$$

As you can see, we have a remainder of 1. In the past, we would have written this as r1 or $\frac{1}{5}$. However, sometimes it's more helpful to represent these remainders using decimals. Because our place-value system is based on 10, we can use the *same rules for dividing the remainder as we do for whole numbers*. We simply add a decimal point to show we're now dealing with tenths, add a zero to the dividend, and keep dividing!

$$
\begin{array}{r}
9.2 \\
5\overline{)4\,6.0} \\
-\,4\,5 \\
\hline
1\,0 \\
-1\,0 \\
\hline
0
\end{array}
$$

If we buy a package of 5 for $46, then each DVD costs $9.20.

Note that we added a 0 after the 9.2 to make it 9.20; remember, adding zeros to the right of the decimal point does not change the meaning, as $\frac{2}{10}$ is equivalent to $\frac{20}{100}$. It's common to add a zero when working with dollars if we have 0 hundredths so that it's easier to quickly assess the cost. We know $9.20 means 9 dollars and 20 cents.

Rounding

Let's say we spent $1,189 to make 42 quilt racks and want to find the price per quilt rack. How much did we spend per rack?

```
        28.309
   42)1,189.000
      - 84
        349
      - 336
         130
        -126
           400
          -378
            22
```

Okay, we've divided a *lot* of digits, and we still have a remainder. Often when dividing numbers, the answer keeps going on, not expressing evenly as a fraction of 10, 100, 1,000, etc., for a long time, if at all.

The good news is that we don't typically need that exact of an answer. For most purposes, we can round our answers after a certain number of decimals. In this case, we'll round to the nearest cent. $28.309 rounds to $28.31. So we spent $28.31 per rack.

Remember, when rounding, you **look at the digit to the right of the digit you're rounding.** If it is 5 or higher, you round up; if it is less than five, you round down.

- **1.25** rounds to **1** (if rounding to ones) or to **1.3** (if rounding to tenths).
- **1.23** rounds to **1** (if rounding to ones) or to **1.2** (if rounding to tenths).
- **1.527** rounds to **2** (if rounding to ones) or to **1.5** (if rounding to tenths) or to **1.53** (if rounding to hundredths).

> Unless otherwise specified in this course, you can **round all answers to two decimal places (hundredths)**. Since we only go to hundredths in money (100 pennies equals $1), rounding to the hundredths makes sense whenever dealing with money.
>
> Note that in order to round to the hundredths place, you will need to keep dividing through the thousandths place. That way you will be able to look at the thousandths-place digit to determine if you should round up or round down.

We have to round to simplify problems. Our need to round is another reminder that we can't keep track of everything. Unlike God, we are limited in what we can handle.

Finding an Approximate Answer

In real life, if all we're looking for is an approximate answer, we'll frequently round to the nearest whole number.

For example, if you were buying items at the store, you might want to know about how much you were committing to spend, but you might not need to know the exact amount. Approximating your answer will typically be enough to let you know if you have enough cash for your purchase.

$$\$5.99 + \$7.99 + \$1.98 \approx \$6 + \$8 + \$2 = \$16$$

However, we need to use our judgment with rounding. For instance, if you're asked to find how many shirts a certain yardage of fabric can make and the answer comes back 4.75 shirts, you cannot round and assume the fabric will yield 5 shirts. Even though 4.75 rounds to 5, if you don't have enough fabric to finish the fifth shirt, you'll only be able to make 4 shirts. **Always make sure your answer makes sense.**

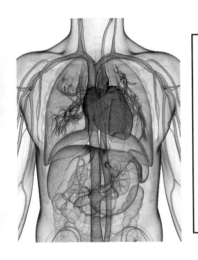

Keeping Perspective — 60,000 Blood Vessels

In this lesson, we used the decimal notation to help us record the answers to division . . . including divisions that have remainders. As we learn these skills, keep in mind that the skills we learn can help us in real life . . . including in exploring and better appreciating God's creation. In the corresponding worksheet in your *Student Workbook*, you are going to use division and decimals to explore the blood vessels in your body. Have fun using math to get a fresh glimpse of how we truly are fearfully and wonderfully made (Psalm 139:14)!

7.5 Conversion and More with Decimals

It's time to dig just a little deeper into decimals. While you may already know some of these techniques, take advantage of the opportunity to better understand why the techniques work.

Conversion

We've already seen how to convert between fractions with a denominator of 10, 100, 1,000, etc., and decimals — we just put the numerator after the decimal point.

$$\frac{4}{10} = 0.4$$

$$\frac{40}{100} = 0.40$$

$$\frac{400}{1,000} = 0.400$$

But often, our denominator cannot be renamed into a denominator of 10, 100, 1,000, etc. Take $\frac{1}{7}$ — how can we convert it?

While you likely already know how to convert this fraction (by dividing the numerator by the denominator), let's think about why this is the case. Fractions represent division, so it would make sense to convert them to decimals simply by completing the division!

$$\frac{1}{7} = 7\overline{)\begin{array}{l} 0.142 \\ 1.000 \end{array}}$$
$$\begin{array}{r} -\ 7 \\ \hline 30 \\ -\ 28 \\ \hline 20 \\ -\ 14 \\ \hline 6 \end{array}$$

If we keep dividing, we'll end up with 0.14285714285. For our purposes, though, let's just round the answer to the nearest hundredth: 0.14. The decimal equivalent of $\frac{1}{7}$ is approximately 0.14.

Dividing a Decimal Number by a Decimal Number

The "rule" for dividing a decimal number by a decimal number is to **count the number of digits to the right of the decimal point in the divisor and move the decimal point that number of digits to the right in the dividend.** Then divide as usual.

$$8.23\overline{)3.776} \text{ changes to } 823\overline{)377.6}$$

The decimal point in both the divisor and the dividend moved two spaces to the right.

$$2.686\overline{)4.67} \text{ changes to } 2686\overline{)4670.}$$

The decimal point in both the divisor and the dividend moved three spaces to the right; notice that to move the dividend three spaces, we had to add a 0.

$$3.41\overline{)5.893} \text{ changes to } 341\overline{)589.3}$$

The decimal point in the dividend and the divisor moved two spaces. It is okay to have a decimal in the dividend, just not in the divisor.

Any guesses why this rule works? Look at a problem written as a fraction, remembering that the fraction line is a way of representing division.

$$0.86\overline{)4.6} = \frac{4.6}{0.86}$$

Notice how if we were to multiply this fraction by $\frac{100}{100}$, we would end up removing the decimal from the divisor (0.86).

$$\frac{4.6}{0.86} \bullet \frac{100}{100} = \frac{460}{86}$$

7. DECIMALS 167

This fraction could now be written as $8\overline{6)4\,6\,0}$.

When we move the decimal point in the dividend and divisor in a division problem, we're really multiplying both the dividend and the divisor by a fraction worth 1, which doesn't change the value (yet another application of the identity property of multiplication!).

> ## Keeping Perspective
>
> Remember that you're learning all these mechanics so that you'll be equipped to use decimals in everyday life. Decimals help us when shopping, designing greeting cards, reading temperatures, comparing distances — the list could go on and on. Since decimals give us a way to represent partial quantities as part of our base-10 place-value system, they are an incredibly useful tool.

7.6 Chapter Synopsis

Decimals, and the rules for working with decimals, serve as useful tools we can use while depending on God and joyfully doing the work He has given us. Writing partial quantities in the decimal system lets us work with them with the same ease as we can whole quantities.

- Our **decimal system** extends to include partial quantities. Partial quantities written in the decimal system (i.e., **decimals**) have assumed denominators of 10, 100, 1,000, etc., with each digit to the right being $\frac{1}{10}$ of the previous one.

- Because decimals are part of the same place-value system we use for whole numbers, we can **add and subtract using the same method** we do for whole numbers, **being careful to correctly line up the digits**. We can also use the **same methods for multiplying and dividing**, using simple rules to deal with the decimal points. Always remember that rules in math are typically a shortcut for working with some consistency God created and sustains.

- When dividing with decimals, some numbers go on and on. We often round, as we rarely need that precise of an answer. **Unless otherwise specified, in this course we are rounding to the nearest hundredth.**

- Because fractions represent division, we can **convert fractions to decimals by simply dividing the numerator by the denominator.**

[CHAPTER 8]

Ratios and Proportions

8.1 Ratios

Now that we've delved into how to express and work with whole numbers, fractions, and decimals, it's time to apply what we learned to ratios and rates. And just what are ratios and rates? Let's take a look.

Understanding Ratios

Suppose a recipe called for 3 cups of flour and 1 cup of sugar. How can we easily express the relationship between flour and sugar?

One way is to express the **ratio**. A ratio is "the relative size of two quantities expressed as the quotient of one divided by the other." [1] In other words, "ratio" is a fancy name for using division to compare quantities! So the ratio of flour to sugar for the above example would be the quotient (result) of 3 divided by 1. Or, if we wanted to list the sugar first, the ratio of sugar to flour would be the quotient (result) of 1 divided by 3.

Throughout our study of fractions, we dealt with many ratios. When we expressed dividing 1 pie between 6 people as $\frac{1}{6}$, we were really expressing the ratio, or relationship expressed via division, between pie and people.

A ratio is simply a comparison between two quantities using division.

Expressing Ratios

We can (and have been!) using a variety of notations to represent ratios. Since fractions are a way to represent division, they're often used to represent ratios. It is also commonly accepted to use a colon.

3 cups flour | 1 cup sugar

Ratio = relationship found by dividing one quantity by the other

Ratio of flour to sugar: 3:1 or $\frac{3}{1}$ | Ratio of sugar to flour: 1:3 or $\frac{1}{3}$

It doesn't matter what notation we use to express it: if we're expressing the relationship between two quantities by *dividing* one by the other, we're working with a ratio.

Understanding Rates

You've likely heard the term "rate" before. A **rate** is a specific type of ratio. The exact definition for what makes a ratio a rate varies, but the important thing is to know that all rates are ratios — comparisons of two quantities using division. Whenever you hear the term "rate," know that you're dealing with a ratio.[2]

Reading Ratios

When we know a fraction represents a ratio, or relationship between two quantities, we read it differently. We would read $\frac{3}{1}$ or 3:1 as "3 to 1" or "3 per 1" and $\frac{1}{3}$ or 1:3 as "1 to 3" or "1 per 3."

When solving real-life problems, watch for situations that use the words *to* and *per*. Miles *per* hour, growth *per* day, euros *per* dollar, food *per* person, girls *to* boys, adults *to* children — all are ways of describing ratios.

How we read ratios is just a convention to emphasize we are talking about 3 compared to 1 (or 1 compared to 3).

Comparing Ratios

When trying to see how ratios compare with one another, we'll often complete the division in each ratio and compare the results.

Example: Compare $\frac{8}{9}$ and $\frac{2}{3}$.

$$\frac{8}{9} = 0.89 \quad \frac{2}{3} = 0.67$$

It's easy to see at a glance how much greater $\frac{8}{9}$ is than $\frac{2}{3}$ once we complete the division and look at the results as decimals. Notice how all the different notations you've been learning work together!

Keeping Perspective

Ratios (and **rates**!) are just a way of comparing two quantities with division. They can be written as fractions and operated on the same way as fractions, so there aren't a lot of new rules to learn. Instead, it's time to put the skills you know to use.

The corresponding worksheet in your *Student Workbook* will help you explore a ratio called the "golden ratio" that we encounter throughout creation, including in the seed arrangement in sunflowers. Along the way, you'll discover how ratios can help you in art!

8.2 Proportions

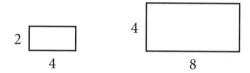

Notice how the ratio between the height and the length of these rectangles ($\frac{2}{4}$ and $\frac{4}{8}$) both reduce to the same fraction: $\frac{1}{2}$. Even though these rectangles are different sizes, the *ratios* between their sides are the same. In both rectangles, one side is half the size of the other.

We call two equal ratios a **proportion**. Both ratios in a proportion are equivalent fractions — they represent the same relationship between two quantities.

$$\frac{2}{4} = \frac{4}{8}$$

The word **proportion** is sometimes used interchangeably with **ratio**, but in math, we'll use it to mean two equal ratios.

> **Term Time**
> As you may have noticed, mathematicians like to give names to different parts of equations. Names help us communicate clearly. It is a lot easier to say, "Go to Susan's Grocery Store" than to say, "Go to the place where they sell food that is down the street and to the left." In a similar way, it is a lot easier to use specific words to describe parts of equations. The terms **extremes** and **means** are the names typically used to refer to different parts of a proportion.
>
>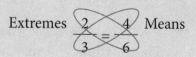

Solving Problems

Proportions often help us solve real-life problems. Say you were planning a party. According to the package, you need 2 quarts of lemonade for every 8 people. If you expect 96 people to attend the party, how many quarts of lemonade should you make?

Here you know the ratio of lemonade to people for 8 people. You want to keep that *same ratio* with 96 people. So you really know three parts of a proportion and need to find the fourth — the number of quarts of lemonade to make in order to keep the same ratio of lemonade to people.

$$\frac{2 \text{ quarts}}{8 \text{ people}} = \frac{? \text{ quarts}}{96 \text{ people}}$$

So how do you find the number of quarts to make? One way is to think through what to multiply or divide by to **finish creating an equivalent fraction** (see 5.3).

Notice that our denominators are 8 and 96.

$$\frac{2 \text{ quarts}}{8 \text{ people}} = \frac{? \text{ quarts}}{96 \text{ people}}$$

So what would we have to multiply or divide 8 by to get to 96?

$$8 \bullet ? = 96$$

$$96 \div 8 = 12$$

If we'd have to multiply 8 by 12 to get 96, then we'd have to *multiply the entire first ratio* by $\frac{12}{12}$ to form an *equivalent ratio* with 96 in the denominator.

$$\frac{2 \text{ quarts}}{8 \text{ people}} \bullet \frac{12}{12} = \frac{24 \text{ quarts}}{96 \text{ people}}$$

If we have 96 people coming and want to keep the same ratio of lemonade to people, we need to make 24 quarts of lemonade.

$$\frac{2 \text{ quarts}}{8 \text{ people}} = \frac{24 \text{ quarts}}{96 \text{ people}}$$

Let's take another example. Suppose we want to find the missing number in this proportion.

$$\frac{21}{3} = \frac{7}{?}$$

How can we figure it out? Again, let's **think through what to multiply or divide by to finish creating an equivalent fraction**. This time, we know both the numerators rather than the denominators. Our numerators are 21 and 7.

$$\frac{21}{3} = \frac{7}{?}$$

So what would we have to multiply or divide 21 by to get 7? We can figure this out by thinking about what times 7 would equal 21.

$$7 \cdot \,? = 21$$

$$21 \div 7 = 3$$

If we'd have to divide 21 by 3 to get 7, then we'd have to divide the entire first ratio by $\frac{3}{3}$ to form an equivalent ratio with 7 in the numerator.

$$\frac{21}{3} \div \frac{3}{3} = \frac{7}{1}$$

We've now completed our proportion.

$$\frac{21}{3} = \frac{7}{1}$$

Comparing Ratios

If a recipe calls for 2 cups of spinach and 7 cups of water, and someone put in 9 cups of spinach and 23 cups of water, did he keep the ratio between spinach and water the same?

It's easy to tell — just write the ratios between spinach and water for both the recipe and the actual, and then see if they form a proportion.

Ratio of spinach to water in recipe: $\frac{2}{7}$

Ratio of spinach to water in actual: $\frac{9}{23}$

No, these do not form a proportion, as $\frac{2}{7} \neq \frac{9}{23}$.

Notice that we could also have seen that these two ratios are not equal by **converting each one to a decimal**:

$$\frac{2}{7} = 0.29 \qquad \frac{9}{23} = 0.39$$

$$0.29 \neq 0.39$$

It doesn't really matter what notations we use — the important thing is to see if the two ratios are equal. Equal is equal, regardless of the notation.

Keeping Perspective

Ratios and **proportions** are tools we're going to be using over and over and over again. As we continue our mathematical explorations, we'll use them to draw scale drawings, convert between units of measure, and even find the height of a tree!

8.3 Ratios and Proportions Containing Decimals

Sometimes, we'll have fractions or decimal numbers as part of a ratio or proportion. For instance, if a recipe called for $3\frac{1}{2}$ cups of flour and $1\frac{1}{4}$ cups of sugar, our ratio of flour to sugar would look like this:

$$\frac{3\frac{1}{2}}{1\frac{1}{4}} \quad or \quad 3\frac{1}{2} : 1\frac{1}{4} \quad or \quad \frac{3.5}{1.25} \quad or \quad 3.5 : 1.25$$

For another example, let's say we wanted to compare the sides of the rectangle below.

$$\frac{1}{4} \quad \boxed{} \\ \frac{3}{4}$$

Any of the methods below express the ratio of length to width — notice the fractions and decimals!

$$\frac{\frac{3}{4}}{\frac{1}{4}} \quad or \quad \frac{3}{4} : \frac{1}{4} \quad or \quad \frac{0.75}{0.25} \quad or \quad 0.75 : 0.25$$

It may seem a little odd at first to have fractions or decimal numbers inside a ratio, but remember that the fraction line represents division — and we can divide partial quantities as well as whole!

We can use and work with decimals or fractions inside of ratios exactly the same way we do with whole numbers inside of ratios. Notice how both these ratios can be simplified to a single decimal number by simply completing the division.

Solving as Decimals	Solving as Fractions
$\dfrac{3.5}{1.25} = 3.5 \div 1.25 = 2.8$	$\dfrac{3\frac{1}{2}}{1\frac{1}{4}} = 3\frac{1}{2} \div 1\frac{1}{4} = 2\frac{8}{10}$ which simplifies to $2\frac{4}{5}$
$\dfrac{0.75}{0.25} = 0.75 \div 0.25 = 3$	$\dfrac{\frac{3}{4}}{\frac{1}{4}} = \frac{3}{4} \div \frac{1}{4} = \frac{12}{4} = 3$

Depending on whether we want a fraction or decimal answer, we might rewrite fractions as decimals in order to solve.

174 | PRINCIPLES OF MATHEMATICS 1

Example: Rewrite $1\frac{1}{5} : 2\frac{2}{20}$ as a single decimal number.

Since we want a decimal answer, we'll rewrite the mixed numbers as decimals.

Rewrite the mixed numbers as decimals: $1\frac{1}{5} = 1.2$ $2\frac{2}{20} = 2.1$

Solve:

$$
\begin{array}{r}
0.571 \\
2.1\overline{)1.2\,0\,0\,0} \\
-1\,0\,5 \\
\hline
1\,5\,0 \\
-1\,4\,7 \\
\hline
3\,0 \\
-2\,1 \\
\hline
9
\end{array}
$$

The answer rounded to the nearest hundredth is 0.57.

Proportions with Decimals or Fractions

Proportions containing ratios made up of decimals or fractions can be solved the same way any other proportion is solved.

Example: If a recipe calls for $2\frac{1}{2}$ cups of chocolate chips per 24 cookies, and we want to make 72 cookies, how many chocolate chips should we put in?

$$\frac{2.5}{24} = \frac{?}{72}$$

$$\frac{2.5}{24} \cdot \frac{3}{3} = \frac{7.5}{72}$$

$$\frac{2.5}{24} = \frac{7.5}{72}$$

We need to include 7.5 or $7\frac{1}{2}$ cups of chocolate chips.

Notice how in the last example we did the math using decimals and then converted back into fractions. **You can freely move between notations to make solving problems easier.** Remember, each notation is just a tool!

Keeping Perspective

With ratios and proportions, we're really just utilizing the notations we already know to help us compare quantities. Don't let the presence of a fraction or decimal inside a ratio fool you — the ratio is still comparing quantities, and you have all the tools you need to understand that comparison.

8.4 Scale Drawings and Models

Now that you're familiar with proportions, it's time to look at two extremely common applications: **scale drawings** and **models**.

When you look at a map, the distances between each place on the map are the same proportion as the distances in real life, only much smaller. For example, if 1 inch on a map represents 100 miles, then places 200 miles apart in real life would be 2 inches apart on the map, as that would form an equivalent ratio (i.e., a proportion).

$$\frac{1 \text{ in}}{100 \text{ mi}} = \frac{2 \text{ in}}{200 \text{ mi}} \qquad \begin{array}{c} 100 \text{ miles} \quad 100 \text{ miles} \\ |\text{\textemdash\textemdash\textemdash}200 \text{ miles}\text{\textemdash\textemdash\textemdash}| \end{array}$$

Since the same reduction was used in drawing everything on the map, it follows that the ratio between any two locations on the map and any two locations in real life will also form equivalent ratios.

For example, if in real life one road is 20 miles from our starting point and another is 5, then on our drawing, if the 20-mile-away road is 0.5 inches away, the other one will be 0.125 inches away, as that would form an equivalent ratio, or proportion.

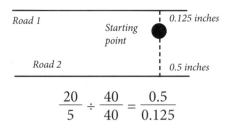

$$\frac{20}{5} \div \frac{40}{40} = \frac{0.5}{0.125}$$

We would say these maps are drawn to **scale**. Maps, model airplanes, drawings in instruction manuals, house blueprints — all these items are drawn or built to scale.

Often, a **scale ratio** will tell us how much an original was reduced or enlarged. If a model airplane has a scale ratio of 1:60, that means that a single unit of the model represents 60 units of the original plane. In other words, the ratio between the model and the original plane is 1 to 60.

Scale drawings and models are what we call **proportional**, or "having a consistent ratio to another quantity."[3] Scale models are basically reduced or enlarged versions of the originals.

Keeping all this in mind, let's take a look at a couple of examples of different ways you might need to explore scale drawings and models. Try to make sure you understand the principle and **avoid the temptation to try to memorize a rule**, as this is a concept you'll use many times, both in math and life, and you need to understand it to the point where you can solve it different ways depending on the situation.

The key principle is that scale drawings or models are proportional to the originals.

Example: Draw a scale drawing of a 6 foot by 8.5 foot room on graph paper.

Graph paper comes with lines already drawn on it, making it easy to use for scale drawings. All we have to decide is how much each box will represent. In this case, it will probably be easiest to make each box represent 1 foot. Thus our model would look something like this:

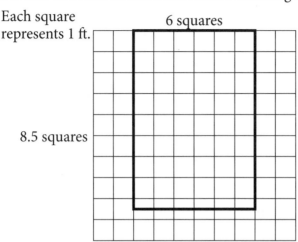

If we needed a smaller drawing, we could make each box represent 2 feet, in which case our scale drawing would look something like this:

Each square represents 2 ft.

3 squares

4.25 squares

Example: If a building is 90 feet tall and 30 feet wide, is a model that's 5 inches tall and 2 inches wide an accurate representation?

How do we tell if the model is an accurate representation? To find out, all we have to do is compare the ratios between the height and width in the drawing with that of the original to see if they are equivalent — that is, if they form a proportion.

$$\text{Ratio between height and width in original: } \frac{90 \text{ ft}}{30 \text{ ft}}$$

$$\text{Ratio between height and width in drawing: } \frac{5 \text{ in}}{2 \text{ in}}$$

Let's convert both of these ratios to decimals to see if they equal.

$$\frac{90 \text{ ft}}{30 \text{ ft}} = 3$$

$$\frac{5 \text{ in}}{2 \text{ in}} = 2.5$$

3 does *not* equal 2.5. These ratios do *not* form a proportion, so the model is not built to scale.

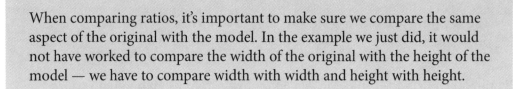

When comparing ratios, it's important to make sure we compare the same aspect of the original with the model. In the example we just did, it would not have worked to compare the width of the original with the height of the model — we have to compare width with width and height with height.

Example: What would the dimensions of a scale model $\frac{1}{60^{th}}$ the size of a building be if the building was 240 feet wide and 600 feet tall?

We've been told that the scale model is $\frac{1}{60^{th}}$ the size of the original. To find the dimensions of the scale model, we just need to multiply each dimension of the original by $\frac{1}{60}$.

Finding the width: $\frac{1}{60} \cdot 240 = 4$

Finding the height: $\frac{1}{60} \cdot 600 = 10$

The model needs to be 4 feet wide and 10 feet tall.

Now, there's often more than one way to solve a problem! Notice that the dimensions in the original and the dimensions for the scale model we just found form a proportion.

$$\frac{240}{600} = \frac{4}{10}$$

We could have found the scale model ratio by dividing our original ratio by $\frac{60}{60}$. This gives us an equivalent ratio that is $\frac{1}{60^{th}}$ the size.

$$\frac{240}{600} \div \frac{60}{60} = \frac{4}{10}$$

There are often several ways to solve problems. Make sure you understand the principle and how different ways work so you'll be equipped to handle any proportion problem you face.

Example: What would the dimensions of a scale picture $\frac{1}{3,000^{th}}$ the size of the Eiffel Tower be if the Eiffel Tower is 125 m wide and 324 m tall? (The *m* stands for meters.)

We can find the dimensions of the scale picture by multiplying each dimension of the original by $\frac{1}{3,000}$.

Finding the width: $\frac{1}{3,000} \cdot 125 = 0.04$

Finding the height: $\frac{1}{3,000} \cdot 324 = 0.11$

The scale drawing would have a width of 0.04 m and a height of 0.11 m.

Alternately, we could have divided the ratio of the original width and height by $\frac{3,000}{3,000}$, giving us an equivalent ratio that is $\frac{1}{3,000^{th}}$ the size.

$$\frac{125 \text{ m}}{324 \text{ m}} \div \frac{3{,}000}{3{,}000} = \frac{0.04 \text{ m}}{0.11 \text{ m}}$$

$$\frac{125 \text{ m}}{324 \text{ m}} = \frac{0.04 \text{ m}}{0.11 \text{ m}}$$

The important thing is that, one way or another, we form the equivalent ratio!

Keeping Perspective

Scale models and **drawings** are examples of using math to model (quite literally!) real life. In working with them, you'll be applying a lot of the different tools you've learned (division, multiplication, fractions, decimals, ratios, proportions, etc.). Have fun playing around and exploring the proportions all around us.

8.5 Mental Math and Decimals

Because our dollar system utilizes the decimal place-value system, we often have to work with decimals mentally. Let's take a quick look.

Paying the Cashier

Say you owe $4.59, and you give the cashier a $5 bill. You would need 1 cent to reach 60 cents, and then 40 cents to reach $5. So your change would be 41 cents.

Did you ever wonder why some cashiers will ask if you have a penny? You can sometimes avoid getting lots of change back from a transaction (and make the cashier's job easier) by paying what it would take to make the change an amount that breaks down into fewer coins. For example, suppose you owe $1.26. If you give $2, you'll get $0.74 back (2 quarters, 2 dimes, and 4 pennies). But if you give $2.01, you will get $0.75 back (3 quarters), which results in fewer coins.

Think about what we just did. We saw that our total was 1 penny over an even quarter amount (our total was $1.26, and the nearest even quarter amount was $1.25), so we paid 1 penny more. That resulted in negating that extra penny, leaving our change in even quarter increments.

Likewise, if our total had been $1.61 and we gave $2.01, it would change the amount of change from $0.39 to $0.40, which can be given with fewer coins.

Multiplying and Dividing Decimals

It's also quite helpful to be comfortable with multiplying and dividing numbers with decimals mentally. For example, you may want to find the cost of buying 3 pounds of a vegetable at $2.99 a pound. While you can often round to find an approximate answer, you may sometimes need a more exact one. Here are three approaches you might find helpful:

1. Round, and then subtract the appropriate number of cents to reflect what you rounded.

    Example: Solve 3 • $2.99

    Round the $2.99 to 3, giving you 3 • 3, or 9.

    You rounded up a penny when you multiplied, so now subtract 3 pennies (1 penny times the 3 times we multiplied the rounded number) from the answer, giving you $8.97.

2. Multiply the dollars and then the cents, adding them together at the end.

    Example: Solve 4 • $2.50

    4 • $2 = $8

    4 • $0.50 = $2

    $2 + $8 = $10

3. Think in terms of converting to whole dollars.

    Example: You have 13 quarters. How many dollars do you have?

    4 quarters equals $1, so 12 quarters equals $3, as 12 ÷ 4 = 3. You have 13 quarters, so you would have $3 + 1 quarter, or $3.25.

Keeping Perspective

Whether we find an answer mentally or on paper, math helps us describe real life. It works because we live in a consistent universe held together by a consistent God.

8.6 Chapter Synopsis

In this chapter, we learned to compare quantities using **ratios** and **proportions**. Along the way, we had an opportunity to practice **mental math** with decimals.

Below are a few key takeaways:

■ **Ratios** are a way of using division to compare quantities. Comparing quantities with division often helps us solve real-life problems and learn about the way different aspects of creation relate. Because a ratio can be expressed as a fraction or decimal number, we can work with them the same way we do fractions or decimals.

■ **Rates** are just specific types of ratios.

■ A **proportion** is the name given to two equal ratios. We can use what we know about equivalent fractions to find missing numbers in a proportion. **Means** and **extremes** are names given to describe different parts of a proportion.

We'll revisit ratios (including rates) and proportions extensively later in this book when we look at unit conversion. For now, though, just familiarize yourself with them and begin using them to explore God's creation.

[CHAPTER 9]

Percents

9.1 Introducing Percents

In this unit, we're going to look at yet another agreed-upon notation for expressing ratios: **percents**. While you are likely already somewhat acquainted with percents, let's take a deeper look at them and their function.

Exploring Percents

Say you're charged $2 in tax on an item that cost $200. We could represent this as a rate, or ratio, like this: $\frac{\$2}{\$200}$.

Now let's say you cross over a state line and end up paying $0.25 in tax on a $20 purchase. We could represent this as a rate like this: $\frac{\$0.25}{\$20}$.

To compare the tax rate of these two states we need to compare these fractions. Any idea of how we could do that? We could convert them to fractions with the same denominator. Since multiples of 10 are easy to work with, let's express them both as fractions of 100.

Sales tax rate in first state: $\frac{\$2}{\$200} \div \frac{2}{2} = \frac{\$1}{\$100}$

Sales tax rate in second state: $\frac{\$0.25}{\$20} \bullet \frac{5}{5} = \frac{\$1.25}{\$100}$

Now it is easy to see that, while the tax *amount* was less on our purchase in the second state, the *rate*, or ratio, between the tax and the items' cost was actually *higher*. Expressing both quantities as fractions of $100 makes them a lot easier to compare!

Percents are an even shorter way to compare ratios as portions of 100. The word *percent* is "short for Latin *per centum*, by the hundred."[1] *Per* actually means "by," and *centum* means "hundred."[2] Thus, 1 percent means "1 per 100."

To save ourselves from writing the whole word "percent," we typically abbreviate it. In the past, *percent*, or *per cent*, has been expressed ꝑ cento, per ꞔ, per ÷, and other various forms.[3] Today, these are shortened to just %. This symbol is a shorthand way to let us know a number is representing a portion of a hundred. Notice how there are two 0s in the % sign, just as there are two 0s in 100.

Sales tax rate in first state: $\dfrac{\$2}{\$200} = \dfrac{\$1}{\$100} = 1$ percent $= 1\%$

Sales tax rate in second state: $\dfrac{\$0.25}{\$20} = \dfrac{\$1.25}{\$100} = 1.25$ percent $= 1.25\%$

> ### Tax Rates
>
> Notice that we refer to the amount charged for tax as a rate. A 5% tax rate results in a tax of $5 *per* every $100. The rate, or ratio, between the tax and dollars spent is $\dfrac{\$5}{\$100}$.

Using Percents

Percents are an extremely useful notation for describing ratios. They help us describe interest, sales tax, discounts on products, tips left at restaurants, and more. You'll also frequently encounter percents in the news expressing how a certain opinion or quantity relates to the whole — that is, what portion of people plan to vote for which candidate, approve of the current president, etc.

As you encounter percents, remember that **percents are a shorthand way to express ratios as fractions of 100. Thus, they do not give the exact quantities, but rather the correct relationship.** For example, saying 60% of the people must agree in order to pass a law is the same thing as saying 60 out of every 100 people must agree. This does *not* necessarily mean there are 100 people and 60 agree. It means that the *ratio* between people who agree and the total people must be equivalent to 60 out of 100. There may only be 50 people total, in which case 30 must agree ($\dfrac{30}{50} = \dfrac{60}{100}$).

Converting between Percents and Decimals

If you keep in mind that a percent means "by a hundred," it's easy to express numbers as percents.

Since a percent is a way of describing a portion of 100, all we have to do to convert decimals to percents is to multiply the decimal by 100 and add the percent sign.

Example: Express 0.125 as a percent.

Multiply by 100: 0.125 • 100 = 12.5

Add a percent sign: 12.5%

Note that the percent sign, like the fraction bar, represents division. It means "per 100," or divided by 100. Even though we multiplied our decimal number by 100, we didn't change its value, as we then added a sign that means it must be divided by 100.

It follows that to convert a percent to a decimal, we simply divide by 100.

Example: Express 12.5% as a decimal.

Divide by 100 and remove the percent sign: 12.5 ÷ 100 = 0.125

Converting between Percents and Fractions

To express a fraction as a percent, we can either convert the fraction to a decimal and then to a percent, or rewrite it as a fraction of 100 and then convert to a percent by simply removing the denominator and adding the percent sign.

Example: Express $\frac{4}{10}$ as a percent.

Rewrite as a Decimal First	**Rewrite as a Fraction of 100 First**
$4 \div 10 = 0.40$	$\frac{4}{10} \cdot \frac{10}{10} = \frac{40}{100}$
$0.40 = 40\%$	$\frac{40}{100} = 40\%$
Answer: 40%	Answer: 40%

To express a percent as a fraction, just replace the percent sign with a denominator of 100, and then simplify if needed.

$$40\% = \frac{40}{100} = \frac{4}{10} = \frac{2}{5}$$

> A percent expresses a quantity — whatever it is — as a portion of 100. You can switch back and forth from fractions, decimals, and percents easily if you remember this.

We mentioned in 7.3 that, because of place value, multiplying by 100 is only a matter of moving the decimal point two digits to the right.

$$0.1\underset{\smile}{2}\underset{\smile}{5}$$
$$\times\ 1\ 0\ 0$$
$$\overline{1\ 2.5}$$

We also mentioned that dividing by 100 is only a matter of moving the decimal point two digits to the left.

$$1\ 2.5$$
$$\div\ 1\ 0\ 0$$
$$\overline{0.1\ 2\ 5}$$

---

9. PERCENTS | 185

Keeping Perspective

Fractions, decimals, and percents — they are all useful ways of recording ratios and quantities. Just as different names are used to refer to animals (a bird can be referred to as a oiseau [French], pájaro [Spanish], vogel [German], птица [Russian], etc.),[4] so ratios and quantities can be expressed differently. Which way to express ratios and quantities depends on what we're trying to accomplish.

God did not just give us the ability to think of one way to record ratios — He created us capable of using many different valuable notations. We're able to be creative with ratios because God gave us this ability.

0.40	40%	$\frac{40}{100}$
0.25	25%	$\frac{25}{100}$
0.30	30%	$\frac{30}{100}$
0.23	23%	$\frac{23}{100}$
0.125	12.5%	$\frac{12.5}{100}$

9.2 Finding Percentages

In this lesson, we're going to put percents to use! As we do, we will add a few new terms to our understanding of percents.

Completing the Proportion

No doubt you've been to the store and seen a 20% off sign . . . or gotten a coupon in the mail. How do you figure out how many *dollars off* you'll get, say, on an $86 purchase with a 20% off coupon?

Remember how we mentioned in the first lesson of this chapter that a percent expresses a ratio in terms of *per hundred*? It tells us the ratio, or relationship, but doesn't specifically tell us the actual quantity in a situation. To find that, we have to find an equivalent ratio.

Here, we're looking to find the number that will represent the same part of $86 as 20 does of 100. We're really trying to find the missing number in a proportion!

$$\frac{20}{100} = \frac{?}{86}$$

86 divided by 100 is 0.86. So if we multiply $\frac{20}{100}$ by $\frac{0.86}{0.86}$, we'll find our missing numerator: 17.20.

$$\frac{20}{100} \cdot \frac{0.86}{0.86} = \frac{17.20}{86}$$

$$\frac{20}{100} = \frac{17.20}{86}$$

20% of $86 is $17.20. So, a 20% discount on an $86 purchase comes to $17.20.

Using Names

When working with percents, there are some common terms you may encounter.

$$\text{Rate} \longrightarrow \left(\frac{20}{100}\right) = \frac{17.20 \; \longleftarrow \text{Percentage}}{86 \; \longleftarrow \text{Base}}$$

Rate is a portion per hundred — in other words, our percent (although often rewritten in fractional or decimal notation).

Base is the number off which we want to find the percent.

Percentage is often used synonymously with percent. However, to keep it easy to tell what we're referring to, we'll use the term "rate" or "percent" if we mean the rate or ratio, and the term "percentage" if we mean the **result we'd get if we found the portion of a base that the rate (i.e., %) represents.**

Using this lingo, we could describe the proportion to find a percentage like this:

$$\textbf{Rate} = \frac{\textbf{Percentage}}{\textbf{Base}}$$

where "Rate" equals a percent — that is, *a portion of 100*.

Keeping Perspective

Because so many sales and news items are reported in terms of percents, you will encounter them often. Being familiar with how to find a **percentage** is an extremely useful tool . . . and it's no different than what you've already been doing in finding the missing number in a proportion!

9.3 More Finding Percentages/ Multiplying and Dividing Percents

In the last lesson, we looked at how to use a proportion to find a certain percentage. There's actually another way of finding percentages. To understand it, we need to back up and review what we learned about multiplying fractions. We saw in 6.1 that it was helpful to think of multiplications with partial quantity multipliers as finding that portion *of* the other number. For instance, when we multiplied $\frac{1}{4}$ by 2, we were really asking what $\frac{1}{4}$ *of* 2 was.

9. PERCENTS 187

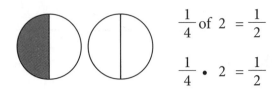

This same concept applies to percents. After all, $\frac{1}{4}$ and 25% represent the same quantity. So 25% *of*, or times, 2 equals $\frac{1}{2}$.

$$25\% \cdot 2 = \frac{1}{2}$$

When we multiply any number by a percent, we're really finding a specific portion *of* that number. Likewise, when trying to find a percentage, we're really looking for a portion *of* a base. For example, when finding how many dollars off an $86 purchase we would get if we got a 20% discount, we really want to find 20% *of* $86. Thus, rather than writing out a proportion, we **could use multiplication to find the percentage of a number**. If we take the rate times our base, we'll find our percentage.

$$\text{Rate} \cdot \text{Base} = \text{Percentage}$$

Using this method, we'd find 20% of $86 by finding the product of 20% • $86. Of course, to use this method, we have to know how to multiply percents.

Multiplying & Dividing Percents

We can multiply (or divide) percents simply by **converting the percent to either a decimal or a fraction.**

Example: Find 20% • $86

Converting to a Decimal	Converting to a Fraction
$0.2 \cdot \$86 = \$17.20$	$\frac{20}{100} \cdot \$86 = \frac{\$1,720}{100}$
Answer: $17.20	Answer: $17.20

Note that we could use division to check our answer.

Check: $17.20 ÷ 20%

Converting to a Decimal	Converting to a Fraction
$\$17.20 \div 0.2 = \$86$	$\$17.20 \div \frac{2}{10} = \$17.20 \cdot \frac{10}{2} = \frac{\$172}{2} = \$86$
Answer: $86	Answer: $86

As you can see, the decimal was the easier way to complete the multiplication and the division. **In general, to multiply or divide a percent, convert it to a decimal.**

Finding Percentages

Notice that multiplying 20% by $86 gave us *the same answer* ($17.20) that we found in the last lesson when we set up a proportion. Whether we find the answer through a proportion or via multiplication, we're finding the same thing: the amount *of* $86 that represents the same ratio as 20 *per* 100.

There is a mathematical reason why both methods (setting up a proportion and multiplying any number by a percent) arrive at the same answer, but we'll have to wait for Book 2 to examine it. For now, just get comfortable with using either method to find the percentage of a base.

Keeping Perspective — Remembering Why

In this lesson, we saw that we can multiply or divide percents by simply converting them to decimals or fractions and then multiplying or dividing as normal. Remember, we can perform operations on paper like this and expect them to hold true in real life only because of God's consistency in holding all things together. Operations with percents, decimals, and fractions all depend on the same consistencies — the consistencies don't change, even though we express them in different creative ways. God's truths do not change.

For ever, O LORD, thy word is settled in heaven (Psalm 119:89).

Just as there is more than one way to find a percentage, there are often multiple ways to go about solving a problem. Avoid the temptation to just memorize the mechanics of one method. Instead, make sure you really understand the process. Different methods come in handy at different times!

9.4 Adding and Subtracting Percents/ Finding Totals

We encounter percents all the time when shopping. You've already been practicing finding a percent off a total, but let's build on this knowledge to go one step further.

Suppose you're out shopping. If you have a 15% off coupon and use it on a $20 item, you'll save $3, as 0.15 • $20 = $3.

You now know your savings, but how much will you actually end up paying for the item?

There are a couple of different ways to find the answer. But to understand one of them, we have to first look at how to add and subtract percents.

Adding and Subtracting Percents

Any guesses how to add or subtract percents? If you guessed converting to decimals or fractions, you're correct: that is definitely one way to add and subtract them.

But, since percents are basically numerators of 100, we can take a shortcut. Let's take a look.

9. PERCENTS | 189

Example: Solve 49% + 2%

Adding as fractions: $\frac{49}{100} + \frac{2}{100} = \frac{51}{100} = 51\%$

Adding as decimals: $0.49 + 0.02 = 0.51 = 51\%$

Notice there really was no need to rewrite the percents as decimals or fractions; since all percents are numerators of the same denominator (100), we could have just added them together as they were!

Adding as percents: 49% + 2% = 51%

Likewise, to subtract percents, all we need to do is subtract them. After all, they are just numerators of 100.

Example: Solve 51% – 49%

Subtracting as fractions: $\frac{51}{100} - \frac{49}{100} = \frac{2}{100}$

Subtracting as decimals: $0.51 - 0.49 = 0.02 = 2\%$

Subtracting as percents: 51% – 49% = 2%

Since percents are all numerators of the same denominator (100), adding and subtracting is just a matter of adding and subtracting the percent amount.

Math Builds

Once again, we're applying what we already know about addition and subtraction to yet another notation: percents. You'll find that for each new notation we learn, we'll revisit the basic operations to see how they can be done in the new notation. And each time, we'll build on the basic assumption that addition and subtraction are consistent. We can rely on this assumption because we can rely on God to faithfully keep His covenant with the "ordinances" of heaven and earth.

Thus saith the LORD; If my covenant be not with day and night, and if I have not appointed the ordinances of heaven and earth; Then will I cast away the seed of Jacob and David my servant, so that I will not take any of his seed to be rulers over the seed of Abraham, Isaac, and Jacob: for I will cause their captivity to return, and have mercy on them (Jeremiah 33:25–26).

Finding a Total with Percents Off or Added

Okay, now to return to figuring out your discounted total if you have a 15% off coupon and use it on a $20 item. There are a couple of different ways you could figure this out.

1. You know the original cost and you know the percent of the discount, so you could just find the discount amount and then subtract that from the original cost.

   15% • $20 = 0.15 • $20 = $3

   $20 − $3 = $17

   You'll end up paying $17 for the item.

2. Or you could approach the problem entirely differently. Rather than finding 15% of $20, you could have viewed the original cost as 100% — full price — and subtracted the 15% discount to find the percent of the original you'll end up paying.

   100% − 15% = 85%

   To find how much you would have to pay, you just need to find 85% of $20!

   0.85 • $20 = $17

   You'll end up paying $17 for the item.

Either way gets us the same answer — they're two different ways to think about the same problem. You'll find both methods helpful at different times.

The same principles apply when adding a percentage to a total as well as when taking a percentage off. For example, if your total purchase was $10, and a 10% sales tax was added to that, you could find the total either of these ways.

1. Finding 10% of $10 and then adding it to $10.

   0.10 • $10 = $1

   $1 + $10 = $11

2. Finding 110% of $10. Here we're just adding together the percents we need to find (100%, which represents our original amount, and our 10% sales tax): 100% + 10% = 110%.

   1.1 • $10 = $11

Keeping Perspective — Percents Everywhere

Percents are a handy tool. Whether calculating the total you'll pay at the store with tax or after a discount, the amount you'll save, the tip you should leave, or the portion of the total number of pages you've completed in a book, percents help us in real life.

But be careful. Percents, like any other math concept, can be used for good and for evil. Unfortunately, percents can be presented in a misleading way. For example, suppose a business claims their new vitamin solution is 96% effective at keeping people healthy. They're claiming that 96 out of 100 people who take their solution will stay healthy. If the business tested their solution only on people who do not normally catch colds, though, then saying the solution is 96% effective would be deceptive![5]

It's also important to note that a 96% effectiveness doesn't mean the company tested the solution on 100 people. They may have tested only 50 people and had 48 stay healthy . . . or 500 out of which 480 stayed healthy. When a result is expressed as a percent, it's sharing the results as a portion of 100. For example, $\frac{48}{50}$ and $\frac{480}{500}$ are both equivalent ratios to $\frac{96}{100}$. Whatever the actual numbers were, a claim of 96% effectiveness just means the ratio between those the vitamin solution kept healthy and those tested was equivalent to 96 out of 100.

As you see percents used in various ways, remember we live in a fallen world. The data behind a percent does not always support a claim, so beware. The Bible teaches us God is the only one we can trust completely.

. . . let God be true, but every man a liar . . . (Romans 3:4).

9.5 Mental Percents

Often (such as when at a store or restaurant), you'll need to find a percentage mentally. When solving percent problems in your head, it's helpful to **round** and **multiply by percents that are easy to calculate.**

Example: What is 20% of $23.75?

Let's round $23.75 to $24.

Rather than finding 20%, let's find 10%. Because our decimal system is based on 10, 10% simply involves moving the decimal point one digit to the left, making it easy to find mentally.

0.10 • $24 = $2.40

Of course, we need 20%, not 10%. But since 20% equals 10% plus 10%, all we need to do is add the amount we found for 10% twice.

10% + 10% = 20%

$2.40 = 10%

$2.40 + $2.40 = 20%

$2.40 + $2.40 = $4.80

20% of $23.75 is approximately $4.80.

The exact answer is slightly less (0.20 • $23.75 = $4.75), but would be harder to find mentally. A mental approximate is close enough for calculating a tip, discount, etc.

Example: What is 15% of $32.50?

Let's round $32.50 to $33.

Now, let's think in terms of percents that are easy to calculate. 15% is 10% plus half of 10% (i.e., 5%). So if we find 10% and then find half of 10% and add the two together, we'll have found 15%.

$$10\% + 5\% = 15\%$$

$$5\% = \frac{1}{2} \cdot 10\%$$

$$0.10 \cdot \$33 = \$3.30$$

$$\frac{1}{2} \cdot \$3.30 = \$1.65$$

$$\$3.30 + \$1.65 = \$4.95$$

15% of $32.50 is approximately $4.95.

To recap, we

■ rounded the dollar amount to the nearest dollar.

■ found 10% by mentally moving the decimal over one place to the left.

■ mentally added the 10% amount we found the number of times needed to find the percentage we wanted to find.

Keeping Perspective

While that may look like a lot of steps on paper, it's quite simple to find an approximate percentage without paper once you get used to it. You can also adjust how closely you round (dollar verses half dollar) in order to get the level of accuracy you need. Once comfortable with finding percentages mentally, you'll be equipped to use percents wherever you go!

9.6 Chapter Synopsis

In this chapter, we explored percents, which give us yet another notation for recording ratios. Here are the highlights from our study:

■ A percent is a shorthand way of representing a ratio as a portion of a hundred.

■ To express a percent as a decimal, we simply divide the percent by 100. To express a percent as a fraction, we put the percent in the numerator with a 100 in the denominator.

■ To multiply or divide percents, we first convert the percent to a decimal or fraction and then proceed as normal. (We usually convert to a decimal, as those are easier to work with.)

■ We can add and subtract percents without converting to a different notation since percents are all numerators of the same denominator (100).

■ We can find the percentage of a number by a proportion or by multiplication.

$$\text{Rate} = \frac{\text{Percentage}}{\text{Base}}$$

$$\text{Rate} \cdot \text{Base} = \text{Percentage}$$

■ Rounding and using easy-to-work-with percents (such as 10%) can help us find percentages mentally.

We'll continue reviewing percents as we move on to other concepts, as it's a frequently used "tool" in everyday life, and one with which you may not have had as many years of experience. Don't be afraid to look back in this book at any point if you need to — the goal is to master the tools so you'll be able to use them for God's glory wherever He calls you.

[CHAPTER 10]

Negative Numbers

10.1 Understanding Negative Numbers

Now that we've looked at the basic operations (addition, subtraction, multiplication, and division) and notations (place value, fractions, decimals, and percents), it's time to apply all we've learned to a different type of number: **negative numbers.**

Understanding Negative Numbers

We often find it necessary to show not just the amount, but also the direction of that amount.

For example, let's say you borrow $5 from your mom while at the store. How would you describe that you *owed* the money instead of that you had it?

$5 we owe $5 we have

Or let's say you were trying to compare the locations of two robots that started at the same point but traveled in *opposite* directions. How can you show that the distances were in opposing directions?

 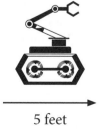

5 feet   5 feet

You could use a **negative sign** (–) and a **positive sign** (+) to differentiate between the amount owed and the amount had, as well as to indicate the direction the robots traveled. The negative sign means *the opposite of*, so –5, read "negative 5," means *the opposite of* 5.

In the following picture, we've used a negative sign to show that we have *negative* $5; in other words, that we have *the opposite of* $5. That is, we owe $5. We used a positive number to show the $5 we had.

– $5   + $5

In the following picture, we've used a negative sign to show that the first robot has gone *negative* five feet, or the opposite of five feet (five feet in the opposite direction), while the other robot has gone *positive* five feet.

– 5 feet   + 5 feet

Negative and positive numbers are perhaps easiest to define and picture on a **number line.** A number line is just a pictorial aid to help us view how numbers relate to each other.

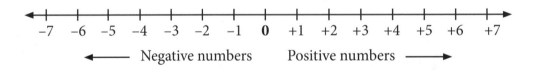

We call numbers to the right of 0 on the number line *positive numbers*, and those to the left *negative numbers*. The negative numbers are the exact same as the positive ones, except they're going in the *opposite direction*. Again, think of a negative sign as meaning *the opposite of*. Negative and positive numbers are simply another convention to help us describe how quantities relate to one another.

Most of the time, we don't bother putting the plus sign in front of a positive number. **If there's no sign in front of a number, we just assume it's positive.** Up until now, you've been dealing with all positive numbers — so you already know how to work with them.

$$4 = +4 \qquad \frac{1}{2} = +\frac{1}{2} \qquad 2.2 = +2.2 \qquad 75\% = +75\%$$

Expanding Negative Numbers

Ratios and partial quantities can be negative. For example, companies will sometimes compare their expenses and what they made in terms of profit margins. A 25% profit margin means they made 25% more than they spent; a –25% profit margin means they spent 25% more than they made. Again, the negative sign means *the opposite of* — rather than making money and having a positive margin, or buffer, between their expenses and income, the company did the opposite: they *lost* money!

Notice that we can put fractional amounts or ratios on a number line, whether written using decimal, fractional, or percent notation.

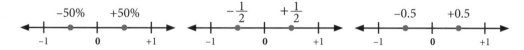

> ## Keeping Perspective
>
> In most real-life situations, we add, subtract, multiply, and divide positive numbers. But there are times where specifying direction, debt, etc., is helpful. Negative numbers, like other notations, prove a useful tool in helping us describe God's creation and work with real-life problems.

10.2 Adding Negative Numbers

Adding negative numbers sounds scary, but I've got good news for you! **You've been adding negative numbers for years.**

That's right. Adding negative numbers is no different than subtracting positive numbers. They're just two different ways of looking at the same thing.

$$4 - 3 = 4 + -3 = 1$$
$$10 - 1 = 10 + -1 = 9$$
$$5 - \frac{1}{2} = 5 + -\frac{1}{2} = 4\frac{1}{2}$$

It's easiest to see how adding a negative number means the same thing as subtraction on a number line. Both $4 - 3 = 1$ (i.e., taking 3 away) and $4 + -3 = 1$ (i.e., adding *the opposite* of 3) mean to move over 3 to the left (the negative side).

$$4 - 3 = 1 \quad \text{and} \quad 4 + -3 = 1$$

What if you have to add two or more negative numbers? Say you owe $5, and then you borrow another $2. Now you have –$5 – $2.

Again, view it as –$5 + –$2. We're already 5 in the negative direction, and we just need to keep moving in the negative direction 2 more spaces. Now we're at –7.

Again, this is easiest to see on the number line.

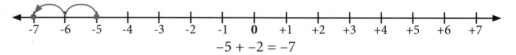

$-5 + -2 = -7$

> When subtracting negative numbers, parentheses are sometimes used to make the problems easier to read, so –$5 – $2 can be expressed as:.
>
> –$5 – $2   or   –$5 + –$2   or   (–$5) + (–$2)
>
> All three mean the same thing. There is nothing within the parentheses to add together, nor is there anything outside the parentheses to first multiply by what's inside. The parentheses are simply being used as visual aids to make the problem easier to take in at a glance.

Making 0 — The Additive Inverse

Notice that if we add a negative and positive that are equally distant from 0, we end up with 0. For example, $5 + -5 = 0$, $6 + -6 = 0$, $1 + -1 = 0$, etc.

Let's take a look at one of these equations: $5 + -5 = 0$. We know that if we walk 5 miles in what we think of as the positive direction and turn around and walk 5 miles in the opposite direction, we'll end up back where we started.

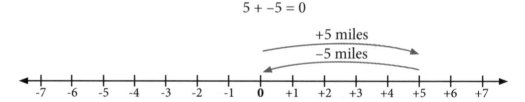

Likewise, if we walk 5 miles in what we think of as the negative direction and turn around and walk 5 miles in the opposite direction, we'll end up back where we started. This equation would be represented as $-5 + 5 = 0$.

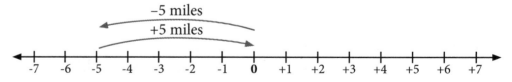

Any number plus its opposite equals 0.

While this seems pretty obvious, it's such an important concept to remember that we have a name to describe the number that, added to a number, equals 0. We call it the **additive inverse** of that number.

In Book 2, we'll use additive inverses to help us solve problems involving unknown numbers. For now, just know that adding the opposite of a number (i.e., its additive inverse) gives us 0.

Problems with Negative Numbers

You might be wondering why you haven't heard people talking about negative numbers in everyday life. **In everyday conversations, we use other terms besides negative numbers.** We would talk about "owing $5" or "traveling 5 miles in the opposite direction" rather than about "negative 5." However, even though we don't use the term "negative number" in standard speech, you need to **understand how to express the idea of opposites mathematically with negative numbers,** as negative numbers will help you solve problems related to these everyday occurrences.

Let's suppose that in the first month of a hypothetical new business, we made $5 and spent $15. We would have lost $10 — in other words, we'd end the month with a negative amount of profit (that is, we'd end with less than we had to start).

$$\$5 - \$15 = -\$10 \quad or \quad \$5 + -\$15 = -10$$

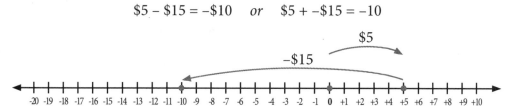

We end with a negative number whenever the number we're subtracting (the subtrahend) is greater than the number we're subtracting from (the minuend). Negative numbers really do apply outside of a textbook!

Now, let's bring in the concept of the additive inverse. When solving a problem with a larger subtrahend than minuend, **it's helpful to think about how much it takes to get to 0**, and then how much further beyond 0 we'd end up.

In the case we just looked at, we started at $5. It took subtracting 5/adding −5 to get to 0 (there's that additive inverse: 5 + −5 = 0).

Since $15 is 10 more than 5, we still have 10 more to go in the negative direction. So our answer will be −$10.

Example: We traveled $10\frac{1}{3}$ miles east down a road, and then turned around and went $20\frac{1}{3}$ miles west (the *opposite* direction). How much farther west from our starting point did we travel?

Using positive numbers for east and negative numbers for west, we could represent what's happening here like this:

$$10\tfrac{1}{3} \text{ mi} + -20\tfrac{1}{3} \text{ mi} = ?$$

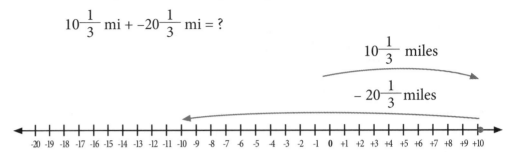

Again, while it's easy to see that we'll end at −10 miles (or 10 miles west) from the number line, note that we could also have figured this out by thinking in terms of what it takes to get to 0. Since we're starting with $10\tfrac{1}{3}$, if we add $-10\tfrac{1}{3}$, we'll get to 0.

We've now added $-10\tfrac{1}{3}$ out of our $20\tfrac{1}{3}$, leaving us with −10 left to add ($10\tfrac{1}{3} + -20\tfrac{1}{3} = -10$). So we need to go 10 beyond 0, leaving us at −10.

Remember, begin thinking of subtraction as adding a negative number — $10\tfrac{1}{3} - 20\tfrac{1}{3}$ means $10\tfrac{1}{3} + -20\tfrac{1}{3}$. So the answer would have to be −10, as we're starting at $10\tfrac{1}{3}$ and moving $20\tfrac{1}{3}$ in the negative direction.

Now, in real life we probably wouldn't express this answer as "negative 10 miles." We'd say "10 miles west."

Bringing It Back to Words — When answering word problems that have a negative answer, **give your answer in words (dollars owed, miles in a specific direction)**, but **show how you obtained that answer using negative numbers**.

Keeping Perspective

As you practice working with negative numbers, remember that negative numbers just mean *the opposite of.* They could represent a different direction, a different electrical charge (with electricity), a debt, or a number of other things.

10.3 Subtracting Negative Numbers

Today, let's continue exploring negative numbers by looking at how to subtract a negative number.

Subtracting by Negative Numbers

Subtracting a negative number is easy if you just remember that the negative sign means *the opposite of*. So if we write – –4, we're really writing "the opposite of the opposite of 4," which is positive 4.

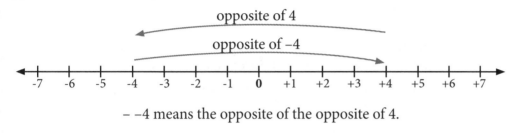

– –4 means the opposite of the opposite of 4.

$$- -4 = 4$$

> Negative numbers might represent the opposite direction or the opposite charge (in electricity) . . . or it might mean doing the opposite of adding (subtracting). But it always means the opposite.

When working with negative numbers, just remember that **every negative sign means *the opposite of*.** It doesn't matter how many negative signs you encounter before a number — each one means *the opposite of*.

An easy way to keep track of the signs is to mentally keep track of whether each one makes the number negative or positive. The first negative sign makes the number negative (i.e., *the opposite of* the number). The second makes the opposite of that, which is positive. Then another makes the opposite of that, which is negative again. Then positive, etc.

You can easily figure this out by saying "negative, positive, negative, positive" as you read off the negative signs. If you end with a negative, the answer is negative; if you end with a positive, the answer is positive.

$$- - - -4 = 4$$
Answer: +4

We can also see this visually on number lines.

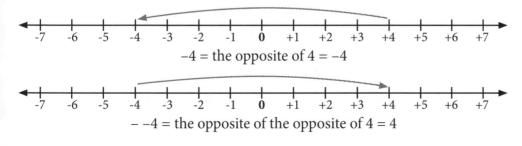

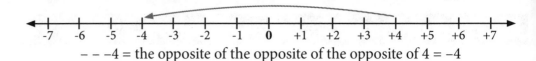
– – –4 = the opposite of the opposite of the opposite of 4 = –4

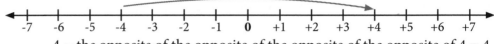
– – – –4 = the opposite of the opposite of the opposite of the opposite of 4 = 4

Applying Subtraction with Negative Numbers to Problems

Now that we've looked at how to handle additional negative signs, let's try dealing with them in an actual problem.

Example: Solve 8 + – –4

We know that – –4 means "the opposite of the opposite of 4," so this means 8 plus "the opposite of the opposite of 4," or 8 + 4!

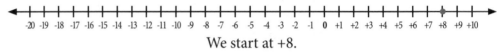

We start at +8.

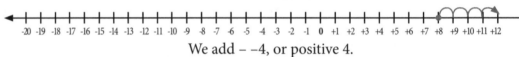

We add – –4, or positive 4.

$$8 + - -4 = 12$$

Note: Our original problem could also have been written 8 – –4. Both 8 – –4 and 8 + – –4 mean the same thing.

Applying Subtraction with Negative Numbers to Fractions

What does $4 + - -\frac{1}{2}$ equal? Don't let the fraction sign fool you — apply the *same* principle as to whole numbers. Each negative sign means *the opposite of*.

Example: Solve $4 + - -\frac{1}{2}$

We know that $- -\frac{1}{2}$ means "the opposite of the opposite of $\frac{1}{2}$," so this means 4 plus "the opposite of the opposite of $\frac{1}{2}$," or $4 + \frac{1}{2}$!

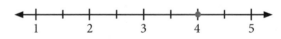

We start at +4.

We add $- -\frac{1}{2}$, or positive $\frac{1}{2}$.

$$4 + - -\frac{1}{2} = 4\frac{1}{2}$$

Note: Our original problem could also have been written $4 - -\frac{1}{2}$. Both $4 - -\frac{1}{2}$ and $4 + - -\frac{1}{2}$ mean the same thing.

Keeping Perspective

I hope you're seeing that we can **use negative numbers to show the opposite of anything**, be that a direction, an amount, a measurement, or something else altogether. If I want to specify the difference between distances above and below the center of a picture, for instance, I could use positive numbers for one direction and negative for the other.

positive

negative

To keep things consistent, we typically use positive numbers for distances *above or to the right*, and negative numbers for distances *below or to the left*, although there's no reason we couldn't reverse that if we needed to.

Keeping in mind that negative signs mean the *opposite of* makes it easy to add and subtract negative numbers. Just remember that the first opposite will be a negative number, the opposite of that would be a positive one, and so forth. So you can "count" the negative signs by saying "negative, positive, negative, etc."

Of course, you may be wondering when we'd really ever subtract negative numbers. While they don't show up often in everyday equations, you'll find yourself using this skill in algebra quite extensively to deal with negatives that turn up as you explore different aspects of God's creation.

10.4 Temperature and Negative Numbers

What's the weather like today? How hot is that soup? Is the turkey done?

We deal with temperature — how hot or cold something is — every day, to one degree or another (pun intended). Since negative numbers help us record cold temperatures, we're going to take this lesson to look at recording temperatures.

If you live in America, you're likely most familiar with the Fahrenheit system for measuring temperature. To better understand this system as well as temperature measuring in general, let's take a little trip through history.

In order to measure how hot something is, we have to have a way of both 1) measuring and 2) describing that heat against some sort of standard. Let's start with measuring.

Sometime in the late 1500s, a man named **Galileo Galilei** invented what could be called the "precursor of accurate thermometers."[1] While there's a whole history behind the development of thermometers, we will look only at the aspect of *describing the heat measured*.

In early thermometers, **each thermometer used its own scale to measure heat**. This made it more difficult to communicate about temperature. To see how, let's give it a try.

When you think about it, we could mark on a thermometer any point as 0, and then measure by whatever increments up and down from there we liked. In the thermometers shown, we've picked two different locations for 0, and made the space between our degrees narrower in one thermometer than in the other (that is, there's less space between each degree). **Notice that we've also used negative numbers to mark points below 0!**

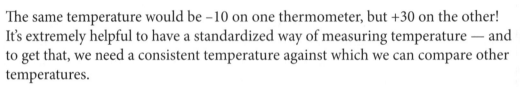

The same temperature would be –10 on one thermometer, but +30 on the other! It's extremely helpful to have a standardized way of measuring temperature — and to get that, we need a consistent temperature against which we can compare other temperatures.

A man named Gabriel Daniel Fahrenheit (1686–1736) "proved that . . . the temperature when ice began to form was always the same."[2] Other men discovered that the temperature at which water boils is always the same. The consistencies of water — created and maintained by God — give us a standard we can use to compare temperatures. Of course, the boiling or freezing point of other liquids or combination of liquids could also work (and have been used) — but water is readily accessible and thus is a logical choice to use as a standard.

Over the years, different scales based on constant temperatures began to emerge. Out of the different scales, two have continued in popular use: the Fahrenheit scale (based on the work of Gabriel Fahrenheit) and the Celsius, or Centigrade, scale (based on the work of Anders Celsius). Both scales are in wide use today, with the Celsius scale used nearly universally outside of the United States and a few other countries. They are shown in their modern form below.

A Galilean thermometer

204 | PRINCIPLES OF MATHEMATICS 1

Fahrenheit Scale
32° = freezing point of water
212° = boiling point of water

Celsius/Centigrade Scale
0° = freezing point of water
100° = boiling point of water

Notice how there are many more degrees between the boiling and freezing points of water in the Fahrenheit scale than in the Celsius scale. This is because each degree in Fahrenheit represents a smaller change in temperature.

Notice that both scales use negative numbers to describe quantities lower than 0 on their scale. Temperature scales are yet one more example of negative numbers in action.

While you should be familiar with the Fahrenheit and Celsius measurement systems in their final forms, in reality, they took time and input from many men to develop. The Fahrenheit scale we use today, for instance, has been modified significantly from Gabriel Fahrenheit's first proposal.

Galileo Galilei

Since Galileo played a key role in developing thermometers, let's take a quick look at his life. His work on thermometers was one invention in a long list of discoveries. You might have heard of him from science, as he promoted the Copernicus theory that the earth circled the sun and helped bring in the scientific method. But, although known for his contributions in science, Galileo was also a professor of mathematics. He used math to help him explore God's creation.

Galileo's "mind was eminently practical."[3] He led a full life of discovery — making one of his "most important discoveries"[4] (that of a pendulum) while observing a lamp swinging in a cathedral at just 18 years of age.

Without going into details about what Galileo believed and taught (which contained a mixture of truth and errors), one thing is clear: Galileo was willing to use math as a practical tool. The universe to him was waiting to be measured and discovered . . . and math was the tool to help him do so.

Galileo Galilei

Keeping Perspective

Here are two important takeaways from our journey through temperature-recording history:

- **Math and science go hand-in-hand**, and both are an exploratory process. Many things you learn in science — such as recording temperature — came through a process that included using math to explore God's creation (in this case, temperature).

- **Discoveries take time and effort** — It's easy to forget that the tools we use today came through much perseverance and work by men who have come before us. Some things take time and perseverance. When God calls you to do something, it can be easy to grow weary if it takes time and effort, but you don't have to let discouragement win. If you feel discouraged, make sure you're doing what God has called you to do and are doing it in His strength, not yours. Then keep looking to Him, trusting Him, and obeying Him.

> *And let us not be weary in well doing: for in due season we shall reap, if we faint not (Galatians 6:9).*

10.5 Absolute Value

Today, we're going to consider a different aspect of negative numbers: their *absolute value*. Let's say you use your house as the starting point. One store is 5 miles to the east of your home, and another is 5 miles (abbreviated *mi*) to the west. If you make your home point 0 and use positive for east and negative for west, you'd have this:

Now, how far apart are the two stores from each *other*? To answer this question, we're *not* interested in the directions the stores are from our home at all — all we care about is the *total distance between them*. We're interested in the total of what we call the *absolute value* of each distance.

Understanding Absolute Value

Absolute value is the fancy name for the value of a quantity regardless of whether it's positive or negative (i.e., its distance from 0). The absolute value of both +5 and –5 is 5. It doesn't matter if that 5 is to the right or left of the 0 on the number line. **Both the 5s are 5 spaces from 0.**

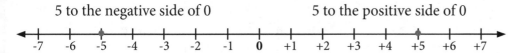

Representing Absolute Value

We represent absolute value with two lines around a number. It's a symbol and term you should become familiar with, as it's used to aid in communication.

$$|-5| = 5$$
The absolute value of negative 5 equals 5.

$$|+5| \text{ or } |5| = 5$$
The absolute value of positive 5 or the absolute value of 5 equals 5.

Applying Absolute Value

In the case of finding how far the two stores are from each other, we only care how many total "spaces," so to speak, they are apart. We can find this by subtracting the two distances from each other and then finding the *absolute value* of our answer. It doesn't matter which store's location we put first if we are careful to watch the negative and positive signs.

Distance between stores?

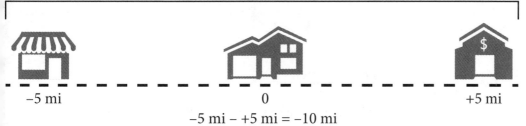

−5 mi − +5 mi = −10 mi

Don't let the positive sign confuse you — it means the same thing as if it were not present. The above could be written −5 mi − 5 mi = −10 mi.

Now, we need to find the absolute value of −10 mi.

$$|-10 \text{ mi}| = 10 \text{ mi}$$

The distance — regardless of the direction we're traveling — between the stores is 10 mi.

Notice that we also could have found the distance by subtracting −5 mi from +5 mi.

$$+5 \text{ mi} - -5 \text{ mi} = +5 \text{ mi} + 5 \text{ mi} = +10 \text{ mi}$$

Remember, two negative signs mean the opposite of the opposite, or a positive.

Again, the absolute value of +10 mi is 10 mi. There's a 10 mi difference between the stores. If we start at the store at +5 mi, we have to travel 10 miles in the negative direction; if we start at the store at −5 mi, we'd have to travel 10 miles in the positive direction. But the distance is the same, either way.

In this case, we could have more easily found the answer by merely adding together the absolute values of each to find the total distance. After all, the absolute values of the individual stores will tell us their distance from zero . . . and since the stores are in opposite directions, if we add up their distances from 0, we'll have their distance from each other.

$$|-5 \text{ mi}| + |+5 \text{ mi}| = 5 \text{ mi} + 5 \text{ mi} = 10 \text{ mi}$$

However, be careful with this method. It works only if the locations we want to find the distance between are in opposite directions of the place we're measuring as 0; it would *not* work if we had two positive or two negative locations. For example, if we had a store +5 mi away and one +10 miles away in the same direction, the two stores are not 15 miles apart. They're 10 mi – 5 mi, or 5 miles apart from each other.

We would need to use the previous method instead, subtracting one distance from the other and then taking the absolute value of the difference.

$$10 \text{ mi} - 5 \text{ mi} = 5 \text{ mi} \quad or \quad 5 \text{ mi} - 10 \text{ mi} = -5 \text{ mi}$$
$$|5 \text{ mi}| = 5 \text{ mi} \quad or \quad |-5 \text{ mi}| = 5 \text{ mi}$$

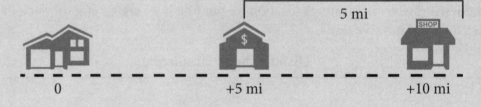

Keeping Perspective

We can use negative numbers to help us describe *direction*, yet we can also use absolute value to help us find the *distance* from a 0 starting point, regardless of the direction. As you apply negative numbers and absolute value to various problems, remember that **absolute value expresses the distance a number is from 0 — it ignores the direction**.

10.6 Multiplying and Dividing Negative Numbers

It's time to learn how to multiply and divide negative numbers. As we do, continue to keep in mind that a negative sign means *the opposite of*.

Multiplying Negative Numbers

Let's say we spend $2 every month for 3 months. If we wanted to track the amount we're spending as a negative (opposite of gaining), we'd have this equation:

$$3 \cdot -\$2$$

What is the product? Well, we've got –$2, and we're taking it 3 times. Since multiplication is repeated addition, this is the same thing as writing –$2 + –$2 + –$2. If we were to add these all together, we'd get a sum of –$6.

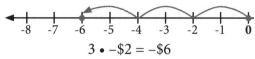

$3 \cdot -\$2 = -\$6$

Taking –2 three times gets us to –6.

Now let's say we want to find out the number of dollars we had 3 months ago. We would add a negative sign on the 3 to stand for 3 months in the past.

$$-3 \cdot -\$2$$

How do we solve this? Well, remember that each negative sign means *the opposite of*. So here we're trying to find the opposite of taking negative $2 three times. We saw that taking negative $2 three times got us –$6, so the opposite of that would be +$6. So 3 months ago, we had 6 more dollars than we do now.

When multiplying negative numbers, all you have to do is remember that **each negative sign means *the opposite of*.** If you have one negative sign involved in the multiplication (–), you'll have a negative answer. If you have two negative signs (– –), you'll have a positive answer. Three negative signs, (– – –) a negative answer, etc.

Notice that this is the same thing we did with subtracting negative numbers. Regardless of the operation, we view each negative sign as *the opposite of*. In fact, we could view multiple negative signs as multiplications by –1.

$$- - - -8 = -1 \cdot -1 \cdot -1 \cdot -1 \cdot 8 = 8$$

Four negative signs — the "opposite of the opposite of the opposite of the opposite of" — give a positive answer.

Dividing Negative Numbers

The same principles we just looked at with multiplication apply to division, too. Since division is the opposite of multiplication, we have to follow the same rules or we won't reverse the multiplication.

Multiplication	Division	
$3 \cdot -2 = -6$	$-6 \div -2 = 3$	$-6 \div 3 = -2$
$-3 \cdot 2 = -6$	$-6 \div 2 = -3$	$-6 \div -3 = 2$
$-3 \cdot -2 = 6$	$6 \div -2 = -3$	$6 \div -3 = -2$

All you have to do to divide negative numbers is to remember that each negative sign means *the opposite of*. If you have one negative sign involved in the division, you'll have a negative answer. If you have two negative signs, you'll have a positive answer. Three negative signs will yield a negative answer.

In short, to divide negative numbers, remember to count the negative signs, remembering that each one means *the opposite of*.

This holds true no matter what type of number you're dealing with. Notice that we can divide fractions, decimals, etc., following this same rule.

$$\frac{1}{2} \div -\frac{1}{3} = -\frac{3}{2}$$

One negative sign — the "opposite of" — gives a negative answer.

$$-5 \div -0.25 = 20$$

Two negative signs — the "opposite of the opposite of" — give a positive answer.

Keeping Perspective

Remember: the negative sign indicates *direction*. Each negative sign means *the opposite of*. If you keep this in mind, you'll be able to add, subtract, multiply, and divide negative numbers with ease!

10.7 Negative Mixed Numbers

Throughout this chapter, you've been working with negative fractions. It's time now to look at negative mixed numbers. There's nothing really special about them, except that it's easier to get mixed up when dealing with them, so they're worth a special mention.

Avoiding a Mixed-Number Mix-Up

When working with negative mixed numbers, keep in mind that a negative sign in front of a mixed number applies to the *entire* mixed number.

$$-2\frac{1}{3} \text{ means the same thing as } -2 + -\frac{1}{3}.$$

It's important to remember this when solving problems. Let's take a look at one.

Example: Solve $5\frac{2}{3} - 6\frac{1}{3}$

Let's start by inserting a plus sign for clarity.

$$5\frac{2}{3} + -6\frac{1}{3}$$

One way to add these two mixed numbers would be to add the whole numbers and fractions separately. But we have to be *very* careful about our negative sign with this approach.

210 | PRINCIPLES OF MATHEMATICS 1

Subtracting the whole numbers gives us −1.

$$5 + -6 = -1$$

Subtracting the fractions gives us $\frac{1}{3}$.

$$\frac{2}{3} + -\frac{1}{3} = \frac{1}{3}$$

Now let's put the two together. Since we have a −1 and a $\frac{1}{3}$, an easy mistake would be to assume the answer is $-1\frac{1}{3}$.

But the $\frac{1}{3}$ part was *positive*, not negative. We have to add $+\frac{1}{3}$ to −1.

$$\frac{1}{3} + -1 = \frac{1}{3} + -\frac{3}{3} = -\frac{2}{3}$$

The correct answer is $-\frac{2}{3}$.

Notice how much easier it would be (and how much less chance there is of a mistake with the negative sign) to simply convert the mixed numbers to improper fractions first.

$$5\frac{2}{3} + -6\frac{1}{3}$$

$$\frac{17}{3} + -\frac{19}{3} = -\frac{2}{3}$$

When solving problems with both negative numbers and mixed numbers, you'll avoid mixed-number mix-ups by **first converting the mixed numbers to improper fractions.**

Keeping Perspective — Negative Numbers and Force

Force is basically the name we give to describe, "loosely speaking, a push or pull on [an] object."[5] When we throw a ball, our arm gives the ball a certain force that sends it flying through the air. It falls eventually, however, because gravity is also pushing/pulling on it, sending it down toward the earth. Understanding and measuring force helps us work with moving objects, design devices to lift objects, and much more.

Negative numbers play a key role in studying and working with force. We can use negative numbers to describe the force being exerted on an object in one direction, and positive numbers to describe the force being exerted in another direction.

For example, when you try to push a box along a carpeted floor, the carpet is exerting resistance against your push. We could represent the resistance as a negative force and your push as a positive force. The picture shows the forces in pounds (abbreviated *lb*).

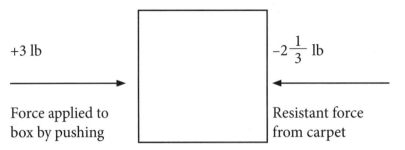

Adding the two forces together will tell you what the resulting force is on the object.

$$+3 \text{ lb} + -2\frac{1}{3} \text{ lb} = +\frac{2}{3} \text{ lb}$$

There's a total of $\frac{2}{3}$ pound of force in the positive direction.

Force is just one example of how negative numbers help us describe God's creation.

10.8 Negative Fractions

So far, we've looked at fractions preceded by negative signs. You know that $-\frac{1}{2}$ means *the opposite of* $\frac{1}{2}$, and how to solve problems with $-\frac{1}{2}$.

$$5 + -\frac{1}{2} = 4\frac{1}{2}$$

$$-\frac{1}{2} \cdot 5 = -\frac{5}{2} = -2\frac{1}{2}$$

But what happens if there's a negative sign *within* the fraction? For example, consider $\frac{-200}{5}$. Is our answer positive or negative?

To find out, think about what it represents. We could rewrite $\frac{-200}{5}$ as $-200 \div 5$, since fractions are a way of representing division. Then, using what we know about dividing negative numbers, we can tell that the answer would be *negative* (−40).

In short, since a fraction represents division, we can figure out if it's negative or positive the **same way we do with division problems** involving negative numbers!

Handling Negative Signs within a Fraction

$\frac{200}{-5} = -40$	$\frac{-200}{5} = -40$	$\frac{-200}{-5} = 40$
$200 \div -5 = -40$	$-200 \div 5 = -40$	$-200 \div -5 = 40$

What happens if there's an additional negative sign outside the fraction *and* one inside?

$$-\frac{-200}{5}$$

This just means *the opposite of* –200 divided by 5. Since –200 divided by 5 equals –40, the opposite of that would be positive 40.

$$-\frac{-200}{5} = --40 = 40$$

Don't let negative signs in fractions fool you. Complete the division inside the fraction following the same rules you normally would, and then deal with any additional negative signs the same way you normally would.

Keeping Perspective

The same principles for working with negative whole numbers apply to fractions containing negative numbers — remember that fractions are a way of representing division.

As you continue studying math, you're going to find that concepts keep building upon one another. The basic principles of addition, subtraction, multiplication, division, etc., are the core ways of describing the consistencies God created and sustains — upper math just keeps applying those same principles to more advanced situations!

10.9 Chapter Synopsis

Well, I hope you've enjoyed our quick tour of negative numbers! The main thing to remember is that a negative sign means *the opposite of*. We can thus use negative numbers to represent travel in an opposite direction, debt, temperatures less than 0, force against an object, and so much more. Knowing how to work with negative numbers proves valuable both in everyday life, and, as we'll discover more in Book 2, in problems where you wouldn't expect negative numbers at all.

Here's a quick review of some of the mechanics we discussed regarding operating with negative numbers:

■ **Adding Negative Numbers** — Adding negative numbers is the same as subtracting positive numbers.

$8 - 9 = 8 + -9$

Answer: –1

Obviously, if we add +4 and –4 together, we get 0. We call the number that, added to a number, equals 0 the **additive inverse** of that number.

10. NEGATIVE NUMBERS | 213

■ **Subtracting Negative Numbers** — Each negative sign means *the opposite of*, so subtracting a negative number is really adding the positive.

8 – –9 = 8 + – –9 (8 plus the opposite of the opposite of 9)

8 + 9 = 17

Answer: 17

If there are more than two signs, we just have to remember that each one means *the opposite of*.

8 – – –9 = 8 + –9 = –1

Look at the first negative sign as making the number negative, then the second making it the opposite of that, or positive, and the third making it negative again.

■ **Multiplying Negative Numbers** — Each negative sign means *the opposite of*, so one negative sign gives a negative answer, two a positive, three a negative, and so forth.

$$-4 \bullet -2 = 8 \qquad -4 \bullet 2 = -8 \qquad 4 \bullet -2 = -8$$

■ **Dividing Negative Numbers** — Each negative sign means *the opposite of*, so one negative sign gives a negative answer, two a positive, three a negative, and so forth.

$$8 \div -2 = -4 \qquad -8 \div 2 = -4 \qquad -8 \div -2 = 4$$

This same principle applies to fractions, as a fraction line represents division.

$$\frac{8}{-2} = -4 \qquad \frac{-8}{2} = -4 \qquad \frac{-8}{-2} = 4$$

We also discussed negative mixed numbers and how it's helpful to convert mixed numbers to improper fractions when there's a negative sign involved (it avoids mix-ups!) and looked at what we call the **absolute value** (i.e., the distance from 0, regardless of direction) of a number and saw how this could be marked like this: |-4|. The absolute value of both +4 and –4 is 4 — they are both 4 spaces from 0 on a number line.

Negative numbers are yet one more way to describe the quantities and consistencies God created and sustains.

214 | PRINCIPLES OF MATHEMATICS 1

[CHAPTER 11]

Sets

11.1 Sets and Venn Diagrams

Imagine for a moment a world without organization. A world where a kitchen drawer has forks, measuring cups, food, and paper all haphazardly mixed together. A world where grocery stores have apples down one aisle, bananas down another, and oranges on the other side of the store. A world where libraries have books with no call numbers or order to their location — just shelves and shelves of randomly piled books.

While those are extreme examples, they illustrate the usefulness of organization. Organization helps us function and refer to things. Even saying "apple" gives us a way of referring to a specific type of fruit, regardless of whether the apple in question is a Golden Delicious or a Granny Smith. We use organization on many levels all the time in life — including in math. Before we take a look at some ways to organize and refer to specific groups of numbers, let's start by understanding the idea of sets and subsets.

Understanding Sets and Subsets

A **set** is simply a collection (i.e., a group), and a **subset** is a set that is a further subdivision of another set. We can (and often do!) make sets and subsets of basically anything.

If you organize your closet by type of clothes (shirts, sweaters, pants, shorts, etc.), we could call each type of clothing a **set**. If you then proceed to organize your shirts by short versus long sleeves, we could call each type of shirt a **subset** of the set shirts — a further organization within the set of shirts.

The beauty of sets is that they can be defined differently. Some people might organize their closet by color instead of type — and that's okay! A set is simply a collection. It can be based on an attribute, or it can even be random. If you went into your closet and grabbed a handful of clothes blindfolded, we could call those clothes a random set. Again, a **set is a collection.**

For example, suppose you wanted to organize recipes into folders. You could view each folder as a set. A recipe for "apple pie" could be categorized under "apples," "pies," "desserts," "autumn," or a number of other sets, depending on the organizational structure you're using.

For another example, suppose you set out to organize your tools. Depending on your organization system, you might put a specific wrench with all your other wrenches in a "wrench" pile, or you might put it in a pile of "plumbing tools."

Now, in all of the above situations, our collections were of physical items, but sets and subsets can be used to organize nontangible things as well. When you write a paper, you organize your ideas into logical (hopefully!) groups, or sets. For another example, we use the term "brown-haired" to refer to all people with a certain characteristic — that is, to a set, or group, of people, even though that set is not geographically located together. Sets help us refer to a grouping or collection as a whole, be that group a physical group or a way of referring to a nontangible grouping.

Defining a Set

Because sets are used extensively in math, there are whole notational conventions regarding how to express a set mathematically.

One way is to simply put brackets { } around the set or its definition. For example, we could represent a set containing all foods that have apples as an ingredient like this:

{all foods with apples as ingredients}

And we could represent the set of all dogs and cats like this:

{dogs, cats}

Picturing Sets and Subsets

Have you ever heard the saying "a picture is worth a thousand words"? Pictures help us process information — even in math!

Venn Diagrams are basically drawings that use circles to show how different sets (collections) relate. Let's use them to better understand sets and subsets.

A Set

In Venn Diagrams, we use a circle to represent a set. The title names the set, and the circle represents the set, or collection.

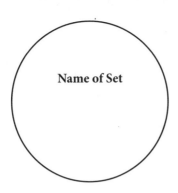

Equivalent Sets

Sometimes sets are really just different names for the same thing. For example, "pop" and "soda" are both names for the same thing (a carbonated drink). Everything that could be called "pop" could also be called "soda." If we think of them as sets, or collections, everything in one set would also be in the other.

We could draw that with a diagram by placing the names for both sets inside the same circle, showing that the same items belong to both sets.

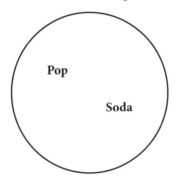

Subsets

As we mentioned, a subset is just a further categorization of a set. So everything inside a subset is also part of the larger set, but not everything inside the larger set is part of the subset. For example, we use "pies" to refer to all pies in general, but "apple pies" refers to a subset of pies — i.e., those with apples. Every apple pie is also a pie, but not every pie is an apple pie, so "apple pie" is a subset of "pies."

We would show this in a diagram by drawing "apple pies" as a circle within the set "pies," thereby showing that everything inside the set "apple pies" is also part of the larger set "pies."

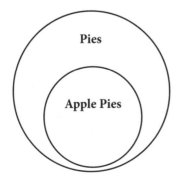

Overlapping Sets

Sometimes sets contain only some items that could be categorized as either set. For example, some desserts are apple dishes and some apple dishes are desserts, but there are desserts that don't contain apples and apple dishes that are not desserts.

We would show this in a diagram by having the circles for each set overlap.

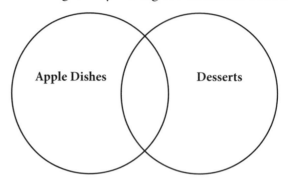

Completely Different Sets

Some sets contain *no* items that could be categorized as a part of either set. For example, if we thought of foods in terms of "gluten-free" and "foods containing gluten," a food item could not be a part of both sets. It would either be gluten-free or it wouldn't; it couldn't be both.

We would draw this by showing no overlapping at all between the two circles.

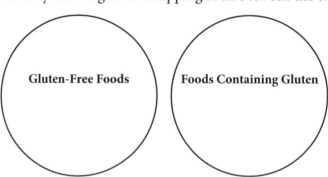

A Glance at History

Venn Diagrams are named after John Venn, whose article helped promote their adoption.[1] Below are some of the diagrams from Venn's book *Symbolic Logic*.[2]

1.   2.   3.   4.   5.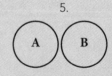

The letters A and B stand for different sets. The circles show how those sets can relate to one another. Notice how they are the same as the drawings we made to show the relationship between different foods, with diagram numbers 2 and 3 both representing subsets (only in 2, A is a subset of B, while in 3, B is a subset of A).

Keeping Perspective

We use **sets**, or collections/groupings, to organize things — whether those things are the clothes in your closet, food, or, as we'll see in the next lesson, numbers with different properties. Some sets are different names for the same things, others are subcategorizations (i.e., **subsets**), some sets overlap and contain some items that could be categorized either way, and other sets are completely different. We encounter sets with all of these characteristics within both life and math.

Remember, sets are just ways of grouping, whether tangibly or by referring to different things as a group. You use the concept of sets all the time without thinking about it when you refer to collections (tangible or not), such as to a team, brown-haired people, or apple pies. Ultimately, sets are yet another way of naming/exploring God's creation — only in sets, we're naming/exploring by groups, or collections, rather than individually. And Venn Diagrams just give us a way to picture how sets relate to one another.

11.2 Number Sets

Numbers can be divided into nontangible sets, or groupings, in order to help us refer to multiple numbers as a group. If you'll recall, back when we learned about fractions, we started referring to nonfractional amounts greater than 0 as **whole numbers**. That was really a number set — a name given to describe numbers with specific characteristics.

Last chapter, we learned about **negative numbers** and **positive numbers** — those were really both number sets!

Let's take a look today at a few common sets used to describe numbers with specific attributes. As you continue to study math and examine in more depth how God causes this universe to operate, you'll find these sets useful in referring to numbers with specific properties. For example, if you need a whole number answer to a problem, you can just say "whole number" rather than saying you need "a nonfractional amount greater than 0."

Positive and Negative

Like we saw in the last chapter, numbers can be thought of in terms of whether they're positive or negative. Notice that a number is either positive or negative — it can't be both. And 0 is in a class all its own . . . it's neither positive nor negative.

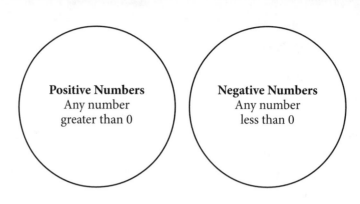

Integers and Its Subsets

Sometimes, we don't care if a number is positive or negative. We only know we need a nonfractional amount. So rather than categorizing numbers by positive or negative, we could categorize them as fractional or whole. But wait a minute — whole numbers are only *positive* — they start at 1! We need to have another name to describe nonfractional quantities in general, regardless of whether they're positive or negative. And we do! We call them **integers**.

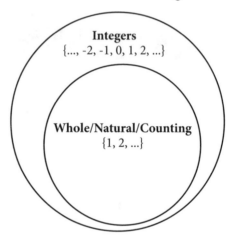

Notice that the sets shown in the Venn Diagram are enclosed with brackets: {1,2 . . .}. **Brackets** are a common convention for representing a set. The **three dots** at the end (. . .) mean that the set, or group, of numbers continues indefinitely in that direction. 4, 5, 6, 7, etc., are also part of the set of whole numbers.

Note that integers can be positive and negative. Positive and negative categorizes numbers based only on direction (positive or negative), while integer categorizes them based on the type of quantity they represent. They're different organizational systems, much as "pies" and "apples" could be different organizational systems for recipes.

On the other hand, *every* whole number is an integer. Whole numbers are a **subset** of integers — they are a smaller grouping within the *same organization system*.

Natural and **counting numbers** are simply different names for whole numbers. They represent the same set. Some people include 0 in the set, and others don't. In this course, we'll treat whole, natural, and counting numbers as starting at 1.

There are still more ways we could subcategorize integers. Some integers can be divided by 2, while others cannot. We call those that can be divided by 2 **even** and those that can't be divided by 2 **odd**. Note that 0 is even, as it *can* be divided by 2 (we saw back in 4.4 that $0 \div 2 = 0$).

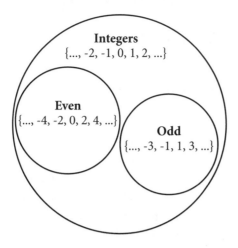

And you already know about **prime numbers**. Prime numbers are whole numbers that can't be divided by any whole number but themselves and 1. Notice that prime numbers are really a subset of whole numbers, which are a subset of integers!

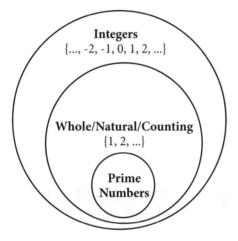

Why Organize Numbers?

Number sets help us **refer to numbers with specific properties.** For example, remember back to when we factored numbers in order to find their least common multiple or greatest common factor? We had to keep factoring until we got to all **prime numbers** — a fancy name for a whole number greater than 1 that cannot be divided evenly by any whole number except itself and 1.

```
         24                    156
         ^                      ^
      12    2               12 x 13
      ^                      ^
    6  x  2               6  x  2
    ^                      ^
  2  x  3               2  x  3
```

24 expressed as a product of its prime factors: 2 • 2 • 2 • 3

156 expressed as a product of its prime factors: 2 • 2 • 3 • 13

Now aren't you glad we could say "prime factors" instead of "factors that are whole numbers greater than 1 that cannot be divided evenly by any whole number except themselves and 1"!

Keeping Perspective

Why are there so many types of numbers? Because God created a complex universe, and it takes a lot of different types of numbers even to begin to describe it!

Biblical principles also give us a framework for understanding why we can organize numbers into sets. For one, God's consistency in holding all things together enables us to have confidence that certain quantities will consistently have specific properties (for example, odd quantities will not be divisible by 2). If quantities did not operate consistently, grouping numbers by their properties would be pretty useless.

Also, if our minds were generated by chance, we would have no reason to suppose we could make logical sets and subsets. But the Bible makes it clear — we are not here by chance. We were carefully fashioned by a loving Creator. We are able to observe the characteristics of numbers because the God who created all things also created us capable of observing and naming His creation.

> *For we are his workmanship, created in Christ Jesus unto good works, which God hath before ordained that we should walk in them (Ephesians 2:10).*

11.3 More on Sets

Remember, sets are a tool to help us refer to a collection as a group. So far, our number sets have been defined based on characteristics, but sometimes we need more specific sets. Let's take a look.

Random Sets

We could group a random collection of numbers together and call them a set. For example, I could say 4, 5, 10, and -1 are a *set* of numbers. We could even call it set "R" for random and represent it as {4, 5, 10, -1}.

Why would we ever do this? Well, random sets of data are often used to test computer programs! Remember, **a set is any collection**, whether random or logical.

Sets Specific to a Situation or Equation

Suppose you need to get greater than a 70 on a test in order to get an A in the course. We could call all your possible scores that would get you an A a *set*.

222 | PRINCIPLES OF MATHEMATICS 1

$$\{71, 72, 73 \ldots 100\}$$

Note: The dots show that 74, 75, etc., are also part of the set. This set assumes you can't get more than 100 in the course, and that you can't get a decimal grade (like 71.5).

Notice how we could have represented this more concretely with symbols.

$$\{\text{integers} > 70 \text{ but} \leq 100\}$$

> Notice we just used a new symbol! $\leq$ **means "less than or equal to."** It's used to show that we need a number less than or equal to 100 (it could be either). Notice that it's the same as the less than sign (<), except it has a line underneath (half of an equal sign: =). There is also a **"greater than or equal to" sign** ($\geq$).

You'll find as you get into upper-level math that we'll often need to specify the numbers that will work for a specific equation. For example, suppose you're ordering CDs in bulk to resell. You've been told it will cost $7 a CD if you order more than 10 but less than 20 CDs, and you want to find your total cost.

Let's put this in math terms:

$$\text{total cost} = 7 \cdot \text{number of CDs}$$

This relationship holds true only when the number of CDs is > 10 but < 20

While we didn't use brackets or mention the word "set," by saying the number of CDs has to be >10 but <20, we essentially just specified a collection, or set, of CD quantities for which this relationship holds true.

For another example, suppose we were dealing with a division. We might need to specify that the number we're dividing by can't be 0 (i.e., $\neq 0$). In other words, we're specifying a set of every number *but* 0!

Whether you see the name "set" or not, you'll encounter sets often in math. In fact, they're a common application of the comparison symbols we looked at back in 1.4: <, >, and $\neq$ (less than, greater than, and does not equal), as well as the variations $\leq$ (less than or equal to) and $\geq$ (greater than or equal to) we briefly mentioned in this lesson.

Sets in Programming

Computer programs are written in languages referred to as code. These languages are partly English and partly math. Quite often, a computer programmer will need to define sets when programming. For example, many online forms require you to fill out a phone number. Some forms will check to make sure that the phone number box actually contains a phone number and not just words. This could be done by defining the set of values allowed for each digit in the box (such as the numbers 0–9) and then programming the computer to check each digit to see if what was entered is a member of the set. Exact details vary based on the specific programming language used and what we're trying to check for, but the principle of sets applies!

11. SETS | 223

> # Keeping Perspective
> Sets are simply collections. They can be collections based on characteristics (such as all positive numbers), randomly chosen, or specific to an equation. They can be written different ways. The point is simply to know that we can specify specific collections, or groups, of numbers as needed.

11.4 Ordered Sets (i.e., Sequences)

You probably learned how to skip count (count by 2s, 3s, 4s, 5s, etc.) at a young age. But did you realize that you were actually counting a specific type of set called a sequence?

> Counting by 2s: 2, 4, 6, 8, 10, 12 . . .
> Counting by 3s: 3, 6, 9, 12, 15, 18, 21 . . .
> Counting by 10s: 10, 20, 30, 40, 50 . . .

Sequence is just a fancy name for describing "an ordered set of quantities."[3] While sequences can have no describable pattern, often numbers within a sequence follow a pattern of some sort. For example, when you count by 2s, each number is 2 greater than the number before it. There's a pattern.

Some patterns are less obvious than the counting sequences you've known for years. Can you identify the pattern in the sequence below?

```
       Skip      Skip      Skip
        ∧         ∧         ∧
4,  5,  7,  8,  10, 11, 13,   ...
    ∨         ∨         ∨
2 numbers  2 numbers  2 numbers
 in order   in order   in order
```

Notice that after two sequential numbers (numbers in order), we skip a number. This patterns repeats. With that in mind, can you figure out what the next two numbers after 13 will be?

The next two numbers will be 14 and 16.

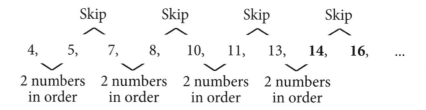

Sequences in Action

We encounter patterns all the time in life. Sequences help us subconsciously categorize these patterns.

Look at this piano. Notice how the black keys are grouped together in a pattern {2, 3, 2, 3, . . .}.

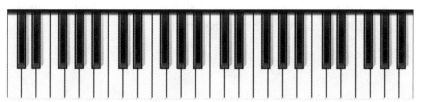

Whether consciously or subconsciously, pianists use the pattern, or sequence, of black keys on the piano to find their way around the keyboard.

Can you figure out the pattern in these numbers?

$$\{0, 1, 1, 2, 3, 5, 8, 13, 21, 34 \ldots\}$$

Notice how this time, *each* number is the *sum* of the two previous numbers (0 + 1 = 1; 1 + 1 = 2; 2 + 3 = 5; and so forth). This particular number pattern (sequence) helps us describe some fascinating order throughout God's creation. In fact, it's so famous it has a name: **the Fibonacci Sequence** (named after the same Leonardo "Fibonacci" Pisano we studied in 2.5).

You'll get a chance in your *Student Workbook* to explore some of the ways this sequence helps us describe God's creation.

> Notice that in {2, 3, 2, 3...} the same numbers (2 and 3) are repeated. Sequences can have repeated numbers. In a set, on the other hand, we wouldn't repeat a number. Since the order doesn't matter in a set, a number is either a part of it or it isn't.

Naming the Relationships

In sequences that follow a pattern, it can be helpful to name the distinguishing aspects of a sequence. For example, in {2, 4, 8, 16, 32 . . .} each number is 2 times the previous number. We would say the **common ratio** is 2.

Think about the name common ratio for a moment. We know a ratio is a comparison via division. If you compare any number in the sequence shown with the one before it as a ratio, you get . . . drum roll please . . . 2! Truly 2 is the common ratio.

$$\frac{32}{16} = 2 \qquad \frac{16}{8} = 2 \qquad \frac{8}{4} = 2 \qquad \frac{4}{2} = 2$$

For a different example, in {2, 4, 6, 8, 10 . . .} each number is 2 plus the previous number. We would say the **common difference** is 2.

Think about the name common difference for a moment. "Difference" is the term we used in subtraction to refer to the result of subtracting one number from another. If you subtract any number in the sequence shown with the one before it, you get . . . another drum roll please . . . 2! Truly 2 is the common difference.

$$10 - 8 = 2 \qquad 8 - 6 = 2 \qquad 6 - 4 = 2 \qquad 4 - 2 = 2$$

Keeping Perspective

Sequences are really just a specific type of set — a group of numbers in which the order matters. Sequences help us describe many aspects of God's creation. In the worksheet that goes with this lesson, you'll explore a sequence that helps us describe the spirals in sunflower seeds, pineapples, a nautilus, and more . . . as well as see how sequences even help us describe a visually appealing ratio used extensively in art.

11.5 Chapter Synopsis

In this chapter, you've had a chance to explore **sets** — a fancy name for collections, be those collections tangible or not. You learned

- that sets can be defined differently depending on our need. For example, the folders we'd use to organize recipes would vary based on how we need to find the recipes, and the nontangible "folders" used to organize numbers can vary based on our need.

- that some sets are **subsets** of other sets — further categorizations of an existing set.

- that **Venn Diagrams** can help us picture how sets relate to other sets.

- that we organize numbers into different sets (including **positive and negative, integers, odd and even, whole numbers, and prime numbers**) in order to easily refer to numbers with different properties.

- that number sets can also be **random** or **specific** to an equation.

- that **sequences** are sets with an order to them. Some sequences have a **common ratio** (a consistent ratio between touching numbers in the sequence), others have a **common difference** (a consistent difference between touching numbers in the sequence), and others follow different patterns or may not have a discernible pattern.

While we don't often talk about sets by name, we use the principle all the time, both in life and in math. In the next chapter, we'll see sets applied in gathering statistics, and in the one after that, you'll explore the sets we use to refer to different shapes.

[CHAPTER 12]

Statistics and Graphing

12.1 Introduction to Statistics

Now that we have some tools under our belt, it's time to take a whirlwind tour through an important field that applies math to explore data: statistics.

Understanding Statistics

The American Heritage Dictionary[1] defines **statistics** two ways:

1. The mathematics of the collection, organization, and interpretation of numerical data; especially, the analysis of population characteristics by inference from sampling.

2. A collection of numerical data.

In other words, the word "statistics" can describe both the process of gathering and working with data *and* the final collection of that data. In this chapter, we'll be taking a broad look at both aspects of statistics. To avoid confusion, we'll use the abbreviation "stat" when referring to the data itself, and "statistic" when referring to the mathematics of collecting, organizing, and interpreting that data.

Let's start our exploration by walking through a simple example of statistics in action: **collecting, organizing,** and **interpreting** numerical data.

Suppose you are a sewing instructor, and you're given information on all the students in your new sewing class. You compile the following list of all their ages:

12	13	12	13	12	8	12
12	13	12	12	14	15	13
12	12	12	12	12	13	

Now you've *collected* the data, but how do you *organize* it to find useful information?

One helpful way to sort data is to look at the **frequency** with which the different **data points** (a fancy name for a specific piece of data within the group of data — each number in the list of ages is a data point) appear. We've done that below in a table.

Frequency Distribution

Age	Frequency
8	1
12	12
13	5
14	1
15	1

Notice how the first column in the table lists all the different ages, and the next lists how many times they appear, or their frequency. For example, the third line down tells us there are 5 students that are 13 years old.

Now let's add a column showing the percent of the total each frequency represents. We have 20 in the class, so having a frequency of 1 is the same as 1 out of 20, or $\frac{1}{20}$ or 5%. We call this percent the **relative frequency.** (We have names for everything in math!) Notice how the relative frequency makes it even easier to see what portion of our class is what age.

Frequency Distribution

Age	Frequency	Relative Frequency
8	1	5%
12	12	60%
13	5	25%
14	1	5%
15	1	5%

Now to *interpret* the data and draw conclusions: notice that 85% of our class is either 12 or 13 (60% + 25% = 85%). What might you conclude off this? Perhaps that it would be logical to target the class to 12 and 13 year-olds!

At a very simple level, we've just done all three steps of statistics: *collected*, *organized*, and *interpreted* data. We took a simple list of data (the ages of our students) and organized it so we could gain information from it.

Now, colleges offer degrees in statistics, so obviously there's often much more to the statistical process. We'll continue to learn about working with data throughout this chapter, as well as in Book 2 of this course.

Keeping Perspective — Useful, but Dangerous

Data collection and interpretation has become increasingly important in our culture. Politicians want to know what people think, businesses want to know what consumers want, stock brokers want to know where they should invest money, sports fans want to know how players are doing — the list goes on. Statistics is a useful application of the math we've been learning.

However, it's important to understand that stats can easily be either knowingly or accidentally presented in a misleading way. For example, suppose a company has stats to show that their sales are up over the last year, and draws the conclusion that their marketing team did a great job. Before we give the marketing team a raise, we'd need to see if it was really the marketing that caused the increase, or something else altogether (a better economy, a new product, etc.). We might also want to check to see if the sale numbers were presented in the same way as those from last year — if the sales were tracked differently, the results might look different when, in reality, they weren't.

Always, always, always (did I say that enough times?) dig a little deeper before just believing a stat you read (and a conclusion reached based on that stat).

The Bible reminds us that we cannot trust ourselves or other men. Only God is entirely trustworthy.

> . . . let God be true, but every man a liar. . . (Romans 3:4).

> But Jesus did not commit himself unto them, because he knew all men, And needed not that any should testify of man: for he knew what was in man (John 2:24–25).

We need to keep this truth in mind as we wade through the stats that bombard us. In this chapter, we'll take a closer look at **collecting** and **organizing** data, along the way exploring how that data can be **interpreted**. Whether or not you ever collect and organize data yourself, what we explore will help equip you to correctly interpret the stats you encounter — and to spot misinterpretations.

To learn more about how stats can be twisted, see Darrell Huff's *How to Lie with Statistics* (New York; W. W. Norton, 1993).

12.2 Collecting Data — Sampling

Collecting data sounds simple enough, but it can be a complicated process. After all, we want to make sure we collect the *right* data — the data we need in order to come to accurate conclusions.

In the previous lesson, we looked at concrete data: the ages of all the students in a specific class. But what if we wanted to find out in general what ages attended sewing classes across the United States? Unless we had a way of collecting data on every single sewing class in the U.S., we'd be forced to look at the ages of just some sewing class attendees and draw conclusions based on them.

> **Term Time**
> A **sample** is "the portion of a population selected for analysis."[2] In other words, a **subset** of the whole population. In our example, the sewing attendees whose ages we look at would be our sample.
>
> By **population**, we mean the whole group (or set!) we're considering — in this case, every single sewing class attendee in the USA.
>
>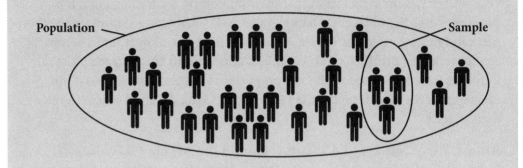

Much of the data you'll encounter in the news and other outlets is based on **sampling** (i.e., taking a sample of a population), so it's an important concept to understand. Let's think through some of the challenges and dangers in basing decisions off a sample.

In order for the data to have meaning, the people in our sample would have to represent the whole accurately . . . and we'd have to interview enough people to have a reasonable picture of that whole.

In our sewing class scenario, if we based a conclusion about ages that attend sewing classes off a sample of just one or two classes, we might end up making a very wrong conclusion. What if those classes were structured to appeal to young girls? Or what if the surrounding area was a retirement community with lots of older women?

For another example, if one section of a congressman's district has voters of a significantly different income, different religious belief, or different type job, taking a sample of that section would not be representative of the whole district.

We need to make sure that the sample we use is a **random** selection that will thus represent the whole population we're trying to find (such as all the different types of sewing classes or voters in the district) — and that our sample is **large enough** to really give us a picture of the whole. (We couldn't survey only 5 people and expect the results to tell us ages of all sewing class attendees in the United States or what all the voters in the district thought.)

When you see a survey result listed or stat quoted, look at **how the information was collected.** Was it based on a random selection? It's also a good idea to look for what are called the **margin of error (MOE)** and **confidence level.** Let's explore these numbers using an actual example.

Before the 2012 presidential elections, Gallup released a report titled "Romney 49%, Obama 48% in Gallup's Final Election Survey." The fine print at the bottom explained that "one can say with 95% confidence that the maximum margin of error is ±2 percentage points."[3]

What exactly does that mean? The margin of error of ±2 percentage points means that the results listed — the 49% and 48% — might vary from total population by 2 percent either direction. In other words, the percent of the whole population in favor of Romney could range from 47%–51%, while the percent in favor of Obama could range from 46%–50%. This means that more people in the population as a whole *could* have actually favored Obama. Statistically, since only a sample was interviewed, we can only expect the results to reflect the whole within a certain percent — calculated here to be within 2%. Up to a certain point after which increasing the size doesn't really change the results, the larger the sample, the smaller the margin of error will be.

The confidence level of 95% means that the survey has a 95% chance of being accurate within the 2% margin of error listed, and a 5% chance of being off by greater than the 2% error listed. In other words, we're 95% sure the results accurately reflect the whole population within 2% either direction . . . but there's a chance that, if we'd tested the whole population instead of a sample, our results would vary by more than 2%.

With results of 49% and 48%, a 2% margin of error, and a 95% confidence level, all we could conclude is that the race is close. The survey didn't show *who* was in the lead. (Plus, it's important to remember that even if the results were more decisive, preelection polls do not necessarily tie with what will happen in the voter box.)

Keeping Perspective

Wow! A lot goes into collecting stats, doesn't it? Wait until we look at organizing them.

While you may never need to sample a population in your lifetime, understanding the sampling process can help you discern through the many stats you'll encounter in life and know if the interpretations offered are reasonable. Remember to look at the process used (was it **random**?), the **margin of error (MOE)**, and the **confidence level.**

12.3 Organizing Data — Understanding Graphs

Have you ever tried looking through pages of numbers? They're not easy to quickly take in! Even the frequency tables we looked at in 12.1 take a moment to process, and they do not always help us see what conclusions we can take away from the data. Seeing a picture or diagram can often help us grasp a concept no amount of words could get across.

For this reason, graphs are an important part of statistics. Graphs pictorially represent the data collected so it can be grasped at a glance. They help us present data in an understandable way.

In this lesson, we're going to look at two different types of graphs: the pie graph and the bar graph. You are probably already familiar with these graphs, as they are often in newspapers or magazines. Hopefully our exploration, though, will give you a greater understanding of the graphs you encounter in everyday life and a little better picture of which graph does what.

Pie Graphs

In a **pie graph**, a circle is used to represent the total (100%), and each shaded section represents the percent of that total a specific number constitutes. **Pie graphs are great for showing how individual numbers relate to a whole.**

In 12.1, we collected data regarding the ages attending a sewing class. Notice how we could graph the relative frequencies of each age on a pie graph, thus visually showing the age composition of the class.

Frequency Distribution

Age	Frequency	Relative Frequency
8	1	5%
12	12	60%
13	5	25%
14	1	5%
15	1	5%

232 | PRINCIPLES OF MATHEMATICS 1

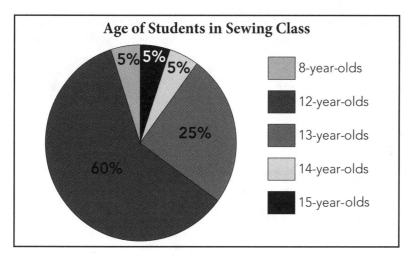

In the pie graph, the entire circle represents 100%. Then 60% of the circle is shaded to represent the portion of 12-year-olds, 25% to represent the 13-year-olds, and the three 5% sections to represent the 8-year-olds, 14-year-olds, and 15-year-olds.

Notice how the pie graph makes it easy to take in at a glance **what portion of a whole** each number represents. It's easy to instantly see what ages make up the majority of the class.

Bar Graphs / Column Graphs

Bar graphs (also sometimes called **column graphs** if the bars are vertical) are another useful way of displaying data. In a bar graph or column graph, bars (surprise!) are used to graphically show the data. Unlike pie graphs, bar graphs don't show how data relates to the whole; instead, they show pictorially **how data compares to other data.**

For example, suppose we had data on how many people three different county parks serve each month.

897	Park A
1,506	Park B
500	Park C

Here, we're not interested in a percent of a whole (that is, a relative frequency) — we're just interested in comparing totals between the parks. We can do that easily on a bar or column graph. Notice that the basic concept of the graph is the same either direction.

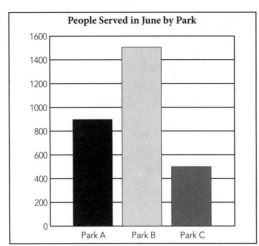 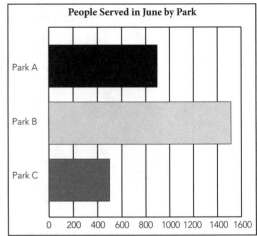

In the park scenario, we are dealing with parks, so we only have one numerical value (the number of people each park served). Sometimes we'll be working with two numerical values . . . and sometimes the data will be grouped together into ranges.

Bar graphs with grouped data are called **histograms**. The histogram shown here displays data regarding the number of houses sold within a specific price range. Notice that the columns shown are right next to each other on a histogram — there's no gap. That's to show that there's no gap in the ranges — each range picks up right where the previous one stops.

There's no gap between the columns.

There's no gap in the ranges. The previous range stopped at $149,999 and this one starts at $150,000.

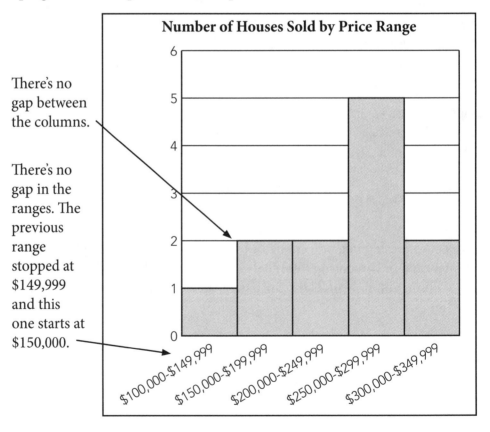

Dr. Brian Ray: Making a Difference with Statistics

Back in 1990, Dr. Brian Ray started an institute to research homeschooling called the National Home Education Research Institute. Over the years, he's researched a lot of data related to home education. His resulting stats have been used in courts to help convince judges that home education is a valid educational option — and shown to many parents wondering whether to home educate. His work reminds us that statistics can make an important positive difference when used for God's glory.

Below is one of his charts, reprinted here with permission. It shows the significant growth in homeschooling since the 1970s. Note Dr. Ray's label to the left that says "Thousands Home Educated." This means that each of the numbers listed vertically represents 1,000 times what's listed. The 200 mark represents 200 • 1,000, or 200,000. Likewise, the 2,000 represents 2,000 • 1,000, or 2,000,000 — that is, 2 million! Abbreviating numbers this way can be helpful when dealing with large numbers.

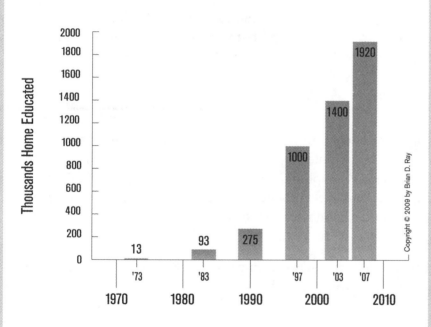

FIGURE 1.
GROWTH OF HOMESCHOOLING IN THE UNITED STATES

Note: Based on a summary of statistics from (a) Bielick, Stacey. (2008, December). 1.5 Million Homeschooled Students in the United States in 2007. Washington, DC: U.S. Department of Education (National Center for Education Statistics). Retrieved December 23, 2008 from http://nces.ed.gov/pubsearch/pubsinfo.asp?pubid=2009030; (b) Lines, Patricia M. (1991, October). Estimating the home schooled population (working paper OR 91-537). Washington DC: Office of Educational Research and Improvement, U.S. Department of Education; (c) Lines, Patricia M. (1998, Spring). Homeschoolers: Estimating numbers and growth. Washington, DC: United States Department of Education, Office of Educational Research and Improvement, National Institute on Student Achievement, Curriculum, and Assessment; and (d) Ray, Brian D. (2008, July 2). Research facts on homeschooling. Retrieved January 7, 2009 from http://www.nheri.org/Research-Facts-on-Homeschooling.html.

Keeping Perspective

In general, **pie** and **bar/column graphs** are most useful with smaller groups of data — once we get more than 6 or so bars or portions to our pie, they get harder to take in at a glance. There are other types of charts for more complex data.

Yet while there are many other types of graphs (and many variations of the ones we looked at), the point is that data can be displayed graphically in order to make it easier to quickly assimilate. Because graphs help summarize data, you'll encounter graphs a lot throughout your life — in magazines, papers, online research, etc. It's important to understand them so you'll be able to see what's being communicated and interpret the graphs correctly.

12.4 Organizing Data — Drawing Bar/Column Graphs

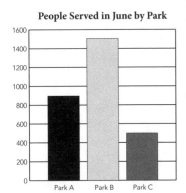

Now that you know how to understand graphs, it's time to draw one! For now, we'll focus on bar/column graphs, but many of the principles we talk about apply to other graphs as well.

The Mechanics

If drawing a bar or column graph by hand, it's easiest to use graph paper. Another option is to use a computer program. Computer programs such as Microsoft Excel have extensive graphing capabilities. Drawing a graph via the computer is obviously easier and likely produces a more professional appearance; however, it's important to understand what the software is doing so you'll know if you're really getting the results you think you are.

Regardless of whether you're drawing it yourself or on a computer, let's take a look at some key things you need to know.

Scale

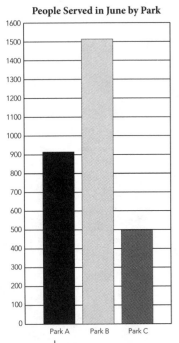

When drawing a bar or column graph, you have to first determine your scale. Often, you'll need to use one square to represent more than one of something. The graph on the top, for example, uses one line to represent 200 people, while the one on the bottom uses one line to represent 100 people.

Note: Most graphs nowadays are made on the computer. Tools like Microsoft Excel automatically calculate a scale when drawing a chart. However, it's important to understand how to choose a scale, as sometimes you'll want to change the scale the computer program selects.

236 | PRINCIPLES OF MATHEMATICS 1

Labels

Since the purpose of a graph is to communicate data visually, it goes without saying that a graph needs to be labeled so viewers can understand what it represents. A graph is virtually meaningless without labels! Be sure to label your graphs. Exactly how you label a graph can vary, but it needs to clearly communicate what each bar represents and the scale used to represent the data.

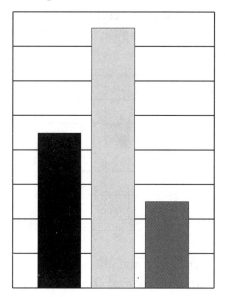

Keeping Perspective

All bar/column graphs don't have to be the same. God's creativity knows no end (just look around at all the different people there are, yet no two are the exact same), and He created man with the ability to think creatively (in a more limited way, of course!). So as you learn to draw graphs, don't be afraid to do so creatively. However, do so with the goal of accurately representing the data.

Always remember that a graph is usually produced to help someone else understand the information. Try to **keep your graph as simple as you can.** Many tools can create fancy-looking graphs — but often the glitter just makes the data harder to understand.

As we seek to use math for God's glory, we'll want to present the facts as accurately as we can, as God is a God of truth. We want our graphs to convey the truth about the data as clearly as possible.

12.5 Coordinates

Before we move on to looking at another type of graph, it's time to learn about the idea of **coordinates**. Coordinates are "each of a group of numbers used to indicate the position of a point, line, or plane."[4] We'll sometimes use coordinates to refer to a region, too.

For example, the shaded square shown below could be described as being 3 spaces to the right horizontally and 4 spaces up vertically. We would refer to "3" and "4" as the coordinates of the box.

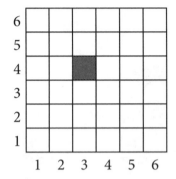

Not surprisingly, there's a standard convention for listing coordinates. The convention is to always **list the horizontal coordinate first and then the vertical coordinate**, separated by a comma and enclosed in parentheses. So the shaded square would be referred to by the coordinates (3,4).

An axis (axes if plural) is "a fixed reference line for the measurement of coordinates."[5] We would call our horizontal reference line our **horizontal axis**, and our vertical reference line our **vertical axis.**

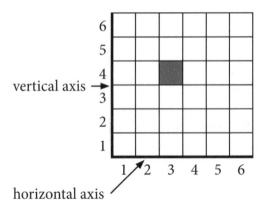

Applying Coordinates

Coordinates are used in many different ways . . . and coordinate systems can vary depending on our needs. Let's take a look at a few examples.

Beginning in Different Places

If you've ever played the game *Battleship*, you used coordinates to describe the different locations on the game board. The game uses letters to describe the distances one direction, and numbers the other. Using this system, we'd describe the box with the shaded circle with the coordinates (1,A).

Notice how we started our counting on the vertical axis in the top left corner. **Coordinate systems can begin wherever we need them to** — the principle is just to use some sort of system to help us refer to locations.

Computer monitors and TVs use coordinate systems to refer to the different pixels (tiny boxes) that make up the monitor/TV. While an actual monitor has hundreds of thousands (or millions) of pixels, the one shown gives the basic idea.

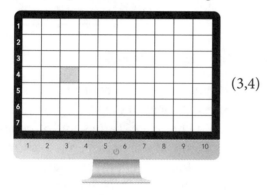

Once again, notice how we started our counting on the vertical axis in the top left corner!

Expanding to Include Negative Numbers

A coordinate system can be expanded to show negative numbers, too. In fact, we could use it on a map, picking any location to be our starting point, and then use a coordinate system to show points east, west, south, and north of that point.

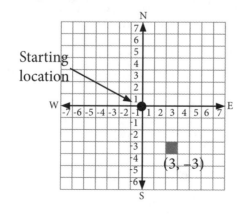

The arrows next to N, E, S, and W are simply reminders that we could continue the graph further.

12. STATISTICS AND GRAPHING | 239

Indicating Points Rather than Squares

While in all the cases so far we used coordinates to refer to a whole "square," so to speak, coordinates are commonly applied to graphs to refer to specific points — the intersections at the edges of the "squares." In fact, there's an extremely useful graphing system known as the Cartesian Coordinate System that helps us with all sorts of math and physics. While we won't be exploring the system in depth until Book 2, notice how in the redone map, we've **used the coordinates to describe points** rather than the entire square. Again, coordinate systems are tools, so we can use them differently depending on our need.

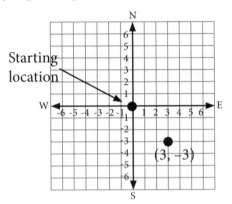

Representing Data and Using Different Scales

We can represent a lot more than locations on coordinate systems! As we'll see more in the next lesson, we can use them to graph data, too. For example, if a plant measured 1 inch after 5 days, 3 inches after 20 days, and 5 inches after 25 days, we could draw a coordinate system and use points on it to show the height by day. We'll use our horizontal axis to represent the time and the vertical the height (we could have graphed it the other way around, but time is typically graphed horizontally, as it's easier to see the growth that way).

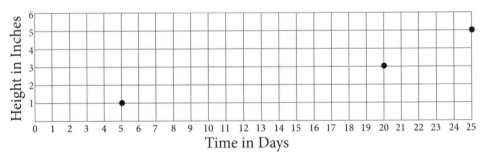

We could make this graph even simpler by using a different scale on our horizontal axis. Rather than using a box to represent 1 day, let's make each box represent 5 days.

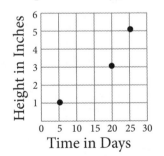

Coordinate systems can be used to represent data and can employ different scales! We'll explore representing data on a coordinate system more in the next lesson.

> ## Keeping Perspective
> Coordinate systems are incredibly useful ways of representing locations, data, shapes (as we'll explore in the next chapter), mathematical relationships (coming in Book 2), and much more. As you play with coordinates, know that they have a variety of uses.

12.6 Organizing Data — Line Graphs

It's time to return to our exploration of data — specifically to organizing it via graphs. Consider the following high temperatures (in Celsius) for Moscow for the week of October 5, 2014, according to Weather.com[6]:

Date	High Temperature in Celsius
5-Oct	5
6-Oct	7
7-Oct	7
8-Oct	9
9-Oct	11
10-Oct	16
11-Oct	14

What graph could we use to graph this data? We *can't* graph it on a pie graph — the data isn't representing a portion of a whole. And while we could use a bar graph, a **line graph** will help us see the general trend, or pattern, in the data much easier.

To draw a line graph, we're going to use coordinates. Let's start by setting up a coordinate system, using the dates one direction and the degrees the other. To make the graph more compact, we'll go ahead and make each box represent 4 degrees.

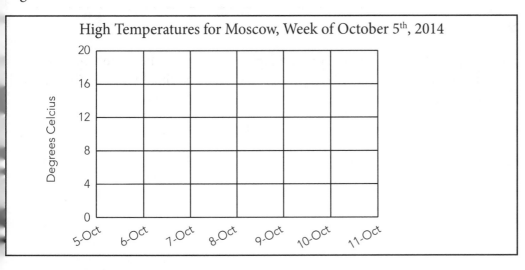

Now we can plot the individual dates and temperatures as coordinates on our graph. We'll show that we had a 5° day on 5-Oct by putting a point at the intersection of 5 degrees and 5-Oct. Since we made each box represent 4 degrees, our point will be just a little above the 4 degree marker. Next we'll put a point at the intersection of 6-Oct and 7, since it was 7° on the 6th. Continuing this way gives us a chart of our data.

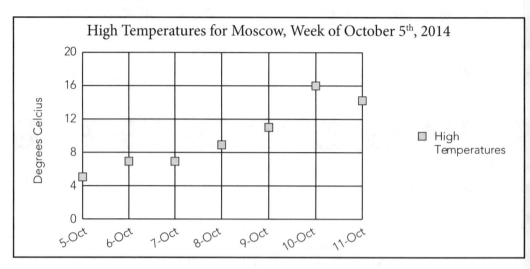

But we're not quite finished yet. Let's connect all the points together with straight lines. The lines will help us see the trend — in this case, that the temperatures have generally risen.

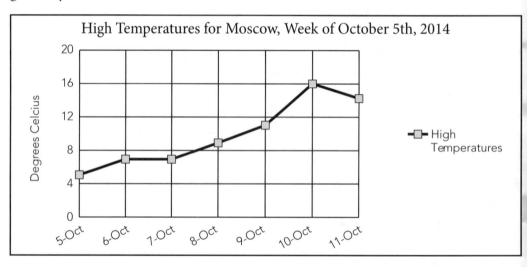

We now have a line graph!

And if we wanted to compare both the highs and lows for the same week (i.e., compare more than one set of data), we could easily do that by just adding another line to the same graph!

Here are the lows for the same week, according to Weather.com[7]:

Date	Low Temperature in Celsius
5-Oct	3
6-Oct	1
7-Oct	0
8-Oct	-4
9-Oct	0
10-Oct	8
11-Oct	9

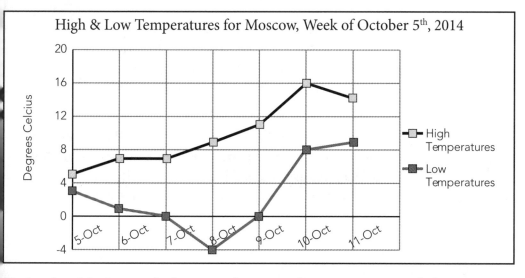

Notice that this time we had to extend our coordinate system to include negative numbers. Also, notice again how line graphs make it easy to visually see the general pattern, or trend. It's easy to see from the graph that, during this week, the high temperature steadily rose, while the low temperature dove and then rose. And again, notice that some of the points fall in between the boxes, as their temperature falls in between the degrees marked.

Sometimes, computer-generated line graphs won't show the actual points (since the purpose of graphing is simplicity, they can get distractive). Below is the same graph with the points removed.

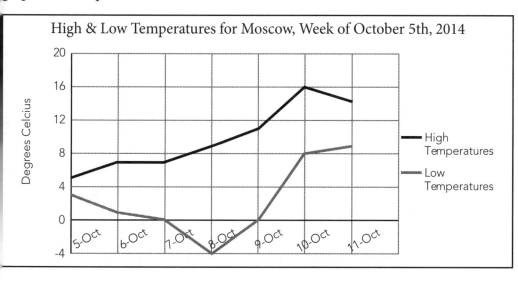

Using Line Graphs

So when should you use a line graph? Consider these helpful summaries of the graphs from a couple of government agencies:

> Line graphs are used to track changes over short and long periods of time. When smaller changes exist, line graphs are better to use than bar graphs. Line graphs can also be used to compare changes over the same period of time for more than one group. — National Center for Educational Statistics[8]

> A line graph displays the relationship between two types of information, such as number and school personnel trained by year. They are useful in illustrating trends over time. — Department of Health and Human Services, Centers for Disease Control and Prevention[9]

In short, line graphs are very useful in showing **trends** and **small changes in data** (like a rise of a degree or two) or comparing **more than one thing** (such as highs and lows). While bar graphs help us compare data when we're looking at *large differences*, a line graph really makes the *small differences* and *patterns* come to life.

Keeping Perspective

Graphs are visual tools to help us describe data — and just as data varies, so do graphs! Familiarizing yourself with these tools will help you better understand the graphs you encounter in life.

12.7 Organizing Data — Averages

Often, it's helpful to summarize data into one number that best represents most of the data. To do this, we look for the **measure of central tendency** — the middle, so to speak, of the data. Let's take a look at how to find the "middle" or "center" of data.

Just Your Average

The most common way of finding the middle of the data is to find the **mean**, which is more commonly called the **average**. We refer to averages all the time, whether in sports (the batting **average**), education (test score **averages**), or business (the **average** consumer).

To find an average, we simply add up all the data points (individual results) and then divide the total by the number of data points. This gives us the number referred to as the average.

244 | PRINCIPLES OF MATHEMATICS 1

For example, according to The Weather Channel,[10] the average daily temperatures in Phoenix, Arizona, are as follows for the first 7 days in the month of October:

$$94 \quad 94 \quad 94 \quad 93 \quad 93 \quad 92 \quad 92$$

To find the average for the first week, we'd

1. Add the daily averages together: 652

2. Divide the total by the number of data points (7 — we have 7 temperatures — one for each day) to find the average: 652 ÷ 7 = 93.1428571429

The average temperature for the first 7 days of October in Phoenix is 93° F (or more precisely, 93.1428571429°). (In other words, if you're heading to Phoenix in October, be sure to pack your shorts!)

Notice that if we were to draw this data on a coordinate graph and then mark the average, the average would be a straight line that evens out the high and low points, thus representing the middle of the data.

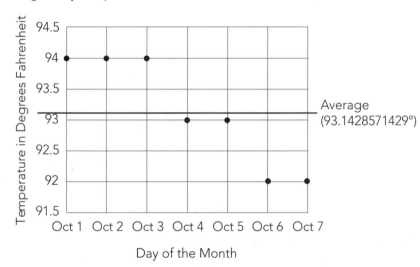

The average is simply a description of the middle of the data, with data above it and below it. **We're essentially summarizing the data with a single number representing the middle of the data.**

Batting averages are an example of this. Over the course of the 22 years he played, Babe Ruth[11] had a batting **average** of 0.342. Let's take a look at what this number means mathematically.

We could (and people do!) track the data every time a player steps up to bat. Each time, that player either hit the ball in a way that got him on base, or else he didn't. So if we tracked his hits to base (which we'll refer to as hits) for each time a hypothetical player went up to bat, we might end up with this at the end of a game:

$$0, \quad 1, \quad 0, \quad 1, \quad 0, \quad 0, \quad 0$$

In other words, the player batted 7 times and got 2 hits.

His batting average for that game would be the sum of his hits, divided by the total number of times he went up to hit.

1. Add the hits he got together: 2

2. Divide the total by the number of data points (we have 7, since he went up to bat 7 times): 2 ÷ 7 = 0.28571428571, which rounds to 0.286. (Batting averages are typically rounded to the nearest thousandth.)

On average, he hit the ball 0.286 of the time (or 28.6% of the time). Notice how on the graph the average line seeks to find a middle number that accurately represents the data. Here again the line evens out the high and low points; the line is closer to the bottom because there are more low points than high points.

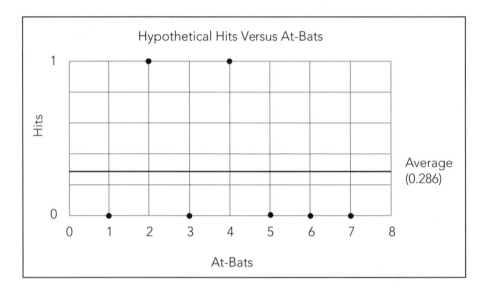

A Shortcut for Batting Average

There's a shortcut for finding a batting average. Let's see if we can come to it with some hypothetical data.

Suppose we have these results for the first 4 games of a season:

Hits at Game 1: 0, 0, 0, 1, 0, 1, 0

Hits at Game 2: 0, 1, 1, 0, 0, 0, 1, 0, 0

Hits at Game 3: 0, 0, 0, 0, 0, 1, 1, 1, 1, 0, 0

Hits at Game 4: 0, 0, 0, 0, 0, 0, 1

If we add all these results together and divide them by the total number of results (i.e, the total number of data points), we get our batting average:

$$\frac{0,0,0,1,0,1,0,0,1,1,0,0,0,1,0,0,0,0,0,0,1,1,1,1,0,0,0,0,0,0,0,0,0,1}{34} = \frac{10}{34} = 0.294$$

Whew! That was hard to add up all those 0s and 1s — and we were only trying to find the average for 4 games! There has to be a shortcut to find the average — and there is.

Rather than adding up all the 0s and 1s and finding the total number of results for all 4 games, let's find it for each game by itself. Notice that since we're collecting data every time the player steps up to bat, the total data points is really the total number of times the player batted (called the **at-bats**). And notice that the sum of the data is really the number of hits we received, since when we don't hit the ball we record a 0 (zero hits).

Let's total up our hits and at-bats.

Game 1

Raw data: 0, 0, 0, 1, 0, 1, 0

Summarized data: 2 hits (data sums to 2); 7 at-bats (7 total data points)

Game 2

Raw data: 0, 1, 1, 0, 0, 0, 1, 0, 0

Summarized data: 3 hits; 9 at-bats

Game 3

Raw data: 0, 0, 0, 0, 0, 1, 1, 1, 1, 0, 0

Summarized data: 4 hits; 11 at-bats

Game 4

Raw data: 0, 0, 0, 0, 0, 0, 1

Summarized Data: 1 hit; 7 at-bats

Now, since our hits are the sum of all the data points, and the at-bats are the number of data points, we can easily add up the hits and divide them by the at-bats to find the average.

$$\text{batting average } = \frac{2+3+4+1}{7+9+11+7} = \frac{10}{34} = 0.294$$

Notice how much easier that was! Most instructions for finding a batting average just list this formula:

$$\text{batting average } = \frac{number\ of\ hits}{number\ of\ at\text{-}bats}$$

But it's important to know that this formula is a shortcut for adding up all the data collected every time a player steps up to bat.

What does the batting average of 0.294 we found mean? It means that the player hit the ball approximately 30% of the time he bats. Does it mean that he'll always hit the ball 30% of the time? Not at all! Some games the player hits the ball more often than others. The average represents the middle of the data — the point that evens out the high and the low points. (In the graph, it looks lower than the middle because there are more points toward the bottom than toward the top.)

12. STATISTICS AND GRAPHING

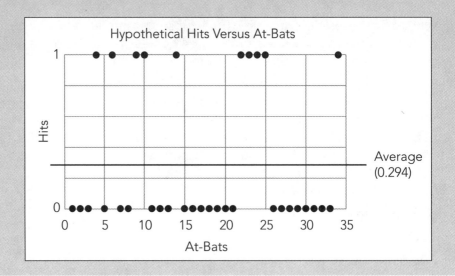

There's Nothing Average About You

Beware of the conclusions you draw off averages, especially ones related to people. As human beings, we tend to compare ourselves with others. When you hear that the average kid scores a certain percent on a test, for example, it's easy to be discouraged if you didn't score that well . . . or to be proud if you scored higher.

The Bible warns us about comparing ourselves with others. God created us all as unique individuals — individuals with different gifts and talents.

> *For we dare not make ourselves of the number, or compare ourselves with some that commend themselves: but they measuring themselves by themselves, and comparing themselves among themselves, are not wise (2 Corinthians 10:12).*
>
> *But unto every one of us is given grace according to the measure of the gift of Christ (Ephesians 4:7).*

The foolishness of comparing ourselves with an average is further magnified when we think about what an average means. The average is a number chosen to summarize a variety of data. It *doesn't* mean that if the average kid does something by age 4, and you didn't do it until age 7, that you somehow were behind . . . or that you're somehow better if you did it by age 3. The "average" is just the middle of the data — most people fall above or below it.

So remember: *each* person is a unique individual. Use your gifts to build others up. You are anything *but* average.

Keeping Perspective

An **average** (also called the **mean**) is a way of summarizing data. Remember, though, that a lot of details get lost in a summary — the real data probably has values both higher and lower than the stated summary number.

12.8 Organizing Data — More on Averages

Sometimes the average we looked at in the last lesson — officially called the **mean** — doesn't give a clear picture of the data. For example, say you did a survey of when people had ice cream for the first time. You got the results below (which are entirely hypothetical!):

$$0.5, 1, 2, 3, 1, 2, 3, 1, 57, 1, 2, 1, 1, 1, 0.5$$

While most people in this survey had ice cream at 1 or 2 years old, if we find the mean/average, the person who didn't have ice cream until 57 really distorts the picture!

$$\text{average} = \frac{\textit{total of data points}}{\textit{number of data points}}$$

$$\text{average} = \frac{0.5 + 1 + 2 + 3 + 1 + 2 + 3 + 1 + 57 + 1 + 2 + 1 + 1 + 1 + 0.5}{15} = 5.13$$

5.13 is hardly representative of when most people we surveyed first had ice cream, but it's the average because of the 57!

Rather than using the mean, we could look for a number called the **median**. A median is a fancy term for the absolute middle number, i.e., the number that would be found if we arrange all the data in order and find the one in the middle.

Data: 0.5, 1, 2, 3, 1, 2, 3, 1, 57, 1, 2, 1, 1, 1, 0.5

Data arranged in order: 0.5, 0.5, 1, 1, 1, 1, 1, 1, 1, 2, 2, 2, 3, 3, 57

Number in the middle:

When there is an even number of data points, we find the median by taking the average (i.e., mean) of the two numbers closest to the middle. For example, if we had an extra 1 in our data, we'd have this:

$$(1 + 1) \div 2 = 1$$

Another option would be to use what's known as the **mode**. The mode is a fancy name for the result that appears the most often. In this case, since the number 1 appears more than any other number, it would be our mode.

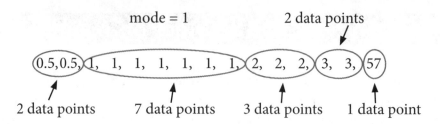

Notice how, in this case, the **mode** or the **median** gives us a much better picture than the mean (i.e., average) did! The mean (average) was 5.13, while the median and mode were 1. The reason the mean was so far off was because of the age of 57 that was so much higher than all the others. We call numbers way out of ordinary "outliers." While the outlier affected the mean, it didn't affect the mode or the median.

While we typically use means/averages much more than either modes or medians, it's wise to be familiar with all three ways of summarizing data.

- **Mean** (average) = *sum* of all the data points divided by the *number* of data points

- **Median** = middle of the data (quite literally — line it all up and find the middle)

- **Mode** = most frequent data point (again, quite literally — figure out which data point appears the most)

Keeping Perspective

Median and **mode** are different ways of summarizing data that in some cases more accurately portray it than a **mean** (average) would. They're other ways of expressing the "measure of central tendency," i.e., the middle of the data.

Always remember, though, that the summary presented is ultimately only as accurate as the data itself. While we haven't looked at how to mathematically determine a sample size needed to draw conclusions, it's pretty obvious that it would be foolish for us to draw *any* conclusions about the entire world's first-tasted-ice-cream-age based on a survey of only 15 people.

The point: question both how stats were gathered *and* how they were summarized to see if the conclusions presented are reasonable.

12.9 Chapter Synopsis

I hope you enjoyed getting a flavor for the different ways we can collect, organize, and interpret data.[12] Since we live in a society where stats bombard us, becoming comfortable with statistical terms and processes will help you examine and draw conclusions for yourself.

Here's what we learned:

- **Collecting Data** — Sometimes data is based on a **sample** of the **population** (the whole). Remember to look at the process used (was it **random?**), the **margin of error (MOE)**, and the **confidence level.**

- **Organizing Data** — **Frequency diagrams** help us summarize data. A variety of **graphs** (including pie graphs, bar graphs/column graphs, histograms, and line graphs) are used to visually display data. (Along the way, we also touched on **coordinates** and **coordinate systems**.) The graph we use depends on what we're trying to display with the data. We can also use **averages (means), modes,** and **medians** to give a numerical summary of the data.

- **Interpreting Data** — Decisions are often based on the data collected. When making a decision, it's important to look at whether the data shows what it's claiming to show. Remember, people are sinful . . . so beware, both when examining stats presented to you and when trying to use statistics yourself.

We'll continue exploring statistics off and on as we continue our mathematic explorations. And keep your eyes peeled for stats in your own life. It just might surprise you how often stats are quoted!

[CHAPTER 13]

Naming Shapes: Introducing Geometry

13.1 Understanding Geometry

It's time now to begin exploring a specific branch of mathematics called **geometry**. Geometry is the branch of math that deals specifically with measuring the earth. In fact, *geo* means "earth" and *metron* means "measure," so *geometry* actually means "earth measure." Geometry is used to survey land, build maps, draw blueprints, design airplanes, and much, much more.

As we explore this useful branch of mathematics, you'll find yourself using the skills we've covered so far, as well as learning additional ones. Let's start our journey through geometry with a little history and some basic terms before we dig into the mechanics of "measuring the earth."

The Great Pyramid of Giza in Egypt, 2560 B.C.

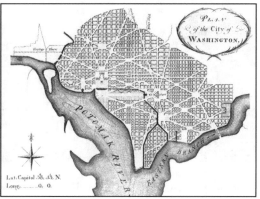
Plan of the City of Washington, March 1792

A Little History

Since the beginning of time, men have been using math to measure the earth. Cities (the Bible tells us Cain built one — Genesis 4:17), pyramids, and

Euclid (Painting is circa 1474.)

other building projects (such as the Tower of Babel) required working with measurements, shapes, and math.

The Greek mathematicians, however, are the ones who formalized geometry. Somewhere around 300 B.C., the Greek mathematician Euclid compiled the *Elements*, a work which taught the current Greek understanding of geometry. To this day, the essence of what is taught in high school geometry is based on Euclid's *Elements*.

Sadly, though, the Greeks had little interest in using geometry to measure the earth. As we'll explore in more depth later in this chapter, they pursued geometry mainly as an intellectual (and even religious) pursuit rather than as a useful tool.

In this course, we'll be looking at the big picture of exploring the shapes around us and using shapes to describe God's creation. We'll see how geometry helps us measure the earth so that when you do take an official geometry course, you'll understand how to apply it outside a textbook. We'll also discover evidences of God's wisdom and care as we go, as we have when studying other concepts.

Some Undefined Terms

Take a look at this honeycomb. Notice that each cell has a shape.

Now look around the room. Everywhere you look, you are looking at shapes. The table, chair, window, books, door, doorknob, etc. — everything you see has a shape and a size. How do we describe these shapes and sizes mathematically?

In order to describe shapes mathematically, we have to start somewhere. In geometry, we start with **undefined terms** — terms that can be described but can't be defined with other already-defined terms. We'll look at three undefined terms:

points, lines, and planes. We can then build our understanding of shapes using these terms as launching pads.

Let's start with what we call **points**. A point refers to an exact location. We represent it with a dot, even though a dot takes up space, and a point technically doesn't (it's just a location). The usefulness of defining a point as an exact location is not to deny the width of points we draw — it's to make it so we can take measurements, draw objects, and learn about objects, without taking into account the width of the marks we make on paper.

Similarly, mathematical **lines** don't have any width. That way, it doesn't matter if we draw a line thick or thin. Euclid described it as "breathless length"[1] — in other words, it can be thought of as a length extending indefinitely in both directions, which we can show with arrows.

Now, of course, physical lines don't really go on indefinitely. But this description makes it so when working with shapes we can extend lines as needed. We'll later use this concept to help find information such as the height of a tree.

*Note: Technically, we call portions of a line a **line segment**, but in this course, we won't be distinguishing between the two.*

While lines could technically be curved, in Euclidean geometry, when you hear the word "line," you can assume a **straight line** is meant.

Straight Line   **Not a Straight Line**

In a mathematical sense, we can think of points and lines as lying in what we call **planes**, or flat, boundless sections of space with a length and a width, but no depth. A more technical definition of a plane is "a flat surface on which a straight line joining any two points on it would wholly lie."[2] By flat surface, we don't necessarily mean a surface we can see. We can think of each side of a toaster oven as lying in a different plane. If you were to lay a large flat surface on top of each side, you could extend that surface indefinitely.

Okay, so why do we need planes? I'm glad you asked! Planes help us break down **three-dimensional (3D)** space into multiple **two-dimensional (2D)** spaces. The toaster, for example, is a 3D figure that has 6 two-dimensional figures (the front, back, top, bottom, left side, and right side).

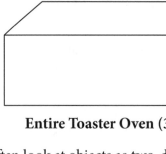
Entire Toaster Oven (3D)

Front of Toaster Oven (2D)

We often look at objects as two-dimensional shapes, dealing with one plane (or surface) at a time. For example, we could look at one side of a quartz crystal at a time (a 2D shape) or look at an entire crystal as a 3D shape.

Quartz crystal

> # Keeping Perspective
>
> The undefined terms we just looked at for points, lines, and planes are the terms typically taught and used — they're part of the geometry system referred to as **Euclidean geometry** (the geometry based on Euclid's *Elements*). However, other geometry systems exist based on using *different descriptions* for these terms. For example, in **spherical geometry**, a line is described as a great circle — which is quite different than the description we looked at today! Yet spherical geometry, like Euclidean geometry, also proves quite useful in describing God's creation. The equator, for example, can be thought of as a line in spherical geometry.
>
> Non-Euclidean geometry is not something you'll be expected to learn. However, it's important to know that other definitions and geometry systems exist. The existence of other systems reminds us that each one is a tool — a way of describing God's universe, which is so complex we need multiple geometries to explore it!

The equator can be thought of as a line in spherical geometry.

13.2 Lines and Angles

It's time to look at more terms and definitions. All of these are ultimately based on the three undefined terms we looked at yesterday (points, lines, and planes). Some of them may be familiar, while others may be new. Either way, keep in mind that these definitions, names, and symbols will help us as we measure and describe different aspects of God's creation.

Understanding Lines and Angles

Notice that we could use **lines** to describe various aspects of real life, from the shape of a mountain in the distance to that of a table to the incline of a tree.

What happens when two lines intersect? They form an angle! An **angle** is a fancy name in Euclidean geometry for the intersection of two straight lines. We call the point at which the lines intersect the **vertex** of the angle. We will commonly mark the angle formed with a little arc.

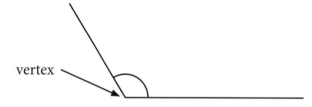

Since lines can be used to describe so many different aspects of creation, so can angles! Just look around you for a moment. You're surrounded by angles! We use angles to describe real-life "intersections," so to speak.

Naming Specific Lines and Angles

Since lines and angles are so useful, it's important to learn to describe them.

We call an angle formed by two lines that intersect to form a perfect capital "L" a **right angle**, and then name the other angles by whether they are wider or narrower than that angle. Angles smaller than a right angle are called **acute angles**, while those larger than a right angle but less than a straight line are referred to as **obtuse angles**.

Notice that we **used half of a box (right angle) to mark a right angle.** You'll find this angle-marking convention helpful as we move forward.

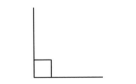

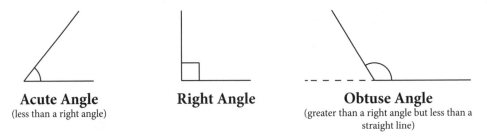

Acute Angle
(less than a right angle)

Right Angle

Obtuse Angle
(greater than a right angle but less than a straight line)

Look around the room for a moment. How many right angles can you find? The edges of books, the intersection of the wall and the floor, the intersection of shelves on a bookshelf — all of these form right angles.

Lines that form a right angle are called **perpendicular**. On the other hand, lines are considered **parallel** if they never intersect — the distance between them is always the same.

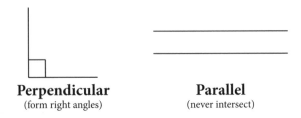

Perpendicular
(form right angles)

Parallel
(never intersect)

Look at the monitor shown. Notice that it has two sets of **parallel lines** that are **perpendicular** to each other, thus forming four **right angles**. We call a shape with these particular properties a **rectangle**.

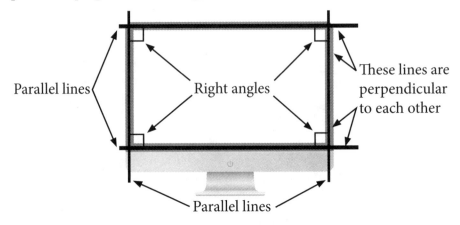

Parallel lines

Right angles

These lines are perpendicular to each other

Parallel lines

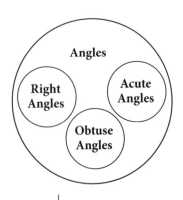

Making the Connection

Remember how we said that the idea of sets is used all the time, even when not mentioned by name? Well, we could think of angles as a set, or collection, and acute, obtuse, and right angles as subsets within that set. Acute, obtuse, and right angles are just ways of further categorizing angles.

258 | PRINCIPLES OF MATHEMATICS 1

Labeling Lines and Angles

As we begin exploring shapes, we need a way to refer to a specific angle or line within a shape. Saying "the angle in the bottom left of the square" gets old fast!

Letters are typically used to label or "name" lines and angles. By placing a letter at the place where lines intersect or end, we have an easy way to name lines and angles.

Labeling Lines

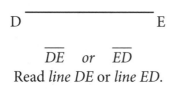

$\overline{DE}$ or $\overline{ED}$
Read *line DE* or *line ED*.

Labeling Angles

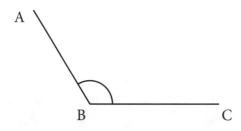

∠ABC or ∠CBA or ∠B
Read *angle ABC* or *angle CBA* or *angle B*.

Notice that the same angle can be labeled different ways. The important thing is that the vertex — the point at which the two lines form an angle (in this case, B) — is the middle letter.

Lines and Angles in a Shape

The same conventions for naming lines and angles apply to lines and angles within a shape.

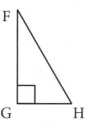

For example, the marked angle would be labeled ∠FGH, ∠HGF, or ∠G. The vertical line would be labeled $\overline{FG}$ or $\overline{GF}$.

Keeping Perspective — Just a Convention

The methods we use to "name" lines and angles are just conventions — agreed-upon ways of communicating. Below are a few other ways angles have been described historically.[3]

Conventions, like English and spelling rules, aid in effective communication. The ones we learned in this lesson will ultimately help us explore and measure the earth.

While we're revisiting the topic of naming, think about what it must have been like for Adam naming the animals back in the Garden of Eden. What delight he must have felt at seeing the amazing creatures God had created!

As you learn names for different lines and angles, don't miss out on the delight of doing so while praising God for His incredible creation, which angles and lines help us describe.

13.3 Polygons

It's time to use the different types of angles and lines we studied in the last lesson to help us categorize shapes. We'll be using names to categorize shapes into sets, or collections, based on specific characteristics (such as their lines and angles). Categorizing shapes will in turn aid us in measuring shapes, finding the right shape for the task, and working with shapes. You may already know a lot of shape names, which is great! It should make this lesson much easier.

Let's start by naming two-dimensional shapes — that is, shapes that lie in a single plane (i.e., they're flat / have no depth). Most of the shapes we'll be looking at in this course are what we call **polygons**.

A **polygon** is the name for closed (i.e., no gap in the shape's outline), two-dimensional figures with straight lines. The prefix *poly* means "many," while *gon* comes from a Greek word that means "angled," so polygon literally means "many angled."[4]

Polygon
(closed, two-dimensional figure with straight lines)

Not a Polygon
(not all lines are straight)

Not a Polygon
(figure is not closed — there is a gap)

Polygons come in all sorts of different varieties — the chart below shows a few you need to make sure you know.

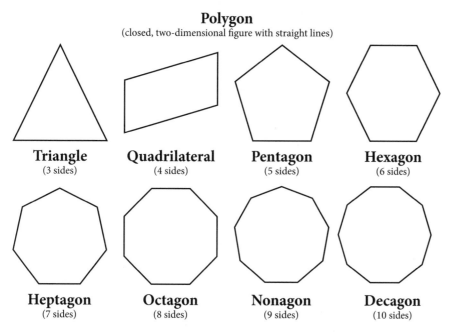

There are polygons with 11, 12, 13, etc., sides — but you will not be expected to learn their names.

Polygons can either be **regular** (all sides and angles are the same) or **irregular** (all sides and angles are not the same). Any polygon can be either regular or irregular. This is just a further way to categorize shapes. **Again, we're basically categorizing shapes into various sets (i.e., groups or collections) and subsets;** this in turn will help us explore them and easily refer to shapes with specific characteristics.

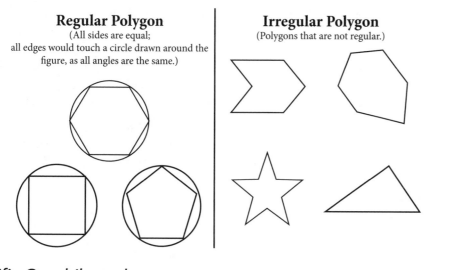

Boosting Your Vocabulary
Learning shape names can also help your vocabulary in other areas! For example, a **tri**athlon has three parts to it, just as a **tri**angle has **three** sides. Any guesses how many years a **quad**ricentennial might be celebrating? Hint: A **quad**rilateral has **four** sides.
(Answer: 400 years)

Specific Quadrilaterals

Quadrilaterals (4-sided polygons) are so common and useful, we further categorize them based on additional characteristics.

13. NAMING SHAPES: INTRODUCING GEOMETRY | 261

Some books define a trapezoid as a quadrilateral with 1, and only 1, parallel sides. Likewise, some define a rhombus/diamond differently. Always remember that definitions can — and do — vary!

Specific Quadrilaterals

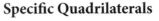
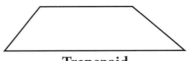

Trapezoid
(quadrilateral with 1 or more pairs of parallel sides)

Parallelogram
(quadrilateral with both pairs of opposite sides parallel)

Rhombus/Diamond
(parallelogram with equal-length sides)

Rectangle
(parallelogram with right angles)

Square
(parallelogram with equal-length sides *and* right angles)

Subcategories (i.e., Subsets)

Imagine if we were trying to design a clothing website. We would need to categorize the different items we wanted to list. We might start with general categories (or sets), such as "women's," "men's," "boys," and "girls." Under each of those categories, we might have other categories (or subsets), such as "shirts" and "pants." Then under those there might be more categories, such as "short-sleeve shirts" and "long-sleeve shirts." A girl's short-sleeve shirt would be a member of the short-sleeve, shirts, and girls' categories.

Similarly, shapes can be called by more than one name (that is, categorized in more than one set or subset). A square, which has equal-length sides and right angles, also meets the definition for a rectangle (right angles), a rhombus (equal-length sides), a parallelogram, a trapezoid, a quadrilateral, and a polygon. It is just a specific type of rectangle, rhombus, parallelogram, etc. When asked to name a shape in this course, **use the most specific name (i.e., call a square a square),** unless otherwise specified.

Keeping Perspective

We're naming shapes to help us as we measure God's creation and complete the tasks He gives us to do. Having a name to describe shapes with certain qualities will make life easier as we seek to explore them — it's much easier to say "parallelogram" than "quadrilateral with both pairs of opposite sides parallel." And knowing what kind of shape we're dealing with often helps us find needed information. As we continue exploring shapes, we'll discover that certain shapes have specific properties. For example, in a parallelogram, the opposite sides and angles will always be equal in measure. Why? Because they have to be in order to form two sets of parallel sides. So if we know a shape is a parallelogram, we also know that its opposites sides and angles are equal in measure. Names prove quite useful!

If you don't already know them, you will need to learn the shape names we looked at, as we'll use them to help us explore shapes. Remember as you study that all we're doing is categorizing and naming shapes based on their characteristics — something we can do because God made us in His image and gave us this ability. Just as Adam named the animals, we can name shapes — and do so while praising the Creator of it all!

The Pentagon, Washington, D.C.

13.4 Circles, Triangles, and Three-Dimensional Shapes

Ready for some more naming of shapes? Let's continue!

Circles

One very common shape that doesn't have straight lines and thus is not a polygon is called a circle.

Using a compass

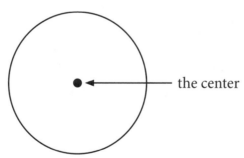

Circle
(closed two-dimensional figure;
each part of the edge is equally distant from the center)

Another way to think of a **circle** is as "the set of points in a plane that are equidistant from a given point."[5] If you have a compass, try drawing a circle. Place

13. NAMING SHAPES: INTRODUCING GEOMETRY | 263

the pointy end of the compass on a piece of paper, and then rotate the end with the pencil around it. Every point on the circle should be equally distant from the point where you placed the pointy end of your compass.

We use circles to represent many different objects. If you look around your house for objects circles can describe, you might be surprised by how many you find. Circles and the relationships between the parts of circles also help us describe various aspects of creation — including gravity.

Triangles

As we saw in the last lesson, **triangles** are polygons with three sides. As we continue exploring shapes, we'll discover that triangles help us describe many different — and often unexpected — aspects of the world. For example, triangles can be used to describe portions of other shapes, as well as the height and the slope of mountains. In fact, back in the 1800s, the geometry book of the time was called *Geodæsia*, and it taught readers how to survey. Triangles were a big part of it! In a more roundabout way, triangles even help us describe sound waves.

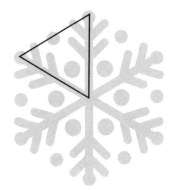

Not surprisingly, we have to have names for triangles with different characteristics, just as we did for quadrilaterals with different characteristics. We subcategorize triangles different ways, based on our needs.

Triangles - Categorized by Length of Sides

 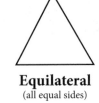 

Isosceles (two equal sides)   **Equilateral** (all equal sides)   **Scalene** (no equal sides)

Triangles - Categorized by Angles

 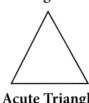 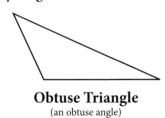

Right Triangle (a right angle)   **Acute Triangle** (all acute angles)   **Obtuse Triangle** (an obtuse angle)

Every triangle can be categorized by both sides and angles, so you'll often see these terms combined.

obtuse scalene triangle   right isosceles triangle

264 | PRINCIPLES OF MATHEMATICS 1

> For the purpose of this course, whenever you're asked to identify a triangle, you should use as many names as you can, unless you're specifically told otherwise.

Labeling Triangles

Just as we can use letters to label lines and angles, we can use letters to label triangles.

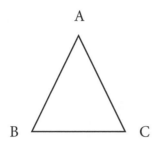

$\triangle ABC$
Read *triangle ABC*.

Three-Dimensional Shapes

While we often look at a three-dimensional shape (a shape with length, width, and height) as several two-dimensional shapes instead (for example, we could view each side of this toaster oven as a rectangle), there are times when we need a name to describe three-dimensional shapes as a whole.

If we wanted to refer to the *entire* toaster oven as a unit rather than just to one side of it, we would call it a *prism*.

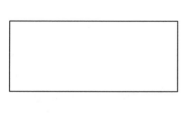 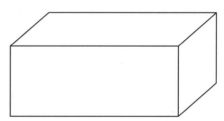 

One side of the toaster oven —
a rectangle

The entire toaster oven —
a prism

A **prism** is a polygon (a square, triangle, hexagon, etc.) with some uniform depth. More specifically, a prism is a solid having two bases that are parallel polygons, and faces (sides) that are parallelograms.

The prism is named after the shape of the bases. A prism with triangular bases is called a triangular prism, one with rectangular bases a rectangular prism, and so forth. So, more properly, the toaster oven shown would be a rectangular prism.

Prism[6]

(A solid with two bases that are parallel polygons and faces [sides] that are parallelograms; the prism is named after the shape of the bases.)

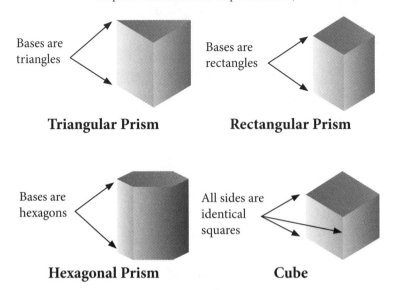

Bases are triangles — **Triangular Prism**

Bases are rectangles — **Rectangular Prism**

Bases are hexagons — **Hexagonal Prism**

All sides are identical squares — **Cube**

A **cylinder** is similar in that it is a circle with depth (or height), but since its bases are not polygons and its sides are not parallelograms, it has it own special name.

Cylinder[7]

(A solid with two bases that are equal parallel circles, having an equal diameter in any parallel plane between them.)

Keeping Perspective

We're surrounded by shapes. Names help us describe and categorize these shapes, as well as other aspects of God's creation. Sometimes it's helpful to "name" just one side of an object — such as the front of the toaster oven — while other times we need to "name" the entire object. Always remember that shapes are tools to help us explore God's creation — tools we can use while praising the Creator.

Let every thing that hath breath praise the LORD. Praise ye the LORD (Psalm 150:6).

13.5 Fun with Shapes

Before we finish our introduction into shapes, it's time to have a little fun with them.

Take a look at the artwork shown. Notice that many of the shapes are identical except for being moved in different ways to form a design. Notice also that some are just smaller versions of others.

Artwork by Florence Feldman, https://picasaweb.google.com/flownow2, used with permission.

Not surprisingly, we have names to describe the different ways we can move shapes around or make them larger or smaller. Remember, names help us communicate with fewer words, keeping the size of math books more manageable. These are terms you need to be somewhat familiar with, as you'll see them again when you study geometry and may even hear them in everyday conversations.

Translate

Move the shape without otherwise changing it.

Translation is how we describe a shape that's simply been *moved*, whether horizontally, vertically, or both.

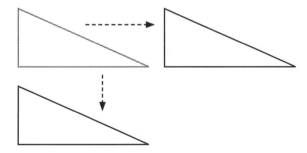

This is perhaps easiest to see on a coordinate graph. We'll use points to represent each vertex (corner) of the triangle, and then connect them with lines.

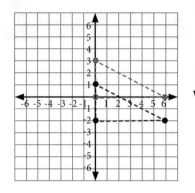

Each point in the bottom triangle is 2 spaces lower vertically than the same point in the top triangle.

Reflect

Make a mirror image of the shape.

When you look at your reflection in the water or mirror, you see an exact mirror image of yourself. Not surprisingly, then, when a shape has been moved so that it forms a mirror image of the first one, we say it's been **reflected** or is a **reflection** of the first one.

Note: The line is included to better show the reflection.

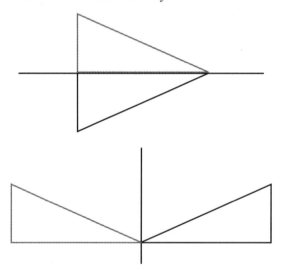

Again, we could also look at reflections on a coordinate graph.

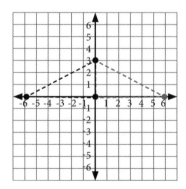

Rotate

Pivot the shape around a point.

When a shape is pivoted around a fixed point, we say it's been **rotated** or is a **rotation**.

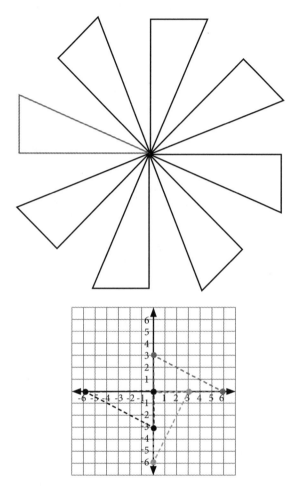

Scale

Keep the proportions the same while changing the size — i.e., make a scaled-up or scaled-down version of the shape.

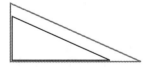

Notice that these two triangles are *not* the exact same size. Instead, one is a proportionally scaled-up version of the other. We call these two shapes **similar**.

Term Time

When two shapes are identical except for their location, we call them **congruent**. Thus a shape that has just been translated, reflected, or rotated would be called congruent. On the other hand, when a shape is a proportional scaled-up or down version of another, we call the shapes **similar**. Scaled shapes are similar. We'll explore these two terms in more depth later, but they're important, so begin familiarizing yourself with them now.

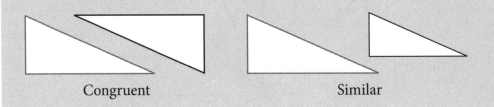

Congruent　　　　　　　　　　Similar

We can also see scaling on a coordinate graph. Notice how the smaller triangle is the same shape, just smaller.

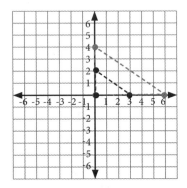

Operations Outside the Box

When you think of the word "operation" in math, you probably think of addition, subtraction, multiplication, and division. But translating, reflecting, rotating, and scaling are also operations. According to the *New Oxford American Dictionary*, an operation is "a process in which a number, quantity, expression, etc., is altered or manipulated according to formal rules, such as those of addition, multiplication. . . ."[8] In each of the processes we looked at in this lesson, shapes were moved in a way we could have described using formal rules!

Just as we did with addition, subtraction, multiplication, and division, we could look at translation, reflection, rotation, and scaling to see whether they are commutative or associative. For example, translation, like addition and multiplication, is a commutative operation — it doesn't matter if we move a shape to the right and then up, or up and then to the right. Either order will get the same end result.

The point? The principles you learn in math end up applying again and again!

Combining the Moves

It often takes more than one word to describe how a shape has changed. Notice how the triangles in the picture below have been both rotated *and* moved (translated).

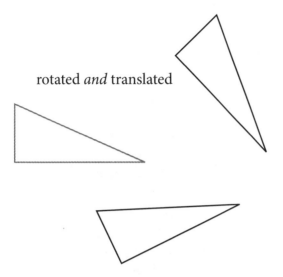

rotated *and* translated

Notice how now we've scaled and translated the triangle.

scaled *and* translated

> ## Keeping Perspective — Shapes All Around Us
>
> We encounter shapes all around us that have been rotated, translated, etc. Here are just a few examples.
>
> **Physics** — When a wheel moves, we can view it as a circle that is being translated and rotated! Each part of the wheel is translating (i.e., moving) *and* rotating around the center point.
>
> When you throw a Frisbee, it rotates around and around a point inside the center of the Frisbee. The Frisbee is also translating (i.e., moving locations) as it flies.
>
> **Graphics** — Artists who work on the computer use terms like "rotate" and "reflect" to tell the computer what to do. In my graphics program, if I want to tilt a picture or shape for a design I'm working on, I know to choose the "Rotate" button. (I can also choose the point around which I want the rotation to occur.) If I want to get a mirror image, I choose "Reflect."

13. NAMING SHAPES: INTRODUCING GEOMETRY | 271

Art — In the artwork shown, the artist, Florence Feldman, has used translation, reflection, rotation, *and* scaling to create her picture. While she probably did it without thinking about the math behind it, math helps us better understand and describe the components of the picture.

Artwork by Florence Feldman, https://picasaweb.google.com/flownow2, used with permission.

Creation — Notice that the daisy's petals are rotated and translated (they've been moved to fit around the center). Sometimes petals are also scaled — that is, some are smaller versions of the others.

Notice also how each half of a butterfly is basically a mirror image (a reflection) of the other half.

Shapes that have been **translated, reflected, rotated, scaled**, or some combination thereof are all used in art . . . including God's beautiful artwork around us!

13.6 Chapter Synopsis

In this chapter, we learned lots of terms used to categorize lines, angles, and shapes. As we move forward, we're going to use the terms we looked at in this chapter, so be sure you are familiar with them.

More importantly than the terms, though, remember that geometry ultimately means "earth measure." What we're learning will help us measure and explore the world around us, from the microscopic to the vast expanses of space.

A Look at History

While their individual philosophies varied, as a whole, the ancient Greek mathematicians viewed math not as a tool to describe God's creation, but as a source of truth. They believed in human intellect absolutely and viewed geometry as something they had created. To them, human reasoning and logic could never be questioned.

In geometry, the Greeks thought they could develop an infallible system of logic and rules. Yet, as we briefly mentioned early in this chapter, other geometry systems exist. For example, we saw that in spherical geometry, a line is defined as a great circle, which is a contrary definition to the definition in Euclidean geometry of "breathless length."[9] The existence of other systems wouldn't make sense if geometry were an infallible truth — if math principles were independent, absolute truths in themselves, how could multiple systems be true and yet contradictory?

A biblical worldview, on the other hand, has no problem with different geometry systems. In fact, multiple "tools" to describe different aspects of creation are exactly what we would expect if we base our thinking on the Bible.

I keep emphasizing that math is a tool because it's so important to view math as one. Humanistic and naturalistic thinking that enthrone reasoning and math as the source of truth often subtly come through in math presentations. I want you to be prepared to recognize and combat this unbiblical thinking.

According to the Bible, our reasoning is fallen, as is the world in which we live. While math and reason can help us *describe* the realities God placed all around us, we can't depend on them as the *source* of truth. God's Word — not our own understanding — is a firm foundation upon which to build our thinking.

> . . . *let God be true, but every man a liar.* . . *(Romans 3:4).*

> *For ever, O* LORD, *thy word is settled in heaven (Psalm 119:89).*

[CHAPTER 14]

Measuring Distance

14.1 Units for Measuring Distance

Imagine for a moment life without measurements. How would we know how much flour to put in a recipe? How would we survey land or record and read maps? How would architects design buildings and articulate that information to builders? How would quantities of food be described on packages? Measurements are important!

God created us with the ability to measure and develop different systems for measuring. To start with, we'll focus on measuring *distance* or *length*.

To measure something, we need a standard we can compare it to — that is, a unit. A **unit** is what we call "a special quantity in terms of which other quantities are expressed."[1]

If we each used our own unit to measure distance, we'd have to go through a lengthy process to communicate to others a specific distance. A builder, for instance, would have to learn a brand-new set of units for every architectural drawing.

To avoid this sort of confusion, there are standardized systems for measurement. Let's take a look at measuring distance using two common measuring systems: the **U.S. Customary System** and the **Metric System**.

The Basic Unit — a Meter

Both the U.S. Customary System and the Metric System base their distance units on a unit called the **meter**.[2] And just how big is a meter?

Its approximate length is easiest to see on a measuring tape. Pull one out and measure 100 centimeters or 39.3701 inches. That's approximately a meter.

$$1 \text{ meter} = 100 \text{ centimeters} = 39.3701 \text{ inches}$$

While the meter is not an actual unit in the U.S. Customary System, the yard is defined in terms of a meter.

I say approximately because, while we typically use devices such as rulers and measuring tapes to measure objects, these devices are not perfect representations (although they're definitely close enough for most practical purposes!). These devices are called **standards** — a "standard" is "a physical realization or representation of a unit."[3]

So what *is* the exact definition of a meter? The meter has actually had various definitions. You might think it would be the length of a specific bar or stick, of which duplicate sticks could be made — and it was at one point in time. The meter is currently defined, though, in terms of "the speed of light in a vacuum" — specifically as "the length of the path traveled by light in a vacuum during an interval of $\frac{1}{299,792,458}$ of a second."[4] This is a repeatable distance (because God causes light to travel consistently!) measurable by devices worldwide rather than a distance that must be measured against one specific bar.

Expanding from the Meter

Now, it would be *quite* difficult to measure some things with a meter. Imagine a stick a meter long with no smaller markings. How would you describe the length of a paper clip? And how many meters would you need to describe a long distance, like the distance between two towns?

While you could describe short distances as portions of a meter and long distances as multiple meters, it would be easier to use longer or shorter units for these distances. The U.S. Customary and the Metric System both have a variety of units to cover both short and long distances.

The U.S. Customary System

In America, the U.S. Customary System of measurement is the system used for most daily purposes.

In terms of a meter, the **yard** is defined as 0.9144 meters. In other words, the yard is just shy of one meter. Pull out that measuring tape again and look for 3 feet or 36 inches — that's one yard. Notice that it's just shy of a meter.

1 yard = 0.9144 meters

1 meter = 39.3701 inches

Each of these numbers marks 1 inch

276 | PRINCIPLES OF MATHEMATICS 1

How about some smaller units to help us measure shorter distances? We've got a **foot**, which is $\frac{1}{3}$ of a yard, and an **inch**, which is technically defined in terms of a portion of a meter but is more commonly known as $\frac{1}{36^{th}}$ of a yard or $\frac{1}{12^{th}}$ of a foot. Notice the foot and inch marks on your measuring tape.

And for longer distances, we have a **mile**, which equals 1,760 yards or 5,280 feet. (Which, as you can imagine, is much too long to show on a measuring tape!)

Below are the units with their common abbreviations, showing how they compare to one another.

12 inches (in) = 1 foot (ft)
3 feet (ft) or 36 inches (in) = 1 yard (yd)
1 mile (mi) = 1,760 yards (yd) or **5,280 feet (ft)**

You will need to memorize the bolded relationships, as you'll need them often in everyday life.

Metric System/SI

The Metric System, or the International System of Units (SI), is a measuring system used around the world, including the United States, especially in scientific fields. We already looked at the definition of a meter. But, like the U.S. Customary System, the Metric System has other units to make it easier to measure shorter and longer distances.

Below are the four most common metric units typically used to express lengths in the Metric System:

1,000 millimeters (mm) = 1 meter (m)
10 millimeters (mm) = 1 centimeter (cm)
100 centimeters (cm) = 1 meter (m)
1 kilometer (km) = 1,000 meters (m)

You will need to memorize these relationships, as you'll need them in everyday life.

As you might guess, millimeters and centimeters, like inches, are helpful for expressing shorter distances (like household objects), while kilometers, like miles, help measure longer distances.

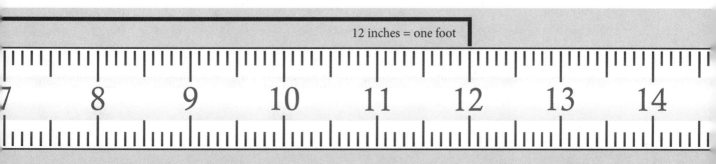

Keeping the Metric Names Straight

The prefix "milli" comes from the Latin word *mīlle*, which means "thousand," and a **milli**meter is $\frac{1}{1,000^{th}}$ of a meter (that is, 1,000 equal a meter). Think of a **milli**on (a million is 1,000 thousands), **mill**ennial ("a span of one thousand years"),[5] etc.

The prefix "centi" comes from the Latin word *centum*, which means "hundred," and a **centi**meter is $\frac{1}{100^{th}}$ of a meter (that is, 100 of them equal a meter). Think of a **centi**pede (they have a lot of legs), a **cent** ($\frac{1}{100^{th}}$ of $1), a **cent**ennial (a 100-year celebration), etc.

The prefix "kilo" is from the Greek *khilioi*, which means "thousand," and a **kilo**meter means 1,000 meters. Any guesses what a **kilo**watt means? Yup — 1,000 watts.[6]

Why both Latin *and* Greek prefixes? While I couldn't find this specifically explained anywhere, it's interesting to note that the derivatives *milli, centi,* and *kilo* are all French, and the Metric System traces much of its roots to France.[7]

The names aren't arbitrary — remembering the meaning of the prefixes can help you keep them straight.

Measuring

Pull out a ruler and a measuring tape. Many of these devices list the U.S. Customary measurements on one side, and the metric measurements on the other, making it easy to measure in both systems.

For now, let's look at the U.S. Customary side — the numbers mark off the inches, and the tiny tick marks in between inches mark fractions of an inch. Each inch is broken into 16 fractional markings. Longer markings mark off $\frac{1}{4}$, $\frac{1}{2}$, and $\frac{3}{4}$ of an inch.

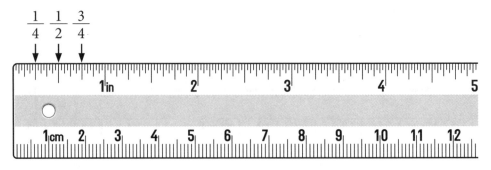

To measure a distance, simply hold your ruler or measuring tape to the distance you're trying to measure. For example, the rock shown is $4\frac{1}{2}$ inches long if we measure all the way to the farthest edge.

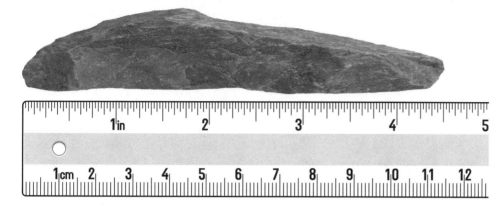

Whose Standard?

When we measure a length, we measure it against a standard — a yard, ruler, etc. — and see how it compares. If we use a faulty standard, we'll get very misleading information.

The same is true spiritually. Many people think of themselves as "good enough" to get to heaven because they see themselves as better than other people. When we die, though, we won't be judged based on how well we did compared to others — we'll be held against *God's standard*. So we'd be wise to look at that standard now and see how we measure up.

Consider just a few of God's commandments from Exodus 20. Be honest with yourself. Are you really a good person when compared with this standard of holiness?

"Thou shalt have no other gods before me." Have you ever loved or desired anything before God?

"Honour thy father and thy mother: that thy days may be long upon the land which the Lord thy God giveth thee." Have you always honored your parents perfectly?

"Thou shalt not kill." Jesus clarified this: *"Ye have heard that it was said of them of old time, Thou shalt not kill; and whosoever shall kill shall be in danger of the judgment: But I say unto you, That whosoever is angry with his brother without a cause shall be in danger of the judgment"* (Matthew 5:21–22). Whoa! Who hasn't been angry without cause?

"Thou shalt not commit adultery." Again, Jesus clarifies: *"Ye have heard that it was said by them of old time, Thou shalt not commit adultery: But I say unto you, That whosoever looketh on a woman to lust after her hath committed adultery with her already in his heart"* (Matthew 5:27–28). The impure thoughts in our hearts and minds make us guilty in God's eyes!

"Thou shalt not bear false witness against thy neighbour." "Bearing false witness" is telling a lie. Have you ever told a lie . . . even a small one? What about when you were little?

See www.LivingWaters.com for more information, both on the gospel and on how to use the 10 Commandments in sharing it with others.

"Thou shalt not covet. . . ." Have you ever wanted something that wasn't yours?

When held against God's standard, we are *all* guilty — we're all idolaters, murderers, adulterers, coveters, and liars. And that's just a few of God's commandments. Plus, in God's eyes, if we're guilty of breaking just *one* of His commandments, we're guilty of breaking them all (James 2:10). The Bible compares even our righteous deeds to filthy rags (Isaiah 64:6). And He warns that **all** sin will be punished — and the punishment for sin is eternal death (Romans 6:23) in a torturous lake of fire (Revelation 20:15; 21:8).

But there's good news! God knew before the world began that man would rebel against Him, and He had a plan. You see, mankind was not always so hopelessly lost in sin — originally, God created us perfect and without any sin at all. But man chose to rebel, bringing sin and death into the world.

Yet God, knowing all the evil you and I would do (and think), chose to come down as a man and claim all that evil as His own, dying on a cross to bear the penalty we deserved. He now offers His righteousness and eternal life in Heaven to all who place their trust in Jesus.

Question: *Have you placed your trust in Jesus, or are you still trying to attain God's standard of perfection on your own? Have you unconsciously raised your own standard to get to Heaven or are you looking at God's standard and His solution? If you haven't trusted Jesus, today is the day! None of us know what tomorrow will hold. If you have trusted Jesus, rejoice in His undeserved righteousness and go tell someone else about what He has done.*

Solving Problems with Measurements

Let's take a look now at solving problems involving measurements.

Example: If I'm $5\frac{1}{2}$ feet tall and I stand on a 6-foot ladder, at what height will my head be?

> We can easily add our measurements together to find the answer.
>
> $$5\frac{1}{2} \text{ feet} + 6 \text{ feet} = 11\frac{1}{2} \text{ feet}$$

Example: If I'm 66 *inches* tall and I stand on a 6-*foot* ladder, how tall will my head be?

> The answer to this question is not as obvious, is it? We cannot simply add 66 inches and 6 feet, or we will get an entirely bogus answer! Since the units of measure are not the same, we need to first reexpress the inches as feet or the feet as inches.
>
> We'll go into more details on how to express distance in different units in the next lesson; for now, I just wanted you to be aware that when dealing with measurements, **you need to make sure you're dealing with the same units!**

280 | PRINCIPLES OF MATHEMATICS 1

Keeping Perspective — A Glance at History

While we focused on the U.S. Customary System and the Metric System units of length in this lesson, throughout history, men have used different units of length. In reading your Bible, you've probably read about the cubit. The cubit was defined as the length from a man's elbow to his middle fingertip. Since this length varies based on the height of a man, this measurement could lead to different standards. (For example, the measurement from Goliath's elbow to fingertip was much different than David's!) The ancient Egyptians used the "Royal Cubit," which was somewhere around 20.6 inches; many other cubits were closer to 18 inches. Can you see why standardization is important?

According to the National Institute of Standards and Technology (NIST), the branch of the Department of Commerce in charge of providing "measurement standards for science and industry" and a "national scientific laboratory in the physical sciences,"[8] our inch, foot, and yard trace their origin back to the cubit, with the Romans influencing some of the smaller units, along with introducing the idea of a mile.[9]

No matter what units we use, the principle is the same: we're using a standard to describe distances. We can do this because God created us with this ability.

14.2 Conversions via Proportions

Say you measured an edge of your garden, and it was 72 *inches* long. You want to buy edging to go along that edge, which is sold by the *foot*. How many *feet* of edging do you need?

In order to figure out how many feet you need, you need to **convert** 72 inches to feet — that is, you need to figure out how to describe 72 *inches* using *feet* as your unit of measurement instead. Any ideas how to do that? There are actually many ways — let's take a look at one of them.

Conversions via Proportions

Think back to what you know about ratios and proportions. As we've seen, a ratio is "the relative size of two quantities expressed as the quotient of one divided by the other,"[10] or basically a fancy name for using division to compare quantities. A proportion is two equal ratios.

Let's use a ratio to compare how inches and feet relate. There are 12 inches in 1 foot, so we could write this ratio as a fraction like this:

$$\frac{12 \text{ in}}{1 \text{ ft}} \quad or \quad \frac{1 \text{ ft}}{12 \text{ in}}$$

It's important to note that this ratio is really worth 1. After all, 12 inches and 1 foot represent the *same distance*, just in different units. This means that we really could substitute either one for the other in the ratio. When we do, we see that the ratio is really worth 1, as any number divided by itself equals 1.

$$\text{substitute 12 in for 1 ft} \quad \frac{12 \text{ in}}{1 \text{ ft}} = \frac{12 \text{ in}}{12 \text{ in}} = 1$$

$$\text{substitute 1 ft for 12 in} \quad \frac{1 \text{ ft}}{12 \text{ in}} = \frac{1 \text{ ft}}{1 \text{ ft}} = 1$$

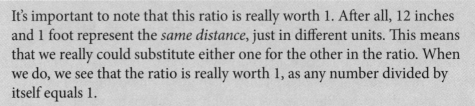

In this course, we'll refer to a ratio between two units that is worth 1 as a **conversion ratio**. It shows us how the two units compare.

Conversion ratio between inches and feet: $\frac{12 \text{ in}}{1 \text{ ft}}$ or $\frac{1 \text{ ft}}{12 \text{ in}}$

Now that we've expressed the relationship between inches and feet as a ratio, it's easy to convert 72 inches to feet. All we need to do is form an equivalent ratio — that is, another ratio that expresses the same distance in both feet and inches, but using 72 inches instead of 12 inches. We can figure out the number of feet to use in the ratio using a proportion!

$$\frac{12 \text{ in}}{1 \text{ ft}} = \frac{72 \text{ in}}{? \text{ ft}}$$

Now we can think through what number would **finish creating an equivalent ratio** (see 5.3). Since $72 \div 12 = 6$, if we multiply by $\frac{6}{6}$, we'll form an equivalent ratio with 72 as our numerator.

$$\frac{12 \text{ in}}{1 \text{ ft}} \cdot \frac{6}{6} = \frac{72 \text{ in}}{6 \text{ ft}}$$

$$\frac{12 \text{ in}}{1 \text{ ft}} = \frac{72 \text{ in}}{6 \text{ ft}}$$

Important! Notice that in both ratios we've put inches in the numerator and feet in the denominator. We could have reversed this ($\frac{1 \text{ ft}}{12 \text{ in}} = \frac{? \text{ ft}}{72 \text{ in}}$), but we have to be consistent. We could *not* put feet in the numerator in one ratio and in the denominator in the other, as we'd no longer be comparing like units to like units.

$$\frac{\cancel{1\text{ ft}}}{12\text{ in}} = \frac{72\text{ in}}{?\text{ ft}} \qquad \frac{\cancel{12\text{ in}}}{1\text{ ft}} = \frac{?\text{ ft}}{\cancel{72\text{ in}}}$$

$$\frac{12\text{ in}}{1\text{ ft}} = \frac{72\text{ in}}{?\text{ ft}} \qquad \frac{1\text{ ft}}{12\text{ in}} = \frac{?\text{ ft}}{72\text{ in}}$$

72 in equals 6 ft.

Feet must be compared with feet and inches with inches.

Watch Your Units

When doing problems, pay attention to the units used, and make sure to **include the units in your answer.** An answer without units when units were involved will be considered partially incorrect, as you've not included what that number represents. Watching your units carefully will serve you well, both in real life and in upper-level sciences.

Keeping Perspective

In math, there's often more than one way to solve a problem. In fact, in the next lesson, you're going to learn yet another way to convert units. Each one will come in handy at different times.

In coming up with these methods, all we're really doing is looking at the notations and skills we've already learned (ratios, proportions, etc.) and seeing if there's a way to apply them to help us in expressing distances. We're then walking away with a "rule" or "method" about how to convert units that simplifies what we discovered. You'll see this process repeated over and over again in math.

14.3 Different Conversion Methods

We saw in the last lesson that we can express the conversion ratio as a fraction: $\frac{12\text{ in}}{1\text{ ft}}$ *or* $\frac{1\text{ ft}}{12\text{ in}}$. We also saw that since both 12 inches and 1 foot represent the *same* length, both these fractions really represent 1.

We then used that knowledge to find the answer via a proportion:

$$\frac{12\text{ in}}{1\text{ ft}} = \frac{72\text{ in}}{?\text{ ft}}$$

$$\frac{12\text{ in}}{1\text{ ft}} \cdot \frac{6}{6} = \frac{72\text{ in}}{6\text{ ft}}$$

Answer: 6 ft

14. MEASURING DISTANCE 283

It's time now to look at two additional methods to convert units. Both these methods prove quite useful, as we'll see.

Conversion via the Ratio Shortcut

Since multiplying by a fraction worth 1 doesn't change the value and since the conversion ratio is worth 1, rather than setting up a proportion, we could *multiply* 72 inches by our conversion ratio instead.

$$72 \text{ in} \cdot \frac{1 \text{ ft}}{12 \text{ in}} = \frac{72 \text{ in} \cdot 1 \text{ ft}}{12 \text{ in}}$$

Remember, 72 in can be thought of as $\frac{72 \text{ in}}{1}$. *Since dividing by 1 doesn't change the value of the number, we ignored the 1, but* **we treated 72 as a numerator,** *as if it were written* $\frac{72 \text{ in}}{1}$.

Dealing with the Units

Now, what does $\frac{72 \text{ in} \cdot 1 \text{ ft}}{12 \text{ in}}$ equal? How do we handle the units? Just as we can divide by $\frac{12}{12}$ to simplify the numbers, we can divide by $\frac{\text{in}}{\text{in}}$ to simplify the units.

$$\frac{72 \text{ in} \cdot 1 \text{ ft}}{12 \text{ in}} \div \frac{\text{in}}{\text{in}} = \frac{72 \cdot 1 \text{ ft}}{12} = \frac{72 \text{ ft}}{12} \div \frac{12}{12} = 6 \text{ ft}$$

Notice that we could have simplified both the units and the numbers as we went. Remember, a fraction line means to divide. Since both the numerator and the denominator have "in", the division will cancel out the multiplication, just as it does with numbers.

$$\overset{6}{\cancel{72 \text{ in}}} \cdot \frac{1 \text{ ft}}{\underset{1}{\cancel{12 \text{ in}}}} = 6 \text{ ft}$$

The "Rule"

We've finally arrived at another "rule," or method, for unit conversions. Once again, this method applies the principles we've learned to a new situation, reducing the amount of thinking we have to do each time.

To convert a unit into another unit, just multiply it by the conversion ratio! Be sure to write the conversion ratio so the unit you're trying to convert will cancel out, leaving you with the unit of measure you need.

We'll refer to this method of converting units as **conversion via the ratio shortcut,** and the proportion method we looked at in the last lesson as **conversion via a proportion.**

Example: Convert 8 miles into feet.

$$8 \text{ mi} \cdot \frac{5{,}280 \text{ ft}}{1 \text{ mi}} = 42{,}240 \text{ ft}$$

Notice that we put the miles as the denominator of our conversion ratio so the miles would cancel out and the answer would be in feet.

Example: Convert 42,240 feet into miles.

$$42{,}240 \text{ ft} \cdot \frac{1 \text{ mi}}{5{,}280 \text{ ft}} = \frac{42{,}240 \text{ mi}}{5{,}280} = 8 \text{ mi}$$

Again, notice that we put the feet in the denominator of our conversion ratio so it would cancel out. **Always write the conversion ratio so the unit you're trying to convert will cancel out.** *Otherwise, you won't succeed in converting to a new unit.*

> The conversion via the ratio shortcut method may seem more involved at first, but it will save you lots of time once you become familiar with it, especially when dealing with multistep conversions.

Conversion via Mental Math

On simple conversion problems, we could convert mentally. Notice that when we converted 72 inches to feet we ended up dividing 72 by 12.

$$72 \text{ in} \cdot \frac{1 \text{ ft}}{12 \text{ in}} = \frac{72 \cdot 1 \text{ ft}}{12} = \frac{72 \text{ ft}}{12} = 6 \text{ ft}$$

If we had needed to find the answer mentally, we could have simply divided 72 inches by 12 inches. When you think about it, we know that 1 foot equals 12 inches. It follows then that if we were to divide the 72 inches by 12, we'd get the equivalent measurement in feet.

If, on the other hand, we'd started with 6 feet and needed to find inches, we could find it using multiplication. If 12 inches equals 1 foot, than 6 feet is going to equal 6 • 12 inches, or 72 inches.

72 inches and 6 feet represent the *same distance*, but in different units.

While it is pretty easy to convert between inches and feet mentally since we're familiar with the units, it's also easy to make a mistake and divide when we need to multiply or vice versa. So be cautious when doing unit conversions mentally.

Also, sometimes the numbers involved make it hard to solve problems completely mentally. For example, converting 83.54 inches to feet would require this division:

$$83.54 \div 12 = ?$$

While that's a little complicated to solve mentally, we could still figure out what we need to multiply or divide mentally and then use paper or a calculator to find the answer.

$$83.54 \div 12 = 6.96$$

On example problems, we'll still show how to set up problems like this one mentally; know, though, that you might need to use paper or a calculator to perform the math.

14. MEASURING DISTANCE | 285

Which Method Is Best?

We've now talked about three different methods for unit conversion:

■ Conversion via a proportion

■ Conversion via the ratio shortcut

■ Conversion via mental math

Which one is best? It depends! In general, it's a good habit to convert via the ratio shortcut, as it makes multistep conversions much easier, as we'll see soon.

Conversion via a Proportion	Conversion via the Ratio Shortcut	Conversion via Mental Math
		We know 1 foot equals 12 inches, so 72 inches divided by 12 will give us our feet
$\dfrac{12 \text{ in}}{1 \text{ ft}} = \dfrac{72 \text{ in}}{? \text{ ft}}$	$\overset{6}{\cancel{72} \text{ in}} \cdot \dfrac{1 \text{ ft}}{\underset{1}{\cancel{12} \text{ in}}} = 6 \text{ ft}$	$72 \div 12 = 6$
Answer: 6 ft	Answer: 6 ft	Answer: 6 ft

Keeping Perspective

Once again, we're continuing to build on what we know to find shortcuts to deal with additional situations. The process of unit conversion we looked at in this lesson is one you'll continue to build on later. As you familiarize yourself with it, remember that each conversion method is ultimately a way of helping us work with real-life distances. It's math in action — a tool to help us measure and describe God's creation.

14.4 Currency Conversions

We often need to convert between a lot more than distance units. The good news is that you have all the skills you need! Let's practice applying these skills to currency.

In America, we use the dollar as our currency, but other countries use other currencies. When traveling, it's often necessary to convert between currencies, exchanging dollars for pounds, euros, etc. The exchange rate is the conversion ratio (also sometimes called the conversion rate; remember, a rate is just a specific type of ratio) between two currencies. If the exchange rate is $\dfrac{1}{3}$, that means that for every 1 unit of one currency, you could receive 3 of the other.

Searching the Internet for the exchange rate between two countries should yield the current rate (although that rate may vary throughout the day, and there may be an additional fee from the vendor who converts the money). Once you know the exchange rate, you can use any of the methods you've learned to convert between the two currencies.

The exchange rate can vary based on a variety of factors, including the economy of the different nations.

Example: If 2 British pounds = 1 U.S. dollar, how much is 8 pounds in U.S. dollars?

We've been given this ratio, or rate: $\dfrac{1 \text{ dollar}}{2 \text{ pounds}}$.

Conversion via a Proportion	Conversion via the Ratio Shortcut	Conversion via Mental Math
$\dfrac{? \text{ dollars}}{8 \text{ pounds}} = \dfrac{1 \text{ dollar}}{2 \text{ pounds}}$	$\overset{4}{\cancel{8 \text{ pounds}}} \cdot \dfrac{1 \text{ dollar}}{\underset{1}{\cancel{2 \text{ pounds}}}} = \dfrac{4}{1} \text{ dollars}$	$8 \div 2 = 4$
Answer: $4	Answer: $4	Answer: $4

Example: While in Britain, you find an item marked 45 pounds. How much will it cost you in U.S. dollars, assuming a conversion rate of 1.602 British pounds to 1 U.S. dollar?

Rate: $\dfrac{1 \text{ dollar}}{1.602 \text{ pounds}}$

Conversion via a Proportion	Conversion via the Ratio Shortcut	Conversion via Mental Math
$\dfrac{45 \text{ pounds}}{? \text{ dollars}} = \dfrac{1.602 \text{ pounds}}{1 \text{ dollar}}$	$45 \, \cancel{\text{pounds}} \cdot \dfrac{1 \text{ dollar}}{1.602 \, \cancel{\text{pounds}}} = \dfrac{45}{1.602} \text{ dollars}$	$45 \div 1.602 = 28.09$
Answer: $28.09	Answer: $28.09	Answer: $28.09

You might need a paper or a calculator to complete the conversion via Mental Math on this one.

Conversion and Missionaries

During tough economic times, missionaries have it extra, extra tough — not only can the dollars they receive decrease because people have less to give, but when the value of the U.S. dollar decreases, as is often the case during an economic depression, the missionaries often get fewer foreign currency per U.S. dollar, meaning their support doesn't go as far. So keep missionaries you support in mind (and prayer) during tough economic times.

Keeping Perspective

The conversion methods you're learning apply in all sorts of situations — including exchanging money in a foreign country. Math can even help you while you're traveling.

14. MEASURING DISTANCE

14.5 Metric Conversions

It's time to dig deeper into the metric measurement system, applying the same conversion methods to metric units.

Understanding the Metric System

While the units we looked at in 14.1 (millimeters, centimeters, meters, and kilometers) are the most commonly known, the Metric System actually contains other units too.

$$10 \textbf{ millimeters (mm)} = 1 \textbf{ centimeter (cm)}$$
$$10 \textbf{ centimeters} = 1 \text{ decimeter (dm)}$$
$$10 \text{ decimeters} = 1 \textbf{ meter (m)}$$
$$10 \textbf{ meters} = 1 \text{ decameter (dam)}$$
$$10 \text{ decameters} = 1 \text{ hectometer (hm)}$$
$$10 \text{ hectometers} = 1 \textbf{ kilometer (km)}$$

Notice how, starting with centimeters, **each unit is worth 10 of the previous unit**. We'll see shortly that this makes the Metric System incredibly easy to work with, since our decimal system is also based on 10.

Conversions within the Metric System

Because each unit in the Metric System is worth 10 of the previous unit, to convert from one unit to the next largest unit, we need only to divide by 10. 1 mm = 0.1 cm (decimal moved over one place to the left, as we divided by 10).

Likewise, to convert from one unit to the next smallest, we need only to multiply by 10.

0.1 cm = 1 mm (decimal moved over one place to the right as we multiplied by 10).

The same methods for conversions that we've looked at so far apply to the Metric System — but because the Metric System is based on 10 and because multiplying or dividing by 10 is simply a matter of moving our decimal point (see 7.3), the math involved is much simpler.

Example: Convert 1 millimeter to centimeters.

Conversion via a Proportion	Conversion via the Ratio Shortcut	Conversion via Mental Math
$\dfrac{1 \text{ mm}}{? \text{ cm}} = \dfrac{10 \text{ mm}}{1 \text{ cm}}$	$1 \text{ mm} \cdot \dfrac{1 \text{ cm}}{10 \text{ mm}} = \dfrac{1 \text{cm}}{10} = 0.1 \text{ cm}$	$1 \div 10 = 0.1$
Answer: 0.1 cm	Answer: 0.1 cm	Answer: 0.1 cm

Example: Convert 3 cm to millimeters.

Conversion via a Proportion	Conversion via the Ratio Shortcut	Conversion via Mental Math
$\dfrac{3 \text{ cm}}{? \text{ mm}} = \dfrac{1 \text{ cm}}{10 \text{ mm}}$	$3 \text{ cm} \cdot \dfrac{10 \text{ mm}}{1 \text{ cm}} = 30 \text{ mm}$	$3 \cdot 10 = 30$
Answer: 30 mm	Answer: 30 mm	Answer: 30 mm

Example: Convert 2 meters to centimeters.

Here we're switching from meters to centimeters — those are more than 1 unit apart. But the math is still quite simple! We know there are 100 centimeters in 1 meter.

Conversion via a Proportion	Conversion via the Ratio Shortcut	Conversion via Mental Math
$\dfrac{2 \text{ m}}{? \text{ cm}} = \dfrac{1 \text{ m}}{100 \text{ cm}}$	$2 \text{ m} \cdot \dfrac{100 \text{ cm}}{1 \text{ m}} = 200 \text{ cm}$	$2 \cdot 100 = 200$
Answer: 200 cm	Answer: 200 cm	Answer: 200 cm

> Since every conversion between metric units is a multiple of 10, conversions within the Metric System can easily be done mentally.

Measuring with Metric

Pull out a ruler and take a look at the metric side of it for a moment. The numbers mark off the centimeters, and the tiny tick marks mark off millimeters. There are 10 millimeters per 1 centimeter.

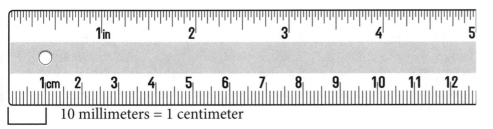

10 millimeters = 1 centimeter

The line shown below is 2 centimeters and 1 millimeter long. We could represent it in centimeters alone as 2.1 cm, since 1 millimeter equals 0.1 cm. (Remember, dividing by 10 just means moving the decimal over to the left.) Notice how easy it is to express the portion representing a part of a centimeter as a decimal!

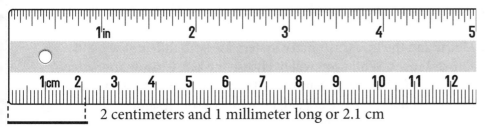

2 centimeters and 1 millimeter long or 2.1 cm

Remember to Convert First!

It's easy to add, subtract, multiply, or divide the wrong numbers and reach an entirely wrong answer if you add unlike units. For instance, if you're asked to add 2 centimeters and 5 millimeters, you can't just add 2 and 5, because they represent different units of measure.

You have to first convert the centimeters to millimeters or the millimeters to centimeters, and *then* add them.

$$2 \text{ cm} + 5 \text{ mm} = 2.5 \text{ cm}$$

$$2 \text{ cm} + 5 \text{ mm} = 25 \text{ mm}$$

See Dr. Jason Lisle's *The Ultimate Proof of Creation: Resolving the Origins Debate* (Green Forest, AR: Master Books, 2009) for more details on how morality only makes sense in a biblical worldview, and on how to use that to challenge other worldviews.

Is Morality Like Measurements?

Most people acknowledge that murder is wrong (God has written His laws upon our hearts — Romans 2:15), but few can explain *why* it's wrong. Without acknowledging a Creator, we have no basis for an absolute standard for right and wrong. Some people will argue that society as a whole determines right and wrong, just as they determine measurement units. But in that case, who is to say that Hitler was wrong for murdering the Jews? His German society didn't view it as wrong!

You see, only the biblical worldview gives us a basis for morality. Morality is not an arbitrary rule like a unit of measure that man can change — it's given to us by God, based on the character of God.

When people tell you they don't believe in God, consider asking them how they explain right and wrong. Point out that by condemning the "Hitlers" of this world, they are acting contrary to their worldview.

The Bible gives us a firm foundation that makes sense out of every area of life — let's share it with others in love!

Keeping Perspective

The Metric and the U.S. Customary System are both different ways to measure distances. While you will likely use the U.S. Customary System more in daily life, becoming familiar with them both is important, as other countries (and many technical fields) use the Metric System.

14.6 Multistep Conversions

Sometimes it takes more than one step to convert between two units. Let's say we need to convert 4 miles into yards, but we can't remember how many yards are in a mile, only that there are 5,280 feet in a mile.

We could look up the ratio between yards and miles, or we could find the answer by breaking this problem down into further steps, converting our miles to feet and *then* to yards.

Conversion via Proportions:

Converting to feet:

$$\frac{4 \text{ mi}}{? \text{ ft}} = \frac{1 \text{ mi}}{5{,}280 \text{ ft}}$$

Answer: 21,120 ft

Now that we have found the feet, we can convert to yards.

$$\frac{21{,}120 \text{ ft}}{? \text{ yd}} = \frac{3 \text{ ft}}{1 \text{ yd}}$$

Answer: 7,040 yd

Conversion via the Ratio Shortcut:

Converting to feet:

$$4 \text{ mi} \cdot \frac{5{,}280 \text{ ft}}{1 \text{ mi}} = 21{,}120 \text{ ft}$$

Converting to yards:

$$21{,}120 \text{ ft} \cdot \frac{1 \text{ yd}}{3 \text{ ft}} = 7{,}040 \text{ yd}$$

Answer: 7,040 yd

Conversion via Mental Math:

Converting to feet:

$$4 \cdot 5{,}280 \text{ ft} = 21{,}120 \text{ ft}$$

Converting to yards:

$$21{,}120 \div 3 = 7{,}040$$

Answer: 7,040 yd

Doing It in One Step

Using the conversion via the ratio shortcut method can greatly simplify multistep conversion problems, as we can do the conversion in a single step by multiplying by more than one conversion ratio until we have the answer in the desired unit.

Notice it's a little challenging to do multistep conversions mentally.

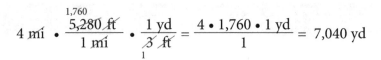

Notice how each unit cancelled out *as we went*!

Also notice that we used $\frac{5{,}280 \text{ ft}}{1 \text{ mi}}$ instead of $\frac{1 \text{ mi}}{5{,}280 \text{ ft}}$ because we needed to have miles in the denominator in order to cancel it out. And we used $\frac{1 \text{ yd}}{3 \text{ ft}}$ instead of 3 ft over 1 yd for the same reason — we needed the feet to cancel.

If you're ever unsure which unit you should put on the numerator and the denominator in a conversion ratio, just **think about what you need to use in order to cancel out the units you don't want and leave only the unit you do.**

Because the conversion via the ratio shortcut makes multistep conversions so much simpler, we'll be emphasizing it throughout this course.

Keeping Perspective

Hopefully, you now see why multiplying by a conversion ratio is such a valuable method to know. Sometimes methods that initially seem silly end up saving time in more complicated situations. The more you learn math, the more you'll realize how different tools combine. Yet all these tools only work because of the inherent consistency God created and sustains. Don't lose sight of the fact that He is the One "…upholding all things by the word of his power…" (Hebrews 1:3).

14.7 Conversions Between U.S. Customary and Metric

Guess what? The same methods you've been using to convert within a system apply for conversions *between* systems. **All we need to know is the ratio between two units (i.e., the conversion ratio).**

Below are the ratios between some common U.S. Customary and Metric units.

$$1 \text{ in} = 2.54 \text{ cm}$$
$$1 \text{ ft} = 30.48 \text{ cm}$$
$$1 \text{ yd} = 0.9144 \text{ m}$$
$$1 \text{ mi} = 1.609344 \text{ km}$$

Now we can convert away! The examples show the conversions using the conversion via the ratio shortcut method, but any of the methods we've looked at would work. We're just using this method because it saves time on multistep conversions, as we saw in the last lesson.

Example: Convert 80 yards to meters.

Ratio between meters and yards: $\dfrac{0.9144 \text{ m}}{1 \text{ yd}}$

Multiply:

$$80 \text{ yd} \cdot \dfrac{0.9144 \text{ m}}{1 \text{ yd}} = 73.152 \text{ m}$$

Example: You drive into Canada and see a sign saying the town you're going to is 100 km away. How many miles is that?

Ratio between kilometers and miles: $\dfrac{1 \text{ mi}}{1.609344 \text{ km}}$

Multiply:

$$100 \text{ km} \cdot \dfrac{1 \text{ mi}}{1.609344 \text{ km}} = \dfrac{100}{1.609344} \text{ mi} = 62.14 \text{ mi}$$

Remember to Think It Through

One common confusion in unit conversion regards which way to express the conversion ratio — should it be $\dfrac{1 \text{ in}}{2.54 \text{ cm}}$ or $\dfrac{2.54 \text{ cm}}{1 \text{ in}}$?

It depends on what unit of measure we want in the end.

If we want to convert from 50 inches to centimeters, we would place the inches on the bottom of the ratio so they would cancel out.

$$50 \text{ in} \cdot \dfrac{2.54 \text{ cm}}{1 \text{ in}} = 127 \text{ cm}$$

If, on the other hand, we wanted to convert 127 centimeters into inches, we would place the centimeters on the bottom of the ratio to cancel them out.

$$127 \text{ cm} \cdot \dfrac{1 \text{ in}}{2.54 \text{ cm}} = \dfrac{127 \text{ in}}{2.54} = 50 \text{ in}$$

Remember, we can write the ratio either way because 1 in and 2.54 cm represent the *same quantity*. Thus, either way, the resulting fraction is worth 1.

$$1 \text{ in} = 2.54 \text{ cm}$$

Substitute 2.54 cm for 1 in:

$$\dfrac{1 \text{ in}}{2.54 \text{ cm}} = \dfrac{2.54 \text{ cm}}{2.54 \text{ cm}} = 1$$

Substitute 1 in for 2.54 cm:

$$\dfrac{1 \text{ in}}{2.54 \text{ cm}} = \dfrac{1 \text{ in}}{1 \text{ in}} = 1$$

> ## Keeping Perspective
>
> Do you see a pattern? We keep applying the same methods to different situations. A lot of math class is expanding on what you know or applying it to new settings. If you ever encounter a problem you don't know how to solve (whether in a textbook or in real life), don't be afraid to try on your own to think through how to apply what you know to it. Chances are you have all the tools you need!

14.8 Time Conversions

It's time to apply unit conversion to a different area altogether: time. Let's start by taking a look at some units we use for measuring time, and then at how to convert between them.

Units of Time

God created time and gave us ways of keeping track of time (day and night, stars, etc.) when He created the universe. Unlike us, God had no beginning — He is outside of time.

> *In the beginning God created the heaven and the earth. And the earth was without form, and void; and darkness was upon the face of the deep. And the Spirit of God moved upon the face of the waters. And God said, Let there be light: and there was light. And God saw the light, that it was good: and God divided the light from the darkness. And God called the light Day, and the darkness he called Night. And the evening and the morning were the first day (Genesis 1:1–5).*
>
> *And God said, Let there be lights in the firmament of the heaven to divide the day from the night; and let them be for signs, and for seasons, and for days, and years (Genesis 1:14).*
>
> *Hast thou not known? hast thou not heard, that the everlasting God, the LORD, the Creator of the ends of the earth, fainteth not, neither is weary? there is no searching of his understanding (Isaiah 40:28).*

Every day, God causes the earth to rotate in a consistent fashion. Whichever side of the earth is tilted away from the sun experiences darkness, while the other side experiences light. As the earth rotates, the stars and moon appear to change their positions overhead. One complete rotation forms a **day**.

Sometimes we need to know specifically how much of a day has passed. If we go outside during the morning, the sun will be in a different spot than if we go outside in the evening. So the sun's position helps us keep track of approximately what time of the day it is. But it's not always possible to see the sun, especially on a cloudy day. Nor is it easy to tell someone exactly when we want to meet using the sun alone. Thus, we use **hours** and **minutes** to refer to portions of a day.

God gave us the 7-day **week** when He created the world in six days and rested on the seventh (Genesis 2:2-3; Exodus 20:8-11). A **year** is based on how long it takes the earth to travel around the sun (approximately 365 days).

Every day and year that passes testifies that God is continuing to hold this world together in a consistent fashion, just as He promised.

> *While the earth remaineth, seedtime and harvest, and cold and heat, and summer and winter, and day and night shall not cease (Genesis 8:22).*

$$\textbf{60 seconds = 1 minute (min)}$$
$$\textbf{60 minutes = 1 hour (hr)}$$
$$\textbf{24 hours = 1 day (d)}$$
$$\textbf{7 days = 1 week (wk)}$$
$$\textbf{365 days = 1 year (yr \textit{or} y)}$$

Conversions between Units

Because we all deal with time every single day, working with and converting between various units of time is essential . . . and you already have all the skills you need! You can convert between units of time the same way you have been between units of distance. Again, we'll focus on the conversion via the ratio shortcut method, as it will save time on multistep problems and is an important method to learn.

Example: How many minutes are in 1 day?

$$1\ \cancel{d} \cdot \frac{24\ \cancel{hr}}{1\ \cancel{d}} \cdot \frac{60\ min}{1\ \cancel{hr}} = 1,440\ min$$

Notice how all we did is set up the ratios so that everything but the unit we wanted — minutes — crossed out. Again, this works because we're multiplying by a value worth 1. $\frac{24\ hr}{1\ d}$ is equivalent to 1, as both the numerator and the denominator represent the same quantity. The same holds true for $\frac{60\ min}{1\ hr}$.

Applications to Time Problems

Because time is such an intricate part of life, we face time problems all the time (pun intended). Some require multiple "tools" from our mathematical toolbox to solve.

For example, let's say that you know you run 8 miles per hour. You want to figure out how far you can run in 2 hours at that pace.

You are probably used to solving this type of problem like this:

$8 \cdot 2 = 16$. You can run 16 miles in 2 hours.

Now that you know about units, it's time to begin including them in the problems. Here is another way to write the problem that shows what's really happening with the units:

$$\frac{8 \text{ mi}}{1 \text{ hr}} \cdot 2 \text{ hr} = 16 \text{ mi}$$

Notice that we put 8 mi over 1 hr: $\frac{8 \text{ mi}}{1 \text{ hr}}$. This is a ratio showing 8 miles per hour. Remember, *per* is a good clue that you're dealing with a ratio.

> **Always include units in problems when units are given**. This practice will keep you from a lot of accidental errors, as well as help you solve more intricate problems. If you do the math correctly, you'll end up in the correct unit. If you don't end up in the correct unit, you'll know you did something wrong.

Example: Jenny can run 8 miles per hour. How far can she run in 90 minutes?

Notice that our speed is given in miles per *hour*, but we're asked how far we can go in 90 *minutes*. In order to get an accurate answer, we have to make sure that we use the same units.

While we could convert the 90 minutes to hours, it's easy enough to **substitute 60 min for the 1 hr** (after all, 60 min = 1 hr), making our speed $\frac{8 \text{ mi}}{60 \text{ min}}$.

Now we can multiply to find the answer.

$$\frac{8 \text{ mi}}{\underset{2}{60} \text{ min}} \cdot \overset{3}{90} \text{ min} = \frac{8 \text{ mi} \cdot 3}{2} = 12 \text{ mi}$$

Example: Jenny runs 8 miles an hour. How far can she run in 20 minutes?

Once again, we'll rewrite our speed as $\frac{8 \text{ mi}}{60 \text{ min}}$ since our time was given in minutes instead of hours.

$$\frac{8 \text{ mi}}{\underset{3}{60} \text{ min}} \cdot \overset{1}{20} \text{ min} = \frac{8 \text{ mi}}{3} = 2.67 \text{ mi}$$

Everyday Time Conversions

While it's important to know how to convert time units on paper, we often also need to convert them mentally. For example, if you know it takes you 25 minutes to drive to a store, and that you need about 35 minutes in the store plus another 10 minutes to get to your 3 p.m. appointment, at what time do you need to leave your house?

To find the answer, start by figuring out how many hours and minutes you need altogether. Mentally add 25 + 35 + 10, which is 70 minutes. Now, you could convert 70 minutes to hours by multiplying by the conversion ratio.

$$\overset{7}{\cancel{70}} \text{ min} \cdot \frac{1 \text{ hr}}{\underset{6}{\cancel{60} \text{ min}}} = \frac{7 \text{ hr}}{6} = 1\frac{1}{6} \text{ hr}$$

Now you could convert the fractional amount back to minutes.

$$\frac{1}{\underset{1}{\cancel{6}}} \cancel{\text{hr}} \cdot \frac{\overset{10}{\cancel{60}} \text{ min}}{1 \cancel{\text{hr}}} = \frac{10 \text{ min}}{1} = 10 \text{ min}$$

Total Time Needed: 1 hr, 10 min

But, that's a long way to go about it for this problem. You can really solve it *all* mentally. You know 60 minutes makes 1 hour, so 70 minutes would be 1 hour and 10 minutes.

Now that you know how long you need, you can figure out when to leave. An hour earlier than 3 p.m. is 2 p.m., and 10 minutes earlier than that is 1:50 p.m.

Keeping Perspective — Don't Get Stuck!

When learning math, it's easy to get stuck on a concept and assume that every problem has to be solved the same way. But in real life, we encounter a variety of concepts all the time. We can't rely on just one mathematical tool — we need to be able to think through what tool to use for each situation.

Since one of this course's goals is to help equip you to use math wherever you may need to (it is, after all, a tool that can be used for God's glory), we'll sometimes throw a problem into a worksheet that requires different tools to solve than those covered in that lesson. Be sure to think through every problem to make sure your answer — and method of solving — makes sense.

14.9 Chapter Synopsis

I hope you had some fun this chapter with measurements! Here's a quick review of what we covered.

■ **Finding standards: units** — To measure, we need a standard, or unit, we can use. We explored the distance units in two measurement systems — the U.S. Customary System and the Metric/SI System.

U.S. Customary System

12 inches (in) = 1 foot (ft)
3 feet or 36 inches = 1 yard (yd)
1 mile (mi) = 1,760 yards (yd) or 5,280 feet (ft)

Metric System/SI

10 millimeters (mm) = 1 centimeter (cm)
10 centimeters = 1 decimeter (dm)
10 decimeters = 1 meter (m)
10 meters = 1 decameters (dam)
10 decameters = 1 hectometer (hm)
10 hectometers = 1 kilometer (km)

■ **Conversions** — We explored various methods for unit conversion (via a proportion, the ratio shortcut, and mental math) and practiced converting distance, currency, and time units. We also learned some common conversion ratios that help convert between the Metric and Customary Systems.

Common Conversions between Systems

1 in = 2.54 cm
1 ft = 30.48 cm
1 yd = 0.9144 m
1 mi = 1.609344 km

Common Conversions of Time

60 seconds = 1 minute (min)
60 minutes = 1 hour (hr)
24 hours = 1 day (d)
7 days = 1 week (wk)
365 days = 1 year (yr *or* y)

As you move on from this chapter, remember that units of measuring distance are simply predefined distances against which we can compare and describe the distances of objects. And methods for unit conversion are just shortcuts that use the consistencies and conventions we know about multiplication, fractions, etc., to easily convert units. Once again, we're using the abilities God gave us to help us more easily name and describe God's creation.

As you continue to learn about and use measurements, ponder the fact that God knows not just the measure of the things easy to measure, but the measure of things we can't possibly measure, such as the dust of the earth, the waters in the oceans, and the stars in the sky. While we can measure some things, our inability to measure so many aspects of creation reminds us again of how much greater God is than we are.

> *Who hath measured the waters in the hollow of his hand, and meted out heaven with the span, and comprehended the dust of the earth in a measure, and weighed the mountains in scales, and the hills in a balance? (Isaiah 40:12).*

[CHAPTER 15]

Perimeter and Area of Polygons

15.1 Perimeter

Now that we've explored distance units and unit conversion, it's time to start measuring!

Suppose you were to take a measuring tape and measure the length of each wall in a room. If you added up those lengths, you would have found the distance around the room.

The mathematical term used to express the distance around a closed shape is **perimeter**.

Finding the perimeter of a polygon is simple — add up all the sides. That will give you the total distance around the shape.

Perimeter of any polygon = sum of the lengths of each side

Example: Find the perimeter of these irregular polygons.

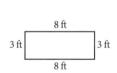 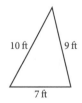 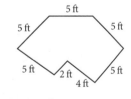

Irregular Quadrilateral
Perimeter =
3 ft + 3 ft + 8 ft + 8 ft
Answer: 22 ft

Irregular Triangle
Perimeter =
10 ft + 9 ft + 7 ft
Answer: 26 ft

Irregular Heptagon
Perimeter =
5 ft + 2 ft + 4 ft + 5 ft + 5 ft + 5 ft + 5 ft
Answer: 31 ft

Example: Find the perimeter of these regular polygons.

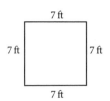 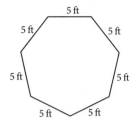

Regular Quadrilateral (i.e., square)
Perimeter =
7 ft + 7 ft + 7 ft + 7 ft
Answer 28 ft

Regular Triangle
Perimeter =
6 ft + 6 ft + 6 ft
Answer: 18 ft

Regular Heptagon
Perimeter =
5 ft + 5 ft + 5 ft + 5 ft + 5 ft + 5 ft + 5 ft
Answer: 35 ft

Making It Simpler

As with anything in math, we're constantly looking for ways to save ourselves needless steps. Let's take a look at some shortcuts we could use to find the perimeter of regular polygons and rectangles.

Regular Polygons

Look at the regular polygons in the example above again. Can you think of a way to simplify finding the perimeter of a regular polygon?

Think back to the definition of a regular polygon. Back in 13.3, we saw that we use the term *regular* in front of a polygon to indicate that we're referring to a polygon whose sides are all the *same lengths*. So, if we know a shape is a regular polygon, rather than *adding* the lengths of all the sides, we could just **multiply the length of one side by the number of sides.**

Perimeter of regular polygon = (number of sides) • (length of each side)

Example: Find the perimeter of the following regular polygons.

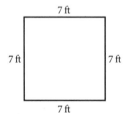 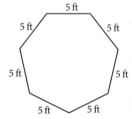

Regular Quadrilateral (i.e., square)
4 Sides
Perimeter = 4 • 7 ft
Answer: 28 ft

Regular (Equilateral) Triangle
3 Sides
Perimeter = 3 • 6 ft
Answer: 15 ft

Regular Heptagon
7 Sides
Perimeter = 7 • 5 ft
Answer: 35 ft

Shortcuts

In Chapter 13, we categorized shapes into sets based on their properties. We just used the knowledge of those properties to think through a "shortcut" for finding the perimeter of the set of *polygons* and the set of *regular polygons*. These "shortcuts" can then help us find the perimeter of real-life shapes, such as a swimming pool, garden, photo, etc.

Rectangles

And what about the perimeter of a rectangle? Even though it's irregular, by definition it has two sets of equal sides. So multiplication could abbreviate our work there, too. To find a rectangle's perimeter, we just need to multiply the length by 2 and the width by 2, and then add the results together.

Perimeter of rectangle = (2 • length) + (2 • width)

Example: Find the perimeter of the rectangle shown.

Perimeter = (2 • 8 ft) + (2 • 3 ft) = 16 ft + 6 ft
Answer: 22 ft

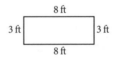

Length and Width: Which Is Which?

You may sometimes wonder which is the length and which is the width. For instance, in this rectangle, which is length and which is width?

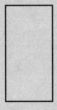

Definitions vary. Some people would call the longest sides the length and the other the width; others would use the word "height" to describe the long sides of this rectangle, and either length or width for the others.

If you were writing a paragraph about an object, the terms you use might matter; however, when it comes to determining length and width in this course, it does not matter which you use. You can use any side of the rectangle to represent the length or the width (or the *height* — you can think of one side as the height and find the perimeter by multiplying 2 times the *height* instead of 2 times the *width*). The point is that we accurately find the perimeter by multiplying both the different-sized sides of the rectangle by two, and then adding the products together so that we've found the distance around all four sides.

Keeping Perspective

Since shapes help us describe real life, knowing how to find the perimeter of a shape is an extremely useful skill. You never know when you'll need to find the distance for fencing around a property, figure out the amount of fabric you need to make a window covering, or need to find the amount of molding needed to go around a room. The worksheet in your *Student Workbook* will give you a chance to practice this helpful skill.

Important Directions for Worksheets

Quite often in real life, we encounter situations where different units come into play. The flooring we want to buy might be sold in yards while our measurements are in feet, a recipe might call for quarts while we can only measure cups, or we might know the acres of some land but need to know the square feet to ascertain how much fertilizer we need.

To help prepare you for using math in real life (as well as in high school science), we'll be including problems in your *Student Workbook* that require you to convert various units **without warning you that the units are different.** Always be sure to read the problem carefully and to check your answer for feasibility — if you didn't convert the units when needed, the answer should seem too high or too low for what you were trying to find.

Your *Student Workbook* problems will also help you expand your problem-solving skills by including more multistep problems that require both arithmetic and geometry to solve. If you have any difficulties, remember to **follow the steps we covered in 2.6 and 4.5.**

15.2 Formulas

In the last lesson, we looked at how to find the perimeter, or distance around, polygons. We also saw how we could simplify the process for many polygons, using multiplication to multiply the sides that are the same lengths.

Now, rather than writing out in *words* the relationship between the sides of a polygon and its perimeter, we could save ourselves time and make the relationship easier to quickly understand at a glance (and easier to work with mathematically) by using *symbols* (such as letters) to stand for the words and operations.

Let's start with the relationship between the perimeter of any polygon and the lengths of its sides. We discussed in the last lesson that the perimeter of a polygon equals the sum of the lengths of each side.

Perimeter of any polygon = sum of the lengths of each side

Now, let's express this relationship using symbols to stand for the words.

$$P = s_1 + s_2 \ldots s_n$$

Here, we used P to stand for perimeter, s_1 for the length of the first side, s_2 for the length of the second side, and $\ldots s_n$ to represent to keep going until we've added up all the sides.

We can also express how to find the perimeter of a regular polygon and rectangle using symbols.

Perimeter of a regular polygon = (number of sides) • (length of each side)

$P = n \cdot s$ or $P = n(s)$ or $P = ns$

P = *Perimeter*
n = *number of sides*
s = *length of a side*

> **Important Notation Note**
> Notice we used the little dot or parentheses to represent multiplication rather than the multiplication sign (x). It's common to avoid using the x when dealing with words or letters, as it can be confused for the letter x.
>
> We also **just put letters next to each other** to show the multiplication ($P=ns$). This is a common way to show multiplication when using letters — it avoids having to write a sign at all! (Do you get the idea we really try to simplify the process as much as possible in math?!)

Perimeter of a rectangle = (2 • length) + (2 • width)

$P = 2 \cdot l + 2 \cdot w$ or $P = 2(l) + 2(w)$ or $P = 2l + 2w$

P = *Perimeter*
l = *length*
w = *width (also sometimes called the height)*

We have now expressed how to find the perimeter of polygons, regular polygons, and rectangles in what we call a **formula** — "a mathematical relationship or rule expressed in symbols."[1] The relationship we're expressing here are the relationships between the perimeter of different polygons and the lengths of their sides.

Using Formulas

To use formulas, all you have to do is substitute the appropriate values for the letters and solve.

Example: Find the perimeter of a regular pentagon when one side is 5 feet long.

The formula for finding the perimeter of a regular polygon is $P = n(s)$.

So all we have to do to find the perimeter is multiply the length of one side by the number of sides in a pentagon (by definition, a pentagon has 5 sides).

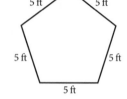

$P = 5(5 \text{ ft})$

Then we can solve to find our perimeter (abbreviated as P in the formula).

$P = 25 \text{ ft}$

Formulas Beyond Geometry

Formulas also apply beyond geometry! In fact, we can use formulas to express all sorts of consistent relationships.

For a simple example, suppose we needed 3 pieces of paper in an art class for every student. We could represent the relationship between paper and students like this:

Total paper needed = 3 • number of students

This can be expressed as a formula using only symbols. We'll use T to represent the total paper needed and s to represent the number of students.

$T = 3 \bullet s$ or $T = 3(s)$ or $T = 3s$

We just wrote a formula for finding the total paper needed — we expressed the relationship using symbols, and we can now use that information to find the total paper needed no matter how many students we end up having in the art class.

This idea of using letters to represent words is not a brand-new concept. When you use letters to represent the relationships in word problems (see 2.6), you are essentially writing a formula!

You'll also discover in science that we use formulas to help us describe the consistent way God holds creation together, from gravity to electricity to the pressure and volume in liquids.

> # Keeping Perspective
>
> In the previous lesson, we thought through in general how a perimeter related to the sides of a polygon, then specifically to regular polygons and rectangles. In this lesson, we expressed those relationships using symbols, making it super easy to just plug the appropriate numbers in to find the perimeter.
>
> Formulas express consistent relationships. They prove useful because of the consistencies all around us. Ultimately, the usefulness of formulas remind us that we live in a consistent universe and serve a consistent, all-powerful God who holds all things together by the power of His Word!

15.3 Area — Rectangles and Squares

Let's continue our exploration of shapes by looking at what we call the **area**, or the space a two-dimensional shape encloses. We'll start our explorations by focusing on rectangles and squares, and then later expand to other shapes.

I know you may already know how to find the area of certain shapes, but let's think through the process together. If you're wondering *when* you'll ever need to find an area, just suppose you need to recarpet your bedroom. You'd need to know the area of the room to figure out how much carpet you need.

Units for Area

Before we look at how to find the area, we need a unit to measure it in. The distance units only measure distance in a single direction, whereas area is expressed in two-dimensional space. So we use what we call a **square unit** to measure area. A square inch is 1 inch wide by 1 inch long (in other words, a square with 1-inch sides), a square foot is 1 foot long by 1 foot wide, a square yard is 1 yard long by 1 yard wide — you get the idea.

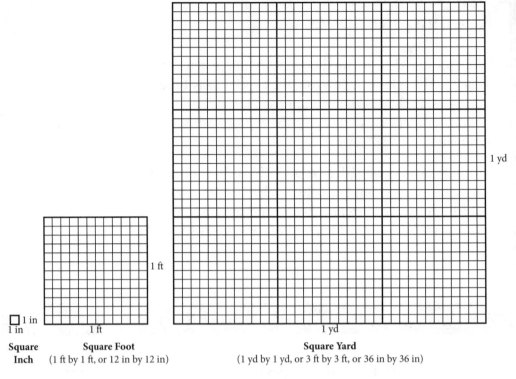

Square Inch   **Square Foot** (1 ft by 1 ft, or 12 in by 12 in)   **Square Yard** (1 yd by 1 yd, or 3 ft by 3 ft, or 36 in by 36 in)

Take a look at the following picture. Notice how we broke up the space inside, or area, of the rectangle into square units. How many squares fit inside the rectangle?

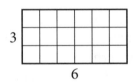

We find that 18 square units fit inside the rectangle above. If each square represented 1 foot wide by 1 foot long, we would say this rectangle has an area of 18 *square feet*. Notice that we use the word "square feet" — the 18 represents 18 1-foot by 1-foot squares.

Finding the Area of a Rectangle

Now that we have a way of expressing area, let's take a look at how to find the area of a rectangle.

Let's say we had a rectangle with a length of 4 units and a height of 6 units. What would the area, or space inside, be?

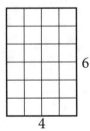

Find out for yourself by counting the squares. You should count 24. This rectangle has an area of 24 square units. If I were to tell you that each unit is 1 square *foot*, you would know that there are 24 square *feet*. However, if each unit is 1 square *inch*, you would have 24 square *inches*.

What would be the area of a rectangle 20 feet long by 16 feet wide/high?

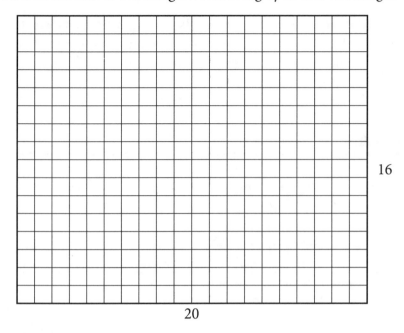

Can you think of a way to find the square units without having to count them? Notice how the length (20) tells us the number of squares in each row, and the width (16) counts the number of rows. So if we **multiply the length by the width**, we'll have counted all the squares! The area of this rectangle is 320 square units, as 20 • 16 = 320.

Notice how this is the same as using multiplication to quickly count any repeated rows — if we lined up chairs so there were 4 to a row and 10 rows, we could add these up by multiplying 10 • 4.

To make things easier, let's represent the area of a rectangle using a formula.

Area of a Rectangle

Area = length • width   or $A = l \cdot w$ or $A = l(w)$ or $A = lw$

A = Area
l = length
w = width (or height)

In other words, **the area of a rectangle equals its length times its width.** It's important to note that sometimes the width of a rectangle is referred to as the height. It doesn't really matter whether you think of the side of a rectangle as the width or the height — the important thing is to know that the area is found by multiplying the two sides together.

Finding the Area of a Square

Since squares are just rectangles (and parallelograms) with all equal sides, we'll be able to find their area by multiplying the length times the width.

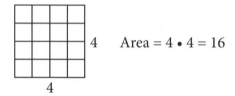

However, since the length and the width are the same, it would be nice to show that all we really have to do is multiply one side by itself.

Area of a Square

Area = side • side   or $A = s \cdot s$ or $A = s(s)$

A = Area
s = side

Application Time

Example: Find the area of a rectangle 10 feet long by 3 feet wide.

> We know that the area equals the length times the width ($A = l \cdot w$), so all we have to do is plug in the numbers and multiply!
>
> A = 10 ft • 3 ft = 30 *square feet* or 30 *sq ft*
>
> Notice that we used square feet to show that the 30 is not just representing a single distance, but rather an *area* measured in 1-foot by 1-foot squares.

Example: Find the area of a square with 5-foot sides.

> We know the area of a square equals the length of a side times itself ($A = s \cdot s$), so all we have to do is plug in the length of our side and multiply!
>
> $A = s \cdot s$
>
> A = 5 ft • 5 ft = 25 *square feet* or 25 *sq ft*

Keeping Perspective — Jesus Used Math!

Since we're learning about measurement, I want to quickly point out that carpenters use measurement and math, and Jesus was a carpenter before He began His public ministry (Mark 6:2–3). Think about that for a minute. The Creator and Sustainer of the universe (including math!) humbled Himself to learn a trade, work with his hands, and likely use math in various ways. Jesus knows what it's like to be a man because He became one — fully man and yet fully God. He totally understands every temptation you face — and He can support you through your temptations.

> *Wherefore in all things it behoved him to be made like unto his brethren, that he might be a merciful and faithful high priest in things pertaining to God, to make reconciliation for the sins of the people. For in that he himself hath suffered being tempted, he is able to succour them that are tempted (Hebrews 2:17-18).*
>
> *For we have not an high priest which cannot be touched with the feeling of our infirmities; but was in all points tempted like as we are, yet without sin. Let us therefore come boldly unto the throne of grace, that we may obtain mercy, and find grace to help in time of need (Hebrews 4:15–16).*

15.4 Area — Parallelograms

Let's expand the information we looked at regarding finding the area of rectangles to parallelograms as a whole.

A rectangle is the name we use to describe a specific type of parallelogram. What if instead of a rectangle, we have the parallelogram below? How do we find its area?

Notice that the parallelogram could be easily rearranged into a rectangle by drawing a perpendicular line and moving the triangle formed to the other side, as shown.

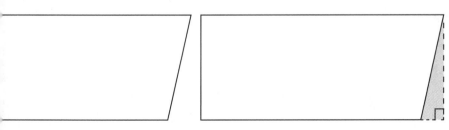

Now, provided we know the length and the height, we could find its area, since we know how to find the area of a rectangle!

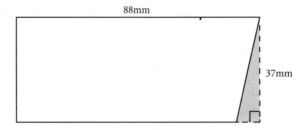

Area = 88 mm • 37 mm = 3,256 sq mm

Rather than redrawing a parallelogram as a rectangle, we could have found the area by simply multiplying the measurements that would form our length and width if we rearranged the parallelogram.

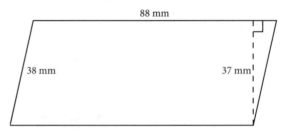

Area = 88 mm • 37 mm = 3,256 sq mm

Notice that we used 37 mm, *not* 38 mm, as our width — 38 mm is the dimension of the slanted side, *not* what would be our width if the parallelogram were rearranged into a rectangle.

To help us remember which sides of a parallelogram to multiply, we're going to use a couple of different terms. We'll call the length the **base** and the measurement of the perpendicular distance from the bottom to the top (i.e., the distance that would be our width if we rearranged the parallelogram to form a rectangle) the **height**.

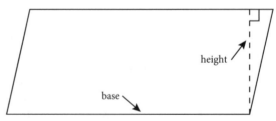

Area of a Parallelogram

   Area = *base* • *height*   or $A = b \cdot h$ or $A = b(h)$ or $A = bh$

   A = Area
   b = base
   h = height

Why Parallelograms?

Now, I know that a parallelogram isn't a shape you run across every day. However, we'll learn in a few chapters how to build on our knowledge of parallelograms to help us find the area of triangles.

Keeping Perspective

You've now learned quite a few formulas. Don't let the multitude of them confuse you. Remember, a *perimeter* is the distance around a closed shape (which is found by adding up the lengths of all the sides), while the *area* is the space inside the closed shape (which is found by finding the number of square units in the shape). The formulas we learned just help us describe the relationships between the perimeter or area and the different parts of polygons.

15.5 Chapter Synopsis and the Basis of Math and Truth

This chapter, we learned how to measure the **perimeter** (outside) of polygons, along with the **area** (enclosed space) of rectangles, squares, and parallelograms. In order to simplify the process so we don't have to think through the problem every time, we used letters to describe the relationships between the perimeter/area and the different parts of polygons (i.e., we used what we call **formulas**).

Here are the formulas we explored:

Perimeter of Any Polygon

> *Perimeter = sum of the lengths of each side*   or $P = s_1 + s_2 \ldots s_n$
>
> *P = Perimeter*
> *s = length of a side*

Perimeter of a Regular Polygon

> *Perimeter = (number of sides) • (length of a side)*   or
> $P = n \cdot s$ or $P = n(s)$ or $P = ns$
>
> *P = Perimeter*
> *n = number of sides*
> *s = length of a side*

Perimeter of a Rectangle

Perimeter = (2 • *length*) + (2 • *width*)   or

P = 2 • *l* + 2 • *w*   or   *P* = 2(*l*) + 2(*w*)   or   *P* = 2*l* + 2*w*

P = Perimeter
l = length
w = width (or height)

Area of a Rectangle

Area = *length* • *width (or height)*   or

A = *l* • *w*   or   *A* = *l*(*w*)   or   *A* = *lw*

A = Area
l = length
w = width (or height)

Area of a Square

Area = *side* • *side*   or   *A* = *s* • *s*   or   *A* = *s*(*s*)

A = Area
s = side

Area of a Parallelogram

Area = *base* • *height*   or   *A* = *b* • *h*   or   *A* = *b*(*h*)   or   *A* = *bh*

A = Area
b = base
h = height

The Basis for Math and Truth

Have you ever thought about how we know something is true? Well, the worldview battle in math is very much a battle for truth. Do we decide truth . . . or does God?

To better understand this battle, let's take a look at two different philosophies regarding the source of truth and at how they apply to math. While there is a lot more that could be said about this topic, hopefully what we cover will give you a glimpse into the worldview battle out there . . . and the solid foundation in God's Word.

Some believe truth is *independent from our experience* — that something is true because it intellectually must be. We call this approach **a priori**, which technically means, "Based on a hypothesis or theory rather than on experiment or experience."[2] In this thinking, math must be true because it makes sense intellectually, regardless of whether it ties with our experience.

Others view truth as something based *completely on experience*, making math true because we experience it to be true. This sort of thinking is known as **a posteriori**, which means "denoting reasoning from facts or particulars to general principles."[3]

Let's bring these terms down to something we discussed this week: the area of a rectangle. How do we *know* the area of a rectangle equals the length by the width?

In *a priori* thinking, we would construct a series of definitions and logical proofs and believe the area of a rectangle equals the product of its sides because we've decided it would and have proven it logically. In *a posteriori* thinking, on the other hand, we would say that experience tells us the area of a rectangle equals the product of its sides, so therefore it must be so.

There's a big problem with both explanations for truth. On the *a priori* side, we have no explanation for why real-life rectangles coincide with what we've "proven" or "decided." On the *a posteriori* side, we can't possibly experience *every* rectangle there will ever be. So how do we know this will always hold true?

Once again, the biblical perspective makes sense out of this difficulty. Even though we cannot experience every rectangle, we know that a consistent, faithful Creator is holding all things together. Thus, we can safely assume that, under the same conditions, rectangles will operate the same way. We also know that the same God created our minds, meaning that we *can* intellectually develop tools to describe His creation and expect them to work in real life. It's so exciting being a Christian — when we start with the Word of God, we find it makes sense out of mysteries that have puzzled scholars for centuries!

As we've discussed before, math is based not on our reasoning or on our experience, but on God. God is truth, and *He* determined and sustains the truth around us.

> *Thy faithfulness is unto all generations: thou hast established the earth, and it abideth (Psalm 119:90).*

> *Jesus saith unto him, I am the way, the truth, and the life: no man cometh unto the Father, but by me (John 14:6).*

[CHAPTER 16]

Exponents, Square Roots, and Scientific Notation

16.1 Introducing Exponents

We discussed in the last chapter how to express area in what we call *square units*. A square foot is a square 1 foot long and 1 foot wide — 9 square feet represents 9 of these square feet.

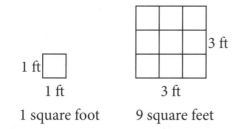

It's time now to take a look at a notation that we can use to help us express these square units more concisely.

Understanding Exponents

Often in math, we deal with multiplying a number by itself multiple times. For example, when finding the area of a square, we're multiplying a number by itself.

Example: Find the area of a square with sides 3 units long.

Answer: **3 • 3 = 9**

We just multiplied 3 by 3.

Exponents are a notation to succinctly write and work with numbers multiplied by themselves. In exponential notation, we use a superscript number to show how many times a number should be multiplied by itself.

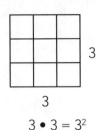

$3 \cdot 3 = 3^2$

Using an exponent, 3 • 3 can be written 3^2. The superscript 2 tells us to multiply the 3 two times.

Reading exponents: We would read 3^2 as "three squared" and refer to finding the answer as "squaring" the number. After all, that's how we would find the area of a square with sides 3 units long!

Extending Exponents

We don't have to stop at squaring a number! Exponents can be used to represent any repeated multiplication. Notice in the chart that **we wrote the number that needs multiplied, and then we wrote a superscript number indicating the number of times that number should be multiplied.**

Notation	Meaning	Read
2^2	2 • 2	Two to the second / two squared *called "squaring" a number*
2^3	2 • 2 • 2	Two to the third / two cubed *called "cubing" a number*
2^4	2 • 2 • 2 • 2	Two to the fourth
2^5	2 • 2 • 2 • 2 • 2	Two to the fifth
2^6	2 • 2 • 2 • 2 • 2 • 2	Two to the sixth
2^7	2 • 2 • 2 • 2 • 2 • 2 • 2	Two to the seventh
2^8	2 • 2 • 2 • 2 • 2 • 2 • 2 • 2	Two to the eighth
2^9	2 • 2 • 2 • 2 • 2 • 2 • 2 • 2 • 2	Two to the ninth
2^{10}	2 • 2 • 2 • 2 • 2 • 2 • 2 • 2 • 2 • 2	Two to the tenth

Term Time
As you continue in math, you may hear people talking about raising a number to a **power**. A power is another name for an exponent — 3 raised to the 4th power is another way of reading 3^4, or 3 • 3 • 3 • 3.

Applying Exponents to Units

When we find an area, we're multiplying a unit by another unit.

    4 ft • **5 ft** = 20 square feet
    3 in • **6 in** = 18 square inches

Now that we have a basic understanding of exponents, we can use them to express square units more simply.

    4 ft • **5 ft** = 20 square feet = **20 ft^2**
    3 in • **6 in** = 18 square inches = **18 in^2**

As with numbers, an exponent to the right of a unit (such as "in^2") represents how many times the unit was multiplied by itself. When we multiply "in" by itself 2 times, we get "in^2".

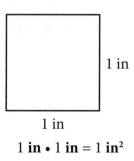

1 in

1 in • 1 in = 1 in²

Using an exponent instead of words to represent square units will help us more easily work with them, as you'll see later this chapter when we explore converting from one square unit to another.

Applying Exponents to Numbers and Units

In 20 ft² and 18 in², the exponent applies only to the units (ft and in), not the numbers (20 and 18). If we wanted to show that 20 ft needed to be squared — that is, 20 ft • 20 ft — we would need to use parentheses, like this: (20 ft)². **The exponent applies to whatever is directly to its left**, whether that's a number, unit, or grouping inside parentheses.

For example, we would find the area of a square 3 ft by 3 ft by finding 3 ft • 3 ft, which would be simplified to (3 ft)². Notice that we put parentheses around 3 ft. The parentheses make it clear that 3 ft needs to be multiplied by 3 ft. Without the parentheses, 3 ft² means 3 square feet, whereas (3 ft)² represents 3 ft • 3 ft, or 9 ft².

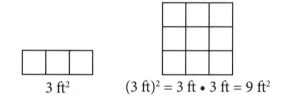

3 ft²   (3 ft)² = 3 ft • 3 ft = 9 ft²

We can also use exponents to simplify formulas.

For example, rather than looking at the area of a square as the area equaling the side times the side ($A = s \cdot s$), we could look at it as the area equaling the side squared ($A = s^2$).

Area of a Square

$Area = side \cdot side$ or $A = s \cdot s$ or $A = s(s)$
$Area = (side)^2$ or $A = s^2$ or $A = (s)^2$

$A = Area$
$s = side$

The exponent is just a shorthand way of representing the multiplication of a side by itself.

Exponents and the Order of Operations

Sometimes, exponents will show up within a problem, such as if we need 20 ft² of carpet for one area, plus we want to carpet an 8-foot square room (8 ft)². Here we have: 20 ft² + (8 ft)².

What do we solve first?

Well, we know we need to multiply before adding, and exponents are really just a way of representing multiplication. We could rewrite the exponent as a multiplication:

20 ft² + 8 ft • 8 ft

And now it's clear what we need to do.

20 ft² + 8 ft • 8 ft = 20 ft² + 64 ft² = 84 ft²

The order of operations are listed here for review. Notice that exponents are listed right after parentheses. This makes sense, since we'd need to simplify the exponents before we can complete the next step: multiplication.

> **Order of Operations**
>
> 1. Solve anything within parentheses first, using the following order (which is the same order we'll follow outside of parentheses):
>
>   a. Solve exponents and roots, from left to right.*
>   b. Multiply or divide, from left to right.
>   c. Add or subtract, from left to right.
>
> 2. Next, **solve exponents** and roots, from left to right.*
>
> 3. Now multiply or divide, from left to right.
>
> 4. Lastly, add or subtract, from left to right.
>
> *There will not be any roots in your problems until they are presented in the next lesson.*

Keeping Perspective

Exponents are a useful way of expressing repeated multiplication. As you focus on learning this part of the mathematical "language," so to speak, don't forget why you're learning it: so you'll be able to describe, understand, and work with real-life problems. The worksheets in the *Student Workbook* for this chapter will give you some simple examples of exponents in action — know that there are thousands of other applications! The more you learn about math and apply it, the more you're going to find yourself working with exponents, as they're an extremely helpful notation.

16.2 Understanding Square Roots

Now that you're familiar with exponents, it's time to take a look at another related notation: square roots.

As we've seen, we call multiplying a number by itself squaring the number.

$$2^2 = 2 \bullet 2 = 4$$

What if we want to find out what number, multiplied by itself, would equal a number? The **square root** (often shown with a $\sqrt{}$ symbol) is a way of concisely asking for the number that, times itself, will equal the number given.

For example, what if we knew the area of a square was 4 ft², but didn't know the length of its sides?

Area = 4 ft²

To find the length of one side of a square when we know the area, we really need to figure out what number times itself will equal the area. This is exactly what the square root of the area will tell us — the number that, times itself, equals the area.

Formula to find the area of a square: *Area = (side)²*   or   *A = s²*

Formula to find the length of each side of a square: $\sqrt{Area}$ *= side*   or $\sqrt{A} = s$

So how do we find the square root of a number? Let's take a look.

> You'll learn in Book 2 how we came up with the formula to find the length of each side of a square. For now, though, just get comfortable with using it.

Finding Square Roots

Think about what we're trying to find when we are finding a square root. We want to find a number that, times itself, equals the given number. In other words, we need to find the *factor* that, times itself, equals the number of which we're finding the square root. Any thoughts as to a technique that can help us find the factors of a number?

Factoring! If we factor a number down to its primes, we can more easily spot the square root.

Example: Find $\sqrt{4}$

Let's start by factoring 4.

$$4$$
$$\wedge$$
$$2 \ \text{x} \ 2$$

Factoring makes it easy to see that the square root of 4 — the number that, times itself, equals 4 — is 2. So the sides of a square whose area is 4 ft² are 2 ft.

16. EXPONENTS, SQUARE ROOTS, AND SCIENTIFIC NOTATION

$$\text{Area} = 4 \text{ ft}^2$$
$$\sqrt{A} = s$$
$$\sqrt{4 \text{ ft}^2} = 2 \text{ ft}$$

Each side is 2 ft. (2 ft • 2 ft = 4 ft²)

Example: Find $\sqrt{36}$

Let's start by factoring 36.

36
3 x 12
3 x 4
2 x 2

36 expressed as the product of its prime factors: 3 • 3 • 2 • 2

This one isn't quite so obvious. But notice that it takes two 3s and two 2s multiplied together to make 36. We're looking for a number that, times itself, equals 36. We can clump prime factors together to find a factor that, times itself, will equal 36. Our answer would be 2 • 3, or 6.

(2 • 3) • (2 • 3) = 36

6 • 6 = 36

$\sqrt{36} = 6$

It's not always necessary to draw a factor tree. Sometimes, you'll know the square root just by thinking through your multiplication facts. In fact, with 4 and 36, you didn't need to draw a factor tree, as you knew 2 • 2 = 4 and 6 • 6 = 36. However, know that factoring can help you find a square root if you get stumped. While we won't show the factoring on the next examples, we could have factored if needed.

Example: Find the length of the side of a square with an area of 64 ft².

We know the square root of the area will equal the length of a side.

$\sqrt{A} = s$

$\sqrt{64 \text{ ft}^2} = s$

$\sqrt{64 \text{ ft}^2} = 8 \text{ ft}$

Each side is 8 feet. (8 ft • 8 ft = 64 ft²)

We can double-check this by finding the area of a square with sides 8 feet long. If we did everything correctly, we should end up with an area of 64 square feet.

area = 64

$A = s^2$

$A = (8 \text{ ft})^2$

$A = 64 \text{ ft}^2$

Example: A square field has an area of 25 square feet. You want to put a fence enclosing the field. This means you need to know the length of each side. What is the length of each side of the square?

Since the length of each side of the square equals the square root of the area (the number that, times itself, will equal the area), we just need to take the square root of 25 to find the length of each side.

$\sqrt{A} = s$

$\sqrt{25 \text{ ft}^2} = s$

$\sqrt{25 \text{ ft}^2} = 5 \text{ ft}$

area = 25

Again, we can double-check this by finding the area for a square with sides 5 feet long. If we did everything correctly, we should end up with an area of 25 square feet.

$A = s^2$

$A = (5 \text{ ft})^2$

$A = 25 \text{ ft}^2$

Note: We could continue on to find the amount of fencing we need by finding the perimeter now that we know the length of each side. If the length of each side is 5 feet and there are 4 sides to a square, the perimeter will be 4 • 5 ft, or 20 ft. So we would need 20 feet of fencing to enclose the square field.

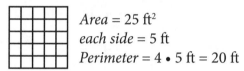

Positive and Negative Square Roots

Technically, the $\sqrt{25}$ could be either 5 *or* -5. We would express this as ± 5. After all, -5 • -5 also equals 25. However, context will often tell us whether we're interested in the positive or negative square root. If there is no context, however, list both.

Example: Find the $\sqrt{16}$

Answer: ± 4

> # Keeping Perspective
>
> As you get into upper-level math, you'll find exponents all over the place. Why? Because *many* real-life consistencies can be represented with repeated multiplication, including interest rates, various aspects of moving objects, and gravity. Because **square roots** reverse squared numbers, they prove quite useful. Most examples of exponents and square roots also involve some algebra and negative numbers, so we'll be holding off on exploring them in more detail until Book 2 . . . and you'll have to wait on other uses until you've learned even more math. But know that you will see exponents and roots again and again. Both exponents and square roots play an important role in our exploration of the consistent way God governs creation.

16.3 Square Unit Conversions

Just as we sometimes need to convert regular units (feet into yards, meters, inches, miles, etc.), we also need to convert squared units. For instance, we might know the area of the tiled sides of a pool in square yards, but need to find it in square feet in order to ascertain how much cleaning agent to use.

Conversions with a Square Conversion Ratio

Squared unit conversions work much the same way as conversions of regular units. If we know the conversion ratio between *squared units*, then any of the methods we've looked at will work.

For example, take a look at this picture of 1 square foot. Notice that it actually contains 144 in². We can use this information to convert between square feet and inches.

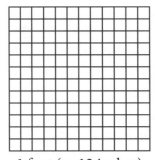

1 foot (or 12 inches)

1 foot (or 12 inches)

1 square foot = 144 in²

Example: Convert 5 square feet to square inches.

Conversion via a Proportion	Conversion via the Ratio Shortcut	Conversion via Mental Math
$\dfrac{5 \text{ ft}^2}{? \text{ in}^2} = \dfrac{1 \text{ ft}^2}{144 \text{ in}^2}$	$5 \text{ ft}^2 \cdot \dfrac{144 \text{ in}^2}{1 \text{ ft}^2} = 720 \text{ in}^2$	$5 \cdot 144 = 720$
Answer: 720 in²	Answer: 720 in²	Answer: 720 in²

Notice that in every method shown, we used a conversion ratio between square feet and square inches. What if we didn't know this ratio? What if all we knew was that 12 inches equals 1 foot?

Rather than memorizing, looking up, or figuring out a whole new set of conversion ratios for square units, let's look at a way to convert square units using the same conversion ratios we use for regular units.

322 | PRINCIPLES OF MATHEMATICS 1

Converting with the Standard Conversion Ratio

If we were dealing with 1 ft instead of 1 *square* foot (1 ft²), we could simply multiply by the conversion ratio ($\frac{12 \text{ in}}{1 \text{ ft}}$) to find our answer.

$$1 \, \cancel{ft} \cdot \frac{12 \text{ in}}{1 \, \cancel{ft}} = 12 \text{ in}$$

As we saw in 14.3, this works since the conversion ratio is worth 1 (both 1 ft and 12 in represent the same quantity) — and multiplying by 1 doesn't change the value of a number (the identity property of multiplication again).

So how can we apply this same thinking to ft²?

Well, think of ft² as ft • ft.

$$1 \text{ ft}^2 = 1 \text{ ft} \cdot \text{ft}$$

Since multiplying by 1 (or a ratio worth 1) doesn't change our value, we could multiply by a *second* conversion ratio, thereby getting a second ft in the denominator and thus canceling out not just feet, but ft • ft.

Example: Convert 1 ft² to square inches

$$1 \text{ ft}^2 = 1 \text{ ft} \cdot \text{ft}$$

$$1 \, \cancel{ft} \cdot \cancel{ft} \cdot \frac{12 \text{ in}}{1 \, \cancel{ft}} \cdot \frac{12 \text{ in}}{1 \, \cancel{ft}} = 144 \text{ in} \cdot \text{in} = 144 \text{ in}^2$$

Do you see the shortcut? We can convert between different square units by simply **multiplying by the appropriate regular conversion ratio twice**.

We didn't have to write out "ft • ft." We just needed to remember ft² means ft • ft, which means we have to multiply by the conversion ratio two times in order to cancel it out.

$$1 \, \cancel{ft^2} \cdot \frac{12 \text{ in}}{1 \, \cancel{ft}} \cdot \frac{12 \text{ in}}{1 \, \cancel{ft}} = 144 \text{ in}^2$$

Example: Convert 121 ft² to square yards.

We need to set up a ratio so the ft² will cancel out. To do this, we need to remember that ft² represents "ft • ft." The ft² will thus cancel out if we multiply by the conversion ratio twice.

$$121 \, \cancel{ft^2} \cdot \frac{1 \text{ yd}}{3 \, \cancel{ft}} \cdot \frac{1 \text{ yd}}{3 \, \cancel{ft}} = \frac{121 \text{ yd} \cdot \text{yd}}{9} = 13.44 \text{ yd}^2$$

Keeping Perspective

The worksheet in the *Student Workbook* will give you a chance to practice with square unit conversions. As you work these problems, think about how, once again, we're continuing to build on what we already know to find simple ways of finding what we need to know. (This is a pattern you'll see over and over and over again in math.) And once again, our method depends on consistencies God created and sustains. For example, if multiplying by 1 (including a fraction worth 1) sometimes changed a quantity's meaning, we couldn't use this method to convert units and expect accurate answers!

16.4 Scientific Notation

Before we close out our exploration of exponents and square roots, let's take a look at an entirely different application of exponents — and one you might not expect! Exponents help us concisely record numbers that would otherwise require an unwieldy number of digits.

For example, the mass of the earth has been calculated at approximately 5,972,200,000,000,000,000,000,000 kg — that's a really lengthy number! Can you imagine trying to write this number often, let alone multiplying or dividing it with other lengthy numbers?

We also encounter lengthy microscopic numbers, such as the rest mass of an electron (approximately 0.000000000000000000000000000000911 kg). Again, imagine trying to keep track of all those zeros!

Thankfully, there's a method for expressing numbers with lots of digits more concisely. This method is called **scientific notation**. Let's take a look at how to describe numbers 1 and greater in scientific notation. (We'll look at how to use it to describe numbers less than 1 in Book 2.)

Expanding the Notation

Before we get started, we need to expand on exponents just a little. So far, we've had exponents such as 2, 3, 4, 5, etc. But exponents can also be other numbers!

Let's take a look at what happens when we use 1 and 0 as exponents.

A number with a 1 as an exponent equals itself — that is, the exponent doesn't change its value at all.

$2^1 = 2 \quad 10^1 = 10 \quad 5^1 = 5$

On the other hand, **a number with a zero as an exponent (^{0}) equals 1.**

$2^0 = 1 \quad 10^0 = 1 \quad 5^0 = 1$

Keep this in mind as we move forward.

Understanding Why

To understand why 0 and 1 give us the answers they do, we need to revisit our old friend the identity property of multiplication again. Remember, multiplying any number by 1 does not change its value. So we could think of any number that's raised to a power as also being multiplied by 1. If we then think of the exponent as how many times we're multiplying 1 by the number to the left of the exponent, we'll have a notation that works for 0 and 1 as well.

$$2^2 = 1 \cdot 2^2 = 1 \cdot 2 \cdot 2 = 4$$

Exponent of 2 means we multiply 1 by the number to the left of the exponent 2 times.

$$2^3 = 1 \cdot 2^3 = 1 \cdot 2 \cdot 2 \cdot 2 = 8$$

Exponent of 3 means we multiply 1 by the number to the left of the exponent 3 times.

Using this train of thought, then,

$$2^1 = 1 \cdot 2^1 = 1 \cdot 2 = 2$$

Exponent of 1 means we multiply 1 by the number to the left of the exponent 1 time.

and

$$2^0 = 1$$

Exponent of 0 means we multiply the number to the left of the exponent 0 times . . . leaving us just 1!

Most people don't bother thinking through why 0 and 1 give the results they do — they just memorize that an exponent of 1 means the number, and an exponent of 0 means 1. However, know that if you were to break down the notation, there's a reason.

Understanding Scientific Notation

To understand scientific notation, we have to review place value. Notice how every place in our place value system can be represented by 10 or 10 raised to a power (i.e., 10 multiplied by itself a certain number of times).

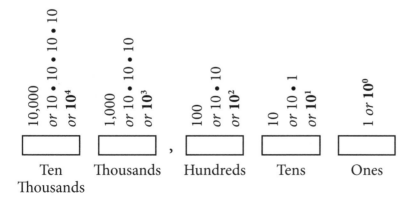

Note: We could represent the decimal places using exponents too, but we'll save that for Book 2.

Knowing this, then, let's take a look at how, rather than using place value, we could represent a number as a multiplication of a power of 10. We'll take 4,000 as a simple example.

When we write 4,000, the place, or location, of the 4 tells us that it represents *thousands*. However, if we didn't want to use place value, we could represent 4,000 as a multiplication: $4 \times 10 \times 10 \times 10$.

We could further simplify this by using an exponent: 4×10^3.

Likewise, 4,900 could be written as $4.9 \times 10 \times 10 \times 10$, which can be expressed with an exponent as 4.9×10^3.

Important! Stop and do the math.

$4.9 \times 10 = 49$
$49 \times 10 = 490$
$490 \times 10 = 4,900$

$4.9 \times 10 \times 10 \times 10 = 4.9 \times 10^3 = 4,900$
4.9×10^3 is representing the **same quantity** as 4,900!

Both 4×10^3 and 4.9×10^3 are written in what we call **scientific notation. Scientific notation is a way to express a number with only one digit to the left of the decimal.** Using it, even numbers with many digits can be expressed with ease. For example, 5,972,200,000,000,000,000,000,000 in scientific notation is 5.9722×10^{24}. Notice that in scientific notation, we don't have to wonder how many zeros there really are or worry we'll forget one — we can easily see the number's place by looking at what power 10 is raised to.

So how exactly do we work with scientific notation? I'm glad you asked! Let's take a look.

Converting to Scientific Notation

To write a number in scientific notation, we just

    1. Move the decimal point so there's only one digit to its left

    2. Multiply by 10 raised to whatever power (i.e., exponent) is needed to put the digits in the appropriate places.

Now that sounds easy enough, but you may be wondering how we know what power (i.e., exponent) of 10 we need to put the digits back in the appropriate places. It's actually quite easy.

When we move the decimal place to the left until there's only one digit to its left, we're really *dividing by a power of 10.*

326 | PRINCIPLES OF MATHEMATICS 1

Dividing by a Power of 10

$$490 \div 10^2 = 490 \div 100 = 4.90$$

4.90.

Moving the decimal to the left 2 places is the same as *dividing by 10²*.

To put the digits in the appropriate place, we need to multiply by the *same power of 10* we divided by when we moved the decimal point to the left. To figure this out, all we have to do is *count how many places we moved the decimal*. If we moved the decimal two places to the left (i.e., divided by 10 two times) we'll have to then multiply by 10 two times — that is, by 10^2. In other words, the power (i.e., exponent) of 10 **will be the same as the number of places we had to move the decimal point.**

$$490 = 4.9 \times 10^2$$

Example: Express 5,972,200,000,000,000,000,000,000 in scientific notation.

Move the decimal point so there's only one digit to the left: 5.9722

Multiply by 10 raised to whatever power (i.e., exponent) is needed to put the digits in the appropriate place values. Since we had to move the decimal point to the left 24 digits (go ahead, count them to see), we need to multiply by 10 raised to the 24th power.

$$5,972,200,000,000,000,000,000,000 = 5.9722 \times 10^{24}$$

Converting from Scientific Notation

To convert a number back from scientific notation to decimal notation, simply complete the multiplication! And again, as we've discussed before, because our place-value system is based on 10, multiplying by 10 or a power of 10 is a matter of moving the decimal to the right.

Multiplying by a Power of 10

$$4.9 \times 10^2 = 4.9 \times 100 = 490$$

4.90.

Moving the decimal to the right 2 places is the same as multiplying by 10^2.

We move the decimal place to the right the *same **number of places as the power of 10!***

Example: Convert 5.9722×10^{24} back to decimal notation.

All we need to do is complete the multiplication by moving the decimal point back 24 places to the right, adding zeros as necessary!

5.9722×10^{24}

$5.9722 \times 1,000,000,000,000,000,000,000,000$

5.972200000000000000000000.

*Important: Notice that we **did not** just add 24 zeros to 5.9722; we **moved the decimal point 24 places**. Since 9722 took up the first four digits, we only added 20 zeros.*

16. EXPONENTS, SQUARE ROOTS, AND SCIENTIFIC NOTATION 327

> # Keeping Perspective
>
> Whew! There sure are a lot of different ways to express quantities, aren't there? Fractions, decimals, percents, ratios, exponents, **scientific notation** — the notations abound. Why? Because God created a complex universe! We find ourselves in need of a variety of different "recording tools" to help us with the different numbers and situations we encounter in life. When we look at incredibly large numbers like the mass of the earth or the distance to the sun, our minds are overwhelmed by the enormity yet intricacy of creation. How much greater is the Creator and Sustainer of it all!

16.5 More Scientific Notation

Before we move on, let's take a quick look at how to compare numbers written in scientific notation, as well as at how to recognize them on the calculator.

Comparing Numbers in Scientific Notation

Return again for a moment to the mass of the earth.

> Earth's mass: 5,972,200,000,000,000,000,000,000 kg

How does that compare to the mass of Venus?

> Venus' mass: 4,867,000,000,000,000,000,000,000 kg

Before you go trying to count all the zeros, let me show you both numbers written in scientific notation:

> Earth's mass: 5.9722×10^{24} kg
> Venus' mass: 4.867×10^{24} kg

Now it's easy to compare them! We can instantly see that the place is the same — they're both being multiplied by 10 raised to the same power (i.e., exponent). Thus, Earth has a greater mass, since 5.9722 is greater than 4.867.

Jupiter, on the other hand, has a mass of approximately 1.89×10^{27} kg. How does its mass compare to earth's?

> Earth's mass: 5.9722×10^{24} kg
> Jupiter's mass: 1.89×10^{27} kg

You might be tempted to think that Jupiter has a smaller mass since 1.89 is less than 5.9722, but first look at the exponents. Notice that 1.89 is multiplied by 10 to the *27th* power, while 5.9722 is only multiplied by 10 to the *24th* power. Thus, 1.89 is representing a number several place values *greater than* 5.9722. This means Jupiter is actually representing a *greater* value!

Jupiter's mass: 1.89×10^{27} kg

$10^{27} = 1,000,000,000,000,000,000,000,000,000$

Jupiter's mass: $1,890,000,000,000,000,000,000,000,000$

Earth's mass: 5.9722×10^{24} kg

$10^{24} = \quad 1,000,000,000,000,000,000,000,000$

Earth's mass: $\quad 5,972,200,000,000,000,000,000,000$

10^{27} represents a greater number than 10^{24}, so Jupiter's mass is actually greater than Earth's.

When comparing numbers in scientific notation, **start by looking at the exponent 10 is raised to**. Remember that a higher exponent of 10 means there are more 0s in the number, making it a greater number.

On the Calculator

Scientific notation isn't just for scientists! Calculators that support scientific notation will automatically convert a number to scientific notation when it runs out of room to show an answer.

For example, if I type 800,000,000 × 90,000 into my calculator, I get 7.2 ^{13}. The "13" off to the right represents the exponent of 10 in scientific notation. In other words, this is the calculator's way of displaying 7.2×10^{13}.

Some calculators or computer programs will put an "E" or "e" (makes sense, since "e" is the first letter in "exponent") before the exponent of 10. On these calculators, 800,000,000 • 90,000 would give a result of 7.2e13 or 7.2E13. Some may even put a positive sign before the exponent (7.2e+13 or 7.2E+13) to show that the exponent is positive (you'll learn about negative exponents in Book 2). Once again, these are all ways of displaying $7.2 \cdot 10^{13}$ on a calculator.

Different Ways to Display 7.2×10^{13}

7.2 ^{13}

7.2e13

7.2E13

7.2e+13

7.2E+13

Scientific Notation and the Distance to the Stars

The closest star to earth, Proxima Centauri, is about 4.3 light years away. Now that may not sound too far, but if we use math to explore this number, we'll discover just how far away that really is — and how great the God who hung this universe together.

In the worksheet in your *Student Workbook*, you'll get a chance to explore more about light years and the size of this universe. To do this, you'll use mainly multiplication, but because you'll be multiplying such large numbers, you'll find your calculator will switch to scientific notation on you!

A light-year is a unit based on the distance we observe light travel in a second here on earth. Although we can see stars billions of light years away, Genesis makes it clear that the earth is only a few thousand years old.

See Appendix A for additional information on starlight and time.

16. EXPONENTS, SQUARE ROOTS, AND SCIENTIFIC NOTATION

329

For example, the distance across the Milky Way Galaxy is estimated to be roughly 100,000 light years. This equates to about 600,000,000,000,000,000 miles, or 6×10^{17} miles in scientific notation.

> I have made the earth, and created man upon it: I, even my hands, have stretched out the heavens, and all their host have I commanded (Isaiah 45:12).

> Behold, the nations are as a drop of a bucket, and are counted as the small dust of the balance: behold, he taketh up the isles as a very little thing (Isaiah 40:15).

How much greater God is than us. Yet this great God knows and cares for each of His children personally.

> I am the good shepherd, and know my sheep, and am known of mine (John 10:14).

By helping us grasp the magnitude of distances like light years, math gives us a better appreciation for the greatness and incomprehensibleness of God, the One who stretched out the distances we cannot even comprehend.

And when we really catch sight of God's greatness, we're forced to fall at His feet as Job did, knowing He knows ever so much more than we could ever attempt to comprehend. Who are we to question His plans?

> Then Job answered the LORD, and said, I know that thou canst do every thing, and that no thought can be withholden from thee. Who is he that hideth counsel without knowledge? therefore have I uttered that I understood not; things too wonderful for me, which I knew not. Hear, I beseech thee, and I will speak: I will demand of thee, and declare thou unto me. I have heard of thee by the hearing of the ear: but now mine eye seeth thee. Wherefore I abhor myself, and repent in dust and ashes (Job 42:1–6).

Keeping Perspective

Scientific notation is just one example of how exponents apply beyond shapes. As you continue to build your mathematical knowledge, you'll learn many others. Exponents are a useful notation in describing and working with many aspects of God's creation.

16.6 Chapter Synopsis

In this chapter, we learned about two new notations: **exponents** and **square roots**. Exponents are simplified ways of writing repeated multiplication, while square roots ask for the number that, times itself, equals the number that's within the square root bracket.

$$2^2 = 2 \cdot 2 = 4$$
$$\sqrt{4} = 2$$

We also explored how to convert between different square units: by following the same principles we use to convert regular units, except **multiplying by the conversion ratio twice** in order to account for the square unit.

$$16 \; m^2 \cdot \frac{100 \text{ cm}}{1 \text{ m}} \cdot \frac{100 \text{ cm}}{1 \text{ m}} = 160,000 \text{ cm}^2$$

And we used exponents to explore yet another way of representing numbers called **scientific notation**. In scientific notation, we represent numbers by placing only one digit to the left of the decimal and multiplying by a power of 10 to show the correct place value.

$$4,900,000,000 = 4.9 \times 10^9$$

We saw that to compare numbers written in scientific notation, we look at the power of 10 first, knowing that a higher power of 10 represents a greater quantity.

$$4.9 \times 10^9 > 6.8 \times 10^8$$

As we learn more and more notations (exponents, square roots, scientific notation, etc.), remember that part of math can be thought of as learning a language — we're mastering agreed-upon notations to help us describe and work with God's creation. As with learning a language, that takes practice! Be patient with yourself and always remember that the reason we need these notations is because we want to be able to effectively use them in the various tasks God brings our way.

Learning math may not always seem important to you right now, but it's an opportunity to practice diligence and faithfulness. Being faithful with the tasks before you today will help prepare you for the tasks before you tomorrow.

He that is faithful in that which is least is faithful also in much: and
he that is unjust in the least is unjust also in much (Luke 16:10).

[CHAPTER 17]

More Measuring: Triangles, Irregular Polygons, and Circles

17.1 Area — Triangles

It's time to return to our study of shapes! Before we took a detour to look at exponents and square roots, we had explored finding the area of rectangles, parallelograms, and squares. Now let's build on that knowledge to find the area of a very common shape: the triangle.

Take a look at this triangle. Any idea how to find its area?

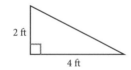

Remember that in math, we try to work from what we know to find what we don't know. Notice how this triangle could be thought of as half of a rectangle.

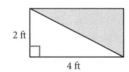

Obviously, each half would have half the area of the whole . . . so our triangle will have half the area of the rectangle!

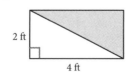

Area of Rectangle: 4 ft • 2 ft = 8 ft²
Area of Triangle ($\frac{1}{2}$ the area of the rectangle):
$\frac{1}{2}$ • 4 ft • 2 ft = 4 ft²   *or*   8 ft² ÷ 2 = 4 ft²

What about triangles like the ones below? How can we find their areas?

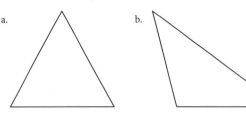

These triangles obviously won't be half a rectangle the same way the right triangle was. But what about other parallelograms besides the rectangle? Or what if the triangle fit inside a rectangle differently? Could these triangles still equal half of a parallelogram? Let's try drawing a parallelogram around each one.

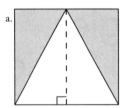
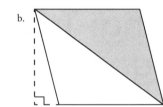
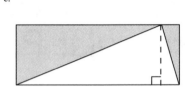

Notice how each triangle is exactly half the parallelogram we drew. While this is easiest to see in diagram b, the same holds true for a and c. (In diagram a and c, the perpendicular line we drew forms two sets of right triangles that are equal to the right triangles in the gray area, which means half the parallelogram is the triangles, and the other half is the gray area.) In all three diagrams, the height of the triangle — the perpendicular line we drew — is the height of the parallelogram.

In Chapter 15, we saw that we can find the area of any parallelogram by multiplying the base times the height.[1]

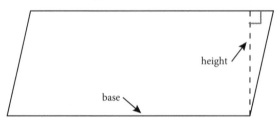

Area of a parallelogram = base • height

Remember, the fraction line means to divide, so
$\dfrac{base \bullet height}{2}$ means base • height *divided by 2*.

So to find the area of these triangles, we just need to find *half* the base times the height, since they represent *half* the parallelogram.

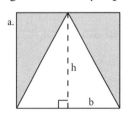
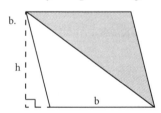
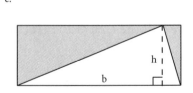

Area of a triangle = $\dfrac{1}{2} \bullet$ **base • height**   or $\dfrac{base \bullet height}{2}$

$A = \dfrac{1}{2} \bullet b \bullet h$ or $A = \dfrac{1}{2}(b)(h)$ or $A = \dfrac{1}{2}bh$

or

$$A = \frac{b \cdot h}{2} \quad \text{or} \quad A = \frac{(b)(h)}{2} \quad \text{or} \quad A = \frac{bh}{2}$$

Note: Remember, when dealing with symbols (including letters), we'll often omit the multiplication sign and write the letters next to each other as a shortcut. We treat letters that are written next to each other or next to a number as a sign to multiply.

> Be careful to use the correct measurement as the height! For height, we need to use the same height we use for parallelograms: the measurement of a line perpendicular to the base going to the top of the shape, *not* the length of a side.

Now, it's important to note that it's not necessary to draw a parallelogram around a triangle to find its area. We just did that to illustrate why taking $\frac{1}{2}$ times the base times the height equals the area of a triangle (because a triangle takes up half the area of a parallelogram with the same base and height). **All we have to do to find the area of a triangle is multiply $\frac{1}{2}$ times the base times the height (i.e. find the base times the height and divide that by 2).**

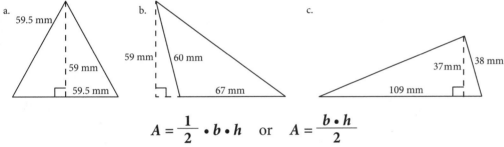

$$A = \frac{1}{2} \cdot b \cdot h \quad \text{or} \quad A = \frac{b \cdot h}{2}$$

Triangle a:

$A = \frac{1}{2} \cdot 59.5 \text{ mm} \cdot 59 \text{ mm} = 1{,}755.25 \text{ mm}^2 \quad \text{or} \quad A = \frac{59.5 \text{ mm} \cdot 59 \text{ mm}}{2} = 1{,}755.25 \text{ mm}^2$

Triangle b:

$A = \frac{1}{2} \cdot 67 \text{ mm} \cdot 59 \text{ mm} = 1{,}976.5 \text{ mm}^2 \quad \text{or} \quad A = \frac{67 \text{ mm} \cdot 59 \text{ mm}}{2} = 1{,}976.5 \text{ mm}^2$

Triangle c:

$A = \frac{1}{2} \cdot 109 \text{ mm} \cdot 37 \text{ mm} = 2{,}016.5 \text{ mm}^2 \quad \text{or} \quad A = \frac{109 \text{ mm} \cdot 37 \text{ mm}}{2} = 2{,}016.5 \text{ mm}^2$

Keeping Perspective — Triangles Everywhere

Even though not too many objects we find in nature look like a triangle, triangles lend themselves well toward helping us explore many aspects of creation. For example, we'll see in the next lesson how they aid in finding the area of various other shapes.

17.2 Area — More Polygons

We can use what we know about the areas of parallelograms and triangles to help us find the area of lots of other shapes! For instance, take a look at the shapes below.

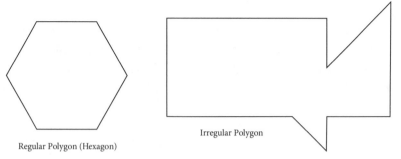

Regular Polygon (Hexagon)   Irregular Polygon

Finding the area of these polygons looks impossible at first glance. However, notice that we could think of both the regular polygon (the hexagon) and the irregular polygon shown as a number of different polygons for which we *do* know how to find the area.

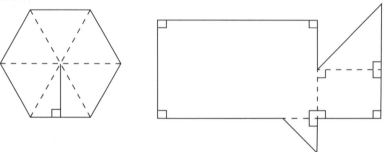

Now we can easily find the area by finding the area of the individual parallelograms and triangles and adding the areas together! Let's try it.

Example: Find the area of this regular hexagon.

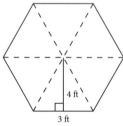

Notice the hexagon, which has 6 sides, can be thought of as 6 triangles. We know the triangles are all the same because we were told the hexagon is regular (to be regular, a hexagon's sides must be the same and its angles must be the same, which would result in the triangles being congruent). All we need to do is find the area of one of the triangles, and then multiply that area by 6 to find the total area.

Area of a Triangle $= \dfrac{1}{2} \cdot b \cdot h$

Area of Triangle $= \dfrac{1}{2} \cdot 3 \text{ ft} \cdot 4 \text{ ft} = 6 \text{ ft}^2$

Area of Regular Hexagon (six sides) $= 6 \cdot 6 \text{ ft}^2 = 36 \text{ ft}^2$

Example: Find the area of this irregular polygon.

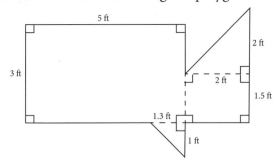

The irregular polygon shown can be thought of as two different parallelograms and two different triangles. We need to find the area of each and then add them together.

Area of large parallelogram = 5 ft • 3 ft = 15 ft^2

Area of small parallelogram = 2 ft • 1.5 ft = 3 ft^2

Area of large triangle = $\frac{1}{2}$ • 2 ft • 2 ft = 2 ft^2

Area of small triangle = $\frac{1}{2}$ • 1.3 ft • 1 ft = 0.65 ft^2

Total Area = 15 ft^2 + 3 ft^2 + 2 ft^2 + 0.65 ft^2 = 20.65 ft^2

Area of Regular Polygons

The same strategy we followed for the regular hexagon works for any regular polygon. To find the area of a regular polygon, multiply the area of the individual triangles formed by the number of those triangles found in the regular polygon. An octagon will have eight triangles, a nonagon nine, a decagon ten, etc.

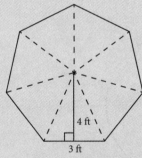
Heptagon (seven sides) = 7 • 6 ft^2 = 42 ft^2

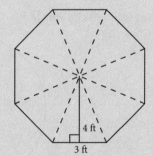
Octagon (eight sides) = 8 • 6 ft^2 = 48 ft^2

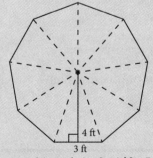
Nonagon (nine sides) = 9 • 6 ft^2 = 54 ft^2

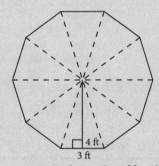
Decagon (ten sides) = 10 • 6 ft^2 = 60 ft^2

Since irregular polygons are irregular, the exact shapes we'll need to use to find the area of each one will vary. The principle is simply that **when we don't know how to find the area of a shape, we think of it as several shapes for which we do know how to find the area.**

Keeping Perspective — When Will I Use This?

Just as some kitchen gadgets (spoons, forks, etc.) are used all the time, while others (the turkey slicer, etc.) might only see use now and then, you'll find yourself using some math concepts frequently and others only now and then. Unless you go into an occupation that requires finding areas regularly, it will probably be one of those "now and then" concepts for you. But that doesn't mean it's not important. Besides knowing how to find an area when needed, you're learning problem-solving skills that will help you in other areas of math (and life) as well.

17.3 Measuring Circles

It's finally time to explore measuring the ever-handy circle. While you may already know some of this information, I think you'll find some interesting new tidbits.

Remembering the Definition

First off, let's review what a circle is.

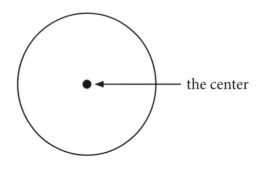

Circle
(closed two-dimensional figure;
each part of the edge is equally distant from the center)

Another way to think of a circle is as "the set of points in a plane that are equidistant from a given point."[2]

Learning the Terms

Because the circle is so unique and useful in describing different aspects of creation, we have different terms to help us describe parts of a circle.

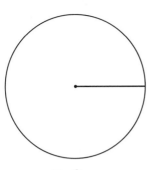 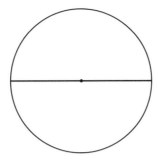

Radius
(distance from the center of a circle to one edge)

Diameter
(distance from one edge of the circle to the other through the center)

With this definition, the **diameter** will always equal 2 times the radius, and the **radius** will always equal $\frac{1}{2}$ the diameter.

$$diameter = 2 \cdot radius \quad \text{or} \quad d = 2 \cdot r$$

$$radius = \frac{1}{2} \cdot diameter \quad \text{or} \quad r = \frac{1}{2} \cdot d$$

You can use this to help you find the radius when you know the diameter, and the diameter when you know the radius! For example, if the diameter is 4 in, the radius will be $\frac{1}{2} \cdot 4$ in or $\frac{4 \text{ in}}{2}$.

With those terms in our tool belt, let's learn how to find the perimeter (called the **circumference** for circles) and the area of circles.

Finding the Circumference/Pi

So how can we find a circle's **circumference**, or the distance around a circle? We can't exactly use a ruler since a circle doesn't have straight edges. But if we *could* stretch the circle into a line, we could then measure the distance.

Try it yourself. Take a piece of string or yarn and use it to outline the circles below, following these steps.

1. Lay string/yarn around each circle, cutting it so it just fits around the circumference. You may need to use tape to help you.

2. Measure each string with a ruler in millimeters. That length is the circumference, or distance around, each circle. Be sure to write down this measurement for each circle. (Measurements: 132 mm, 97 mm, and 162 mm.

Note: Your measurements may not be exactly the same as these, but should be close.)

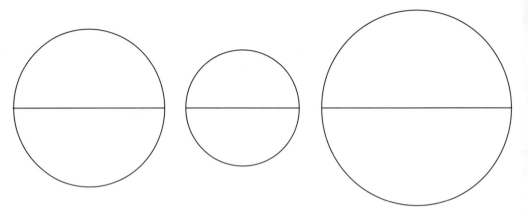

There is a way to find the circumference using math instead of string. To do so, we need to find the exact relationship between the circumference and some other part of the circle that we can easily measure. Let's try seeing if there's a relationship between the circumference of a circle and its diameter.

3. Measure the diameters of the circles above in millimeters. (Measurements: 42 mm, 31 mm, and 52 mm.)

4. Find the ratio between the circumference and the diameter by dividing the circumference you found in Step 2 by the diameter found for that circle in Step 3. While you weren't able to get precise measurements, even the rough ones you took should show that the circumference of a circle is roughly 3 times the diameter. (Ratios: 3.14, 3.13, and 3.12.)

The Greek mathematician Archimedes used triangles, a polygon, angles, and more to help him more accurately find the ratio between the circumference of a circle and its diameter. The picture is one used in his proof.[3] He concluded the ratio was less than $3\frac{1}{7}$ but greater than $3\frac{10}{17}$.

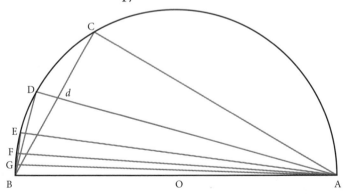

There are complicated formulas and processes used to try to compute the ratio between a circle's circumference and its diameter. All of the attempts fall short, though, to fully expressing this ratio. The actual ratio can't be fully quantified — it's a number that keeps going on and on indefinitely. In other words, the ratio is so precise that we could keep calculating it forever without finding an end! It begins like this:

3.14159265358979323846264338327950288419716939937510582097494459
2307816406286 . . .

Today, we call this number "pi" (pronounced "pie") and use the symbol "π" to represent it. For most practical purposes,[4] we can round pi to 3.14.

$\pi \approx 3.14$

Because we know God is keeping all things together consistently and we know the ratio between the circumference and the diameter of a circle, we can apply that knowledge to help us find the circumference of any circle. We know that if we divide the circumference by the diameter, we get the ratio between the two, which we just saw was π. So if we multiply the diameter by π, we'll find the circumference!

In conclusion, to find the circumference of any circle, we just multiply the diameter of the circle by π. Or, since the radius of a circle is $\frac{1}{2}$ the diameter, we could find the circumference by multiplying the radius by 2 (thus finding the diameter) and then multiplying that by π.

Circumference of a Circle = π • diameter or ***2 • π • radius***

$C = \pi \cdot d$ or $C = \pi(d)$ or $C = \pi d$
$C = 2 \cdot \pi \cdot r$ or $C = 2(\pi)(r)$ or $C = 2\pi r$

Has π Always Been Expressed the Way It Is Today?

Hardly! The symbol π is just a symbol man chose to help express the ratio between the diameter and circumference of a circle — it's actually a Greek letter *p*. It wasn't popularized as a symbol for the ratio between the circumference and the diameter of a circle until the 1700s.[5] On the other hand, the ratio between the circumference and the diameter has been in place since God created the world, just expressed differently over the years.

Example: Find the circumference of a circle with a diameter of 81 feet.

Since the circumference is approximately 3.14 times the diameter, all we need to do is multiply the diameter by 3.14.

$C = \pi \cdot d$

$C = 3.14 \cdot 81 \text{ ft} = 254.34 \text{ ft}$

Example: Find the circumference of a circle with a radius of 20 feet.

$C = 2 \cdot \pi \cdot r$

$C = 2 \cdot 3.14 \cdot 20 \text{ ft} = 125.6 \text{ ft}$

Finding the Area of a Circle

Now that we know how to find the circumference of a circle, let's take a look at how to find the area, or space, inside a circle.

Once again, we need a way to relate the area to some other part of a circle that we can measure. And once again, the work has been done for us. The lengthy process of subdividing the area of the circle into other shapes has already been done, so you won't have to repeat it.[6]

Below is the formula for finding the area of a circle.

Area of a Circle = π • radius • radius

This can be abbreviated to

$A = \pi \cdot r^2$ or $A = \pi(r^2)$ or $A = \pi r^2$

We can use this knowledge of how the area of a circle relates to its radius to find the area of any circle.

Example: What is the area of a circle with a diameter of 6 feet?

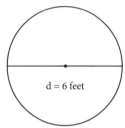
d = 6 feet

Since we only know how the area of the circle relates to its radius, not the diameter, we'll have to first divide the diameter by 2 (i.e., find $\frac{1}{2}$ of it) to find the radius (in this case, 3 ft). Then we can use our formula.

$r = \frac{1}{2} \cdot d$

$r = \frac{1}{2} \cdot 6 \text{ ft} = 3 \text{ ft}$

$A = \pi(r^2)$

$A = \pi(3 \text{ ft})^2 = 28.26 \text{ ft}^2$

Keeping Perspective

Today we explored the ratio between **diameter** and (or **radius** and) the **circumference** (i.e., perimeter) of a circle. We discovered that the circumference is a little over 3 times the diameter but that the actual ratio is a number called **pi** (π) that goes on and on and on. We then applied that knowledge to finding the circumference and area of circles.

We'll learn more about pi in the next lesson, but for now, have fun exploring some real-life circles on the worksheet in your *Student Workbook*. Circles are an incredibly common shape, making exploring them a useful tool!

17.4 Irrational Numbers

Before we move on from our exploration of circles, we need to take a quick detour to look at some additional number sets, or collections, that we use to refer to numbers with different characteristics.

After all, we just learned about pi — a number that, while we abbreviate it 3.14, goes on and on and on and on.

pi = 3.14159265358979323846264338327950288419716939937510582097494459230781640628 6...

This number is different than all the numbers we've dealt with so far. Up until now, all the numbers (including the decimal ones) could be expressed as a ratio of one integer to another. Even 0.33333..., while it goes on and on, could be expressed as $\frac{1}{3}$. But no matter what integers we pick, we could never express pi as a ratio of two integers.

We call numbers that cannot be expressed as a ratio of two integers **irrational numbers**. Irrational numbers never repeat themselves and go on and on for infinity.

In contrast, we call numbers that can be expressed as a ratio between two integers **rational numbers**. All of the other numbers we've looked at are rational numbers. (Notice the "ratio" in **ratio**nal.)

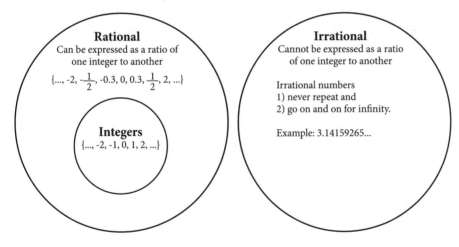

Now I know that the idea of irrational numbers is hard for us to grasp. How does a number go on and on for infinity?

Ultimately, pi and other irrational numbers remind us of our limited knowledge. Why, we can't even describe the exact ratio between a diameter and a circumference! Try as we might, we can never fully calculate pi — it just keeps going and going.

Although our understanding fails, God's understanding is infinite. He knows everything! Psalm 147:5 tells us, "Great is our Lord, and of great power: his understanding is infinite." The infinite nature of irrational numbers gives us a glimpse of our infinite God.

Not only does pi remind us of our limited understanding, but it also points to God's design throughout creation. As you continue exploring math and science, you'll discover that pi proves useful in the most unexpected places, such as with sound waves, general relativity, movements of the heavens, and probability. How can the ratio between the circumference and a diameter help us describe so many aspects of creation? Well, the same God created them all! Just as you can sometimes tell that the same artist painted different pictures, we can see marks of a common designer throughout creation.

Keeping Perspective — A Warning on Pi

Pi is a pretty cool number. It describes the relationship within circles and many other aspects of creation — yet we can't even write it down in its entirety.

Sadly, though, instead of worshiping and standing in awe of the Creator of pi, many people end up standing in awe of the number pi itself. According to one book on pi, "There is almost a cultlike following that has arisen over the concept of π."[7]

It shouldn't surprise us that people would focus on the creation rather than the Creator. The Bible warns us about this in Romans 1:20–23:

> *For the invisible things of him from the creation of the world are clearly seen, being understood by the things that are made, even his eternal power and Godhead; so that they are without excuse: Because that, when they knew God, they glorified him not as God, neither were thankful; but became vain in their imaginations, and their foolish heart was darkened. Professing themselves to be wise, they became fools, And changed the glory of the uncorruptible God into an image made like to corruptible man, and to birds, and fourfooted beasts, and creeping things.*

Let's be sure we praise the Creator when we look at His amazing creation . . . including pi.

17.5 Chapter Synopsis and Pi in the Bible

I hope you had fun this chapter exploring more polygons and circles — and applying your knowledge.

- **Area of Triangles** — We started the week by exploring the area of triangles, discovering that their area is half the area of that of a parallelogram with the same base and height.

$$A = \frac{1}{2} \cdot base \cdot height \quad \text{or} \quad A = \frac{base \cdot height}{2}$$

$$A = \frac{1}{2} \cdot b \cdot h \quad \text{or} \quad A = \frac{b \cdot h}{2}$$

■ **Area of Polygons** — We practiced using what we know about the area of triangles and parallelograms to find the area of other polygons.

■ **Circles** — We looked at the relationships between the **radius, diameter, area,** and **circumference** of a circle. Along the way, we learned about **pi (π)**, the ratio between the diameter and circumference of a circle.

> $diameter = 2 \cdot radius \quad \text{or} \quad d = 2 \cdot r$
> $radius = \frac{1}{2} \cdot diameter \quad \text{or} \quad r = \frac{1}{2} \cdot d$
>
> $Circumference = pi \cdot diameter \quad \text{or}$
> $C = \pi \cdot d \quad \text{or} \quad C = \pi(d) \quad \text{or} \quad C = \pi d$
>
> $Circumference = 2 \cdot pi \cdot radius \quad \text{or}$
> $C = 2 \cdot \pi \cdot r \quad \text{or} \quad C = 2(\pi)(r) \quad \text{or} \quad C = 2\pi r$
>
> $Area = pi \cdot radius \cdot radius \quad \text{or} \quad A = \pi \cdot r^2 \quad \text{or} \quad A = \pi(r^2) \quad \text{or} \quad A = \pi r^2$

■ **Irrational Numbers** — **Pi (π)** introduced us to a whole new type of number — one that goes on and on! Irrational numbers, unlike rational numbers, cannot be expressed as a ratio of two integers.

I hope you had fun in your *Student Workbook* applying the things you learned to various settings, from honeycombs to finding the area of plots of land. As you continue to expand your mathematical knowledge, you'll use these in other situations too.

Pi in the Bible

When describing the molten sea for Solomon's temple, the Bible gives the measurements of both the circumference and diameter of a circular object.

> *And he made a molten sea, ten cubits from the one brim to the other: it was round all about, and his height was five cubits: and a line of thirty cubits did compass it round about (1 Kings 7:23).*

From this text, we see that the diameter (the distance from one brim to the other) was 10 cubits, and the circumference (distance that compassed it round about) was 30 cubits.

C = 30 cubits
d = 10 cubits

Notice that here the circumference is only 3 times the diameter, not 3.14 like we learned in this chapter. Critics of the Bible sometimes point to this and say the Bible got the value for pi wrong, but that's not the case at all.

To start with, engineers round numbers all the time — a precise value is not always needed. Actually, a rounded answer is sometimes considered *more* accurate than a precise one. So the verse could simply contain a rounded number.

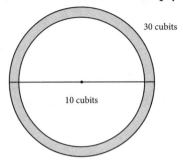

$C = \pi \cdot d$
C = 3.14 • 10
C = 31.4 . . . which rounds to 30 if rounding to the nearest tens place.

Alternately, some argue that the Bible was giving values that adjusted for the thickness of the molten sea, and that these values could show a knowledge of pi to the fourth decimal place, an accuracy that would have been "quite astonishing for ancient times."[8]

You can get this if you assume the sea was 0.225 cubits thick and assume that the Bible gave us the circumference value of the *inside* of the sea and the diameter of the *outside* of the sea. This would make the *inside* diameter 9.55 cubits, as

10 cubits – 0.225 cubits – 0.225 cubits = 9.55 cubits.

Using this value and 3.14 for π, you would get an inside circumference of almost exactly 30 cubits.

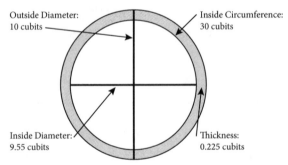

$C = \pi \cdot d$
C = 3.14 • 9.55
C = 29.987, which is nearly exactly 30

Note: The width may have been slightly different than 0.225. In his article "Contradictions: As Easy as Pi," Dr. Jason Lisle uses a value of 0.25 cubits thick on both sides for the sea, making the inside diameter 10 – 0.25 – 0.25, or 9.50 cubits thick. Yet even if you take this larger thickness, the circumference again is nearly 30 cubits.[9]

3.14 • 9.50 cubits = 29.83 cubits

The point? You can trust what the Bible says in every detail. God gave us the information we needed to know within its pages.

[CHAPTER 18]

Solid Objects and Volume

18.1 Surface Area

Now that you know how to find the area of various polygons and circles, let's put that knowledge to use in finding what we call the **total surface area**.

Consider a toaster oven again. What if we want to find the area of *all* the sides of the toaster oven: the front, two sides, top, bottom, and back? We could view *each* side as a rectangle, find the *area* of each rectangle, and *add* all six areas together.

 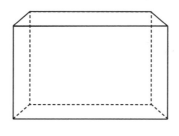

When we are finding the area of all the different sides (i.e., surfaces) of a solid object, we refer to it as finding the **total surface area**.

Total surface area = area of all the sides (i.e., surfaces) of a solid object

Any guesses on how we would find the total surface area of a book? You got it: we would treat each side as a separate rectangle, find the area of each side, and then add up all the areas, remembering to include the sides of the book hidden from view.

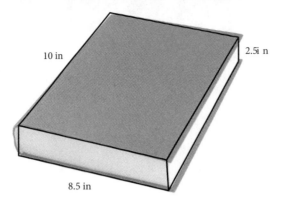

Surface area of top = 10 in • 8.5 in = 85 in^2
Surface area of bottom (same as top) = 85 in^2
Surface area of long side = 10 in • 2.5 in = 25 in^2
Surface area of long side on other side of book = 25 in^2
Surface area of short side = 2.5 in • 8.5 in = 21.25 in^2
Surface area of short side on other side of book = 21.25 in^2
Total surface area =

85 in^2 + 85 in^2 + 25 in^2 + 25 in^2 + 21.25 in^2 + 21.25 in^2 = 262.5 in^2

Note that we could have used multiplication to make this simpler:

2 • 85 in^2 + 2 • 25 in^2 + 2 • 21.25 in^2 = 262.5 in^2

Total surface area is no different than what you have been doing when finding the area, just applied to all the different sides, or surfaces, of an object.

Why Surface Area?

Sometimes, we need to find the surface area of just some sides, or surfaces, of an object. As one example, suppose you were trying to purchase paint for a shed. If you find the surface area of the paintable section (you wouldn't need to include the floor of the barn or the roof, as you don't need to paint those), you'll be able to figure out how much paint to buy.

Whether you find the total surface area or just the area of a few surfaces will depend on what you're trying to do. If you only want to paint the side of a shed, for instance, you would just want the surface area of the side. If, on the other hand, you wanted to paint the entire outside of the shed, you'd need to find the surface area of all the paintable outside areas.

PRINCIPLES OF MATHEMATICS 1

Keeping Perspective — Area and Flight

We've spent a lot of time lately learning how to find the area of shapes. Let's step back for a minute and take a look at how area helps us better appreciate God's design. We'll look specifically at birds and flight.

Grab two sheets of the same type of paper. Crinkle one and leave the other flat. Now drop both sheets. Even though both papers were the same weight, the flat paper should have taken a lot longer to fall to the ground. It took longer because it had a larger surface area.

Now take a look at the hummingbird. The hummingbird has wings with small surface areas, making maneuvering to pollinate flowers possible. Those small areas also means it has to flap its wings continuously to stay aloft, but God gave hummingbirds a heart capable of pumping up to 1,260 times a minute. Not surprisingly, these little birds have an enormous appetite, causing them to pollinate many flowers every day.

On the other hand, God designed vultures and other birds of prey very differently. Their wings have large surface areas, making it possible to soar effortlessly overhead. He also gave them powerful muscles that allow them to get those great wings airborne.

By understanding area, we get a glimpse of the loving design God placed within birds. While we see only a fallen version of His design (there was no death before man's sin, so vultures wouldn't have eaten other animals), we nonetheless see God's incredible care and wisdom. And remember — the same God who designed each bird so perfectly created you.

> *For thou hast possessed my reins: thou hast covered me in my mother's womb (Psalm 139:13).*

You are no accident. Throughout history, God has used different types of people to accomplish His purposes, transforming their weaknesses into strengths by His grace. God didn't make us all the same, just as He didn't make all birds the same. He created each one of us uniquely. And just as He gave birds all they needed to accomplish the tasks He gave them, in Jesus, He's given us all we need. Every day, you have the opportunity to show to the world that God's grace is sufficient to walk victoriously in your set of circumstances as no one else can do, since there is only one you!

> *For as the body is one, and hath many members, and all the members of that one body, being many, are one body: so also is Christ. . . . Now ye are the body of Christ, and members in particular (1 Corinthians 12:12, 27).*

18.2 Volume

It's time to explore a new aspect of measuring shapes: volume. **Volume** is the term we use to describe the space inside three-dimensional, or solid, objects. We'll focus our explorations on prisms and cylinders.

We're going to look at how to find the volume, or space inside, of both prisms and cylinders. Remember, **prism** and **cylinder** are names we use to describe specific three-dimensional, or **solid**, objects as a whole (13.4).

Prism
(bases are polygons; both bases are the same and parallel; sides are parallelograms)

Cylinder
(bases are circles; both bases are the same and parallel)

Before we look at finding the volume, let's take a look at units to describe volume.

Cubic Units

We use what we call a **cubic unit** to measure volume. Unlike a square unit that has two dimensions, a cubic unit has *three* dimensions: length, width, and height.

A cubic unit is 1 unit wide by 1 unit long by 1 unit high.

1 Cubic Unit

So a volume of 4 cubic units would be able to contain 4 of these 1 by 1 by 1 cubes.

4 Cubic Units

Finding Volume

Consider this box, which is a **rectangular prism**.

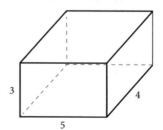

Notice that it's a rectangle with height. We know how to find the area of the rectangular base.

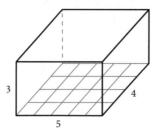

Area of base = 5 • 4 = 20

If we multiply this area by the height, we'll know the volume! Note that to do this, we'll use cubic units (tiny three-dimensional cubes) rather than the two-dimensional squares we used for area.

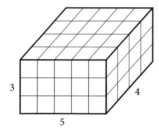

Volume = (4 • 5) • 3 = 60

When we find volume, we are multiplying *three* times and measuring the number of unit *cubes* that fit inside the solid object.

Assuming the dimensions in the prism above are in inches, we would have

5 in • 4 in • 3 in = 60 in • in • in = 60 in³

In other words, this prism holds 60 cubes that are 1 in by 1 in by 1 in. We thus say it has a volume of 60 in³, which is read "60 inches *cubed*."

If, on the other hand, the dimensions of the prism above were in feet, it would have a volume of 60 ft³.

5 ft • 4 ft • 3 ft = 60 ft • ft • ft = 60 ft³

Notice that **we used exponents to abbreviate "in • in • in" to in³ and "ft • ft •ft" to ft³**. Volume is measured by a *three*-dimensional unit (one with length, width, and height) rather than by a two-dimensional unit, and thus we end up multiplying *three* distances to find it.

> ### It's Cubed
>
> While we're looking at cubic units, let's talk about a cube. As we saw in 13.4, a **cube** is basically a prism with sides that are all the same-sized squares. So in a cube, the length, width, and height are all the same. Not surprisingly, a *cubic* unit is a *cube* with sides 1 unit long!
>
>
>
> When we first studied exponents, we learned that 2^3 is read "two cubed." Can you now guess why? The name makes sense, as 2^3 represents the volume of a cube with sides of 2 units.
>
>
>
> Volume = 2 • 2 • 2, or 2^3

A Formula for Volume

Notice that to find the volume, we found the area of the base, and then multiplied that by the height. We can use this process to find the volume of any prism or cylinder.

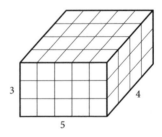

Volume = (4 • 5) • 3 = 60

Volume of a Prism or Cylinder

Volume = area of base • height   or   ***V = B • h***

V = Volume
B = area of the base
h = height

Let's talk about these terms for a minute, as terms can have different meanings in math. The important thing is to understand the *concept(s)* the terms are describing.

Previously, we've used the word **base** to mean the length, or bottom, of a parallelogram or triangle. Now here, we're using it to describe one of the sides of the prism or cylinder that forms the basis for the object (i.e., one of the two identical, parallel sides).

What about height? Well, we're using **height** to describe the height or width for which the shape of the base continues.

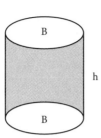
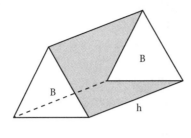

Which Side Is the Base?

Notice that in the case of a rectangular prism, we could think of any side as its base, so long as we view the other dimension as the height.

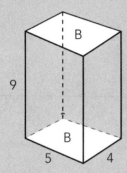
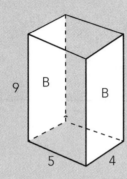
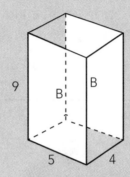

$Volume = (5 \cdot 4) \cdot 9$ $Volume = (9 \cdot 4) \cdot 5$ $Volume = (9 \cdot 5) \cdot 4$
$= 20 \cdot 9 = 180$ $= 36 \cdot 5 = 180$ $= 45 \cdot 4 = 180$

On the other hand, in a triangular prism, we have to view the triangle as the base, or we wouldn't end up being able to calculate the volume using our formula, since the rest of the object would not be the same shape as our base.

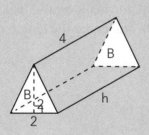
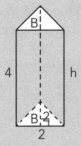

$Volume = (\frac{1}{2} \cdot 2 \cdot 2) \cdot 4 = 8$ $Volume = (\frac{1}{2} \cdot 2 \cdot 2) \cdot 4 = 8$

When picking a base, make sure you pick a side that continues uniformly through the rest of the object.

Example: Find the volume of the rectangular prism shown.

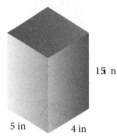

$V = B \cdot h$

In this case, our base is a rectangle, so the area equals the length times the width, or 5 in • 4 in. So our volume equals (5 in • 4 in) • 15 in = 300 in³.

Notice that since our sides are also rectangles, we could have viewed any side as our base. We would get the same answer no matter what base we chose.

(15 in • 5 in) • 4 in = 300 in³

(15 in • 4 in) • 5 in = 300 in³

(4 in • 5 in) • 15 in = 300 in³

Example: Find the volume of a triangular prism whose base is a right triangle and dimensions are shown.

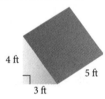

$V = B \cdot h$

In this case, our base is a triangle. We need to first find the area of the triangle. When finding the area of the base, ignore the prism altogether and find the area of the triangle as you normally would. Note that this triangle has a right angle, so its height will equal the length of one of its sides.

Area of base = $\frac{1}{2} \cdot b \cdot h$

Area of base = $\frac{1}{2} \cdot$ 3 ft • 4 ft = 6 ft²

Volume = 6 ft² • 5 ft = 30 ft³

Example: Find the volume of a cylinder whose base has a diameter of 5 feet and that is 15 feet tall.

$V = B \cdot h$

In this case, our base is a circle, so we need to first find the area of the circle.

Area of base $= \pi \cdot r^2$

Area of base $= \pi(2.5 \text{ ft})^2 = 19.63 \text{ ft}^2$

Note that we found the radius by taking $\frac{1}{2}$ of the diameter ($\frac{1}{2} \cdot 5 \text{ ft} = 2.5$).

Volume $= 19.63 \text{ ft}^2 \cdot 15 \text{ ft} = 294.45 \text{ ft}^3$

Remember, π can be approximated as 3.14.

Customizing the Formula

In other books, you may find the formula for volume customized based on the type of base. For example, you might be familiar with finding the volume of a rectangular prism by multiplying the length times the width times the height:

$V = l \cdot w \cdot h$

Basically, these more specific formulas just replace B (which represents the area of the base) with how to find the area of that specific type of base.

Volume of any prism or cylinder:

$V = B \cdot h$

Volume of a rectangular prism:

$V = (l \cdot w) \cdot h$
Replaced B with how we would calculate the area of the base for a rectangle.

Volume of a cylinder:

$V = (\pi \cdot r^2) \cdot h$
Replaced B with how we would calculate the area of the base for a circle.

The first formula applies to any prism or cylinder, while the next one applies only to a *rectangular* prism and the last one only to a *cylinder*. The basic principle is that you find the volume by multiplying the area of the base times the height.

Other Solid Objects

How would we find the area of a pyramid? Notice that we can't just multiply the base times the depth, as the object does not have two parallel bases. A pyramid is not uniform all the way through (it tapers to a point); thus, it is not a prism! You'll learn how to find the volume of this type of solid object later in math; just remember that the method of finding volume you studied in this lesson works only if the object has two parallel bases and a uniform area the whole way through.

18. SOLID OBJECTS AND VOLUME | 355

Keeping Perspective

We've now covered how to find the perimeter (distance around a two-dimensional shape), the area (space inside a two-dimensional shape), and the **volume** (space inside a three-dimensional shape). We use regular units for a single dimension, squared units for two dimensions, and **cubic units** for three dimensions.

Notice that in finding a volume, you're employing other math "tools," such as multiplication and exponents. Once again, we're combining "tools" to help us.

And why are we combining tools and learning to find the volume of shapes? Because finding volume proves quite useful! From a freezer to a shipping box to a fish tank, the skills you're learning can help you find the volume of real-life objects.

18.3 Cubic Unit Conversion

Cubic unit conversion works much the same way as regular or squared unit conversion. Any of the methods we've looked at will work if we know the conversion ratio between *cubed units*.

For example, take a look at this picture of 1 cubic foot. Notice that it actually contains 1,728 in^3!

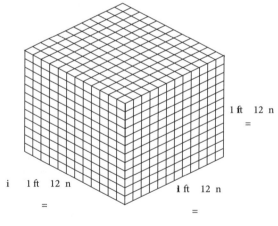

12in • 12 in • 12 in
1 ft^3 = 1,728 in^3

Knowing this, we can easily convert feet to inches or inches to feet. For example, the volume of this prism is 4 ft • 5 ft • 3 ft, which equals 60 ft^3. How many cubic inches does it represent?

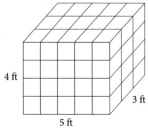

356 | PRINCIPLES OF MATHEMATICS 1

Conversion via a Proportion	Conversion via the Ratio Shortcut	Conversion via Mental Math
$\dfrac{60 \text{ ft}^3}{? \text{ in}} = \dfrac{1 \text{ ft}^3}{1{,}728 \text{ in}^3}$	$60 \text{ ft}^3 \cdot \dfrac{1{,}728 \text{ in}^3}{1 \text{ ft}^3} = 103{,}680 \text{ in}^3$	$60 \cdot 1{,}728 = 103{,}680$
Answer: 103,680 in³	Answer: 103,680 in³	Answer: 103,680 in³

Conversion with Regular Conversion Ratios

Is there a way to convert cubic feet to inches knowing only that 12 in equals 1 ft?

Yes, there is! Remember, ft³ is another way of writing ft • ft • ft. Thus, if we just multiply by $\dfrac{12 \text{ in}}{1 \text{ ft}}$ *three* times, we'll successfully convert our units, as all the ft will cancel out.

$$60 \text{ ft} \cdot \text{ft} \cdot \text{ft} \cdot \frac{12 \text{ in}}{1 \text{ ft}} \cdot \frac{12 \text{ in}}{1 \text{ ft}} \cdot \frac{12 \text{ in}}{1 \text{ ft}} = 103{,}680 \text{ in}^3$$

In the process of canceling out the ft • ft • ft, notice that we ended up with three multiplications of in (we had 12 *in* • 12 *in* • 12 *in*), which gave us an answer in *in³*. Our answer of 103,680 in³ means 103,680 in • in • in!

Just as there's no need to write out in • in • in, there's also no need to write out ft • ft • ft — we just need to know that's what ft³ means and multiply by our conversion ratio three times to cancel it out.

$$60 \text{ ft}^3 \cdot \frac{12 \text{ in}}{1 \text{ ft}} \cdot \frac{12 \text{ in}}{1 \text{ ft}} \cdot \frac{12 \text{ in}}{1 \text{ ft}} = 103{,}680 \text{ in}^3$$

> Note: We know 60 • 1,728 is a lengthy multiplication to do mentally, but the process would work if we could do it.

The basic principle of unit conversion is to multiply by a conversion ratio (or ratios) as many times as needed.
When dealing with squared or cubed units, keep in mind what an exponent means. Then it will be easy to figure out how many times to multiply to properly convert the unit.

Example: Convert 2 ft² to in².

$$2 \text{ ft}^2 \cdot \frac{12 \text{ in}}{1 \text{ ft}} \cdot \frac{12 \text{ in}}{1 \text{ ft}} = 288 \text{ in}^2 \quad or \quad 2 \text{ ft} \cdot \text{ft} \cdot \frac{12 \text{ in}}{1 \text{ ft}} \cdot \frac{12 \text{ in}}{1 \text{ ft}} = 288 \text{ in}^2$$

Remember, ft² = ft • ft.

Example: Convert 2 ft³ to in³.

$$2 \text{ ft}^3 \cdot \frac{12 \text{ in}}{1 \text{ ft}} \cdot \frac{12 \text{ in}}{1 \text{ ft}} \cdot \frac{12 \text{ in}}{1 \text{ ft}} = 3{,}456 \text{ in}^3 \quad or \quad 2 \text{ ft} \cdot \text{ft} \cdot \text{ft} \cdot \frac{12 \text{ in}}{1 \text{ ft}} \cdot \frac{12 \text{ in}}{1 \text{ ft}} \cdot \frac{12 \text{ in}}{1 \text{ ft}} = 3{,}456 \text{ in}^3$$

Remember, ft³ = ft • ft • ft.

> # Keeping Perspective
>
> By now, you might be wondering why unit conversion keeps coming up. Is it really that important to know, especially in today's society, where computers can convert units for us?
>
> First of all, there may be times where you need to convert units without a computer near you. Second, knowing how to do it on your own will help you understand the conversions and make sure you're asking and receiving the data you need. But above all, the skills you're learning in unit conversion are going to serve you well in algebra, which is a very powerful tool for business, science, and just about every other field.

18.4 Measuring Capacity in the U.S. Customary System

Suppose you went to the store and saw this sign:

You'd probably scratch your head and wonder what had happened! Rather than trying to find the base and height of a milk container (which could be difficult to find, especially if it had a handle or odd shape), it's easier to measure a liquid (like milk) based on how it compares to a predetermined unit of volume, such as a gallon, quart, etc. In other words, rather than measuring the space it takes up, we can measure that space based on its **capacity** (i.e., "the maximum amount that something can contain"[1]) as compared to predetermined units of liquid.

Let's take a look at the units for measuring liquid capacity and dry goods capacity in the U.S. Customary System.

U.S. Customary Liquid Capacity Units[2]

The chart shows different units in the U.S. Customary System for measuring liquid capacity. You likely already know many of these units, but take a moment to familiarize yourself with them. The bolded ones are ones you especially need to know, as you'll encounter them often in everyday life.

Historical Note[3]: The gallon we use in the U.S. today comes from what was historically called the "wine gallon"; it represents a volume of 231 in^3.

1 teaspoon (tsp)
3 tsp = 1 Tablespoon (Tbsp)
16 Tbsp = 1 cup (c)
2 c = 1 pint (pt)
2 pt = 1 quart (qt)
4 qt = 1 gallon (gal)

There's one other unit for measuring liquid in the U.S. Customary System that you've likely heard of: the fluid ounce. Take a look at how the fluid ounce compares with the other units for liquid measure. The squiggly equals sign means the comparison is approximate — 2 Tbsp doesn't equal exactly 1 fluid ounce.

2 Tbsp ≈ 1 fluid ounce
8 fluid ounces (fl oz) = 1 cup (c)
16 fl oz = 1 pint (pt)
32 fl oz = 1 quart (qt)
128 fl oz = 1 gallon (gal)

Because we use the word "cup" loosely at times to refer to a drinking cup rather than an official cup, directions will often specify the number of fluid ounces that need to be consumed. For example, when taking some medicines, you need to drink at least 8 fluid ounces of water — depending on the size drinking glass (i.e., "cup") you have, that could be more or less than the "cup" will hold.

> While we'll often refer to fluid ounces as simply "ounces," they are not the same as regular ounces. Fluid ounces are a unit used to measure liquid volume, while regular ounces are a unit used to measure weight.

Units Beware

The British System, while it uses many of the same names, differs in some aspects from the definitions above. For example, the British Imperial Gallon is slightly larger than the U.S. Gallon. Different definitions for units remind us that these names are just tools to help us compare and describe quantities! In this course, you can assume gallon and other units refer to those in the U.S. Customary System, not the British System.

U.S. Customary Dry Capacity Units

Dry goods take up different space than liquids, so it's helpful to have different units for measuring them. And we do!

Notice that we use the units "pint" and "quart" again. It's important to note, though, that a pint represents a larger capacity when it's representing dry goods than when it's representing liquid. The units may have the same name, but they do not represent the same capacity. Unless the problem specifically states otherwise, you can **assume that pint and quart in this course refer to the liquid units.**

2 pints (pt) = 1 quart (qt)
8 quarts = 1 peck (pk)
4 pecks = 1 bushel (bu) / 32 quarts

Unit Conversions

We can convert between capacity units using the same three methods we have been using for other units: via a proportion, via the ratio shortcut, or via mental math. Once again, though, you'll find the ratio shortcut easier than the other two methods when the conversion requires more than one step.

Example: Convert 6 cups to pints.

Conversion via a Proportion	Conversion via the Ratio Shortcut	Conversion via Mental Math
$\dfrac{1 \text{ pt}}{2 \text{ c}} = \dfrac{?}{6 \text{ c}}$	$\overset{3}{\cancel{6} \cancel{c}} \cdot \dfrac{1 \text{ pt}}{\underset{1}{\cancel{2} \cancel{c}}} = 3 \text{ pt}$	$6 \div 2 = 3$
Answer: 3 pt	Answer: 3 pt	Answer: 3 pt

Example: Convert 2 gallons to pints.

Since we don't know the conversion ratio between gallons and pints, we'll have to first convert to quarts, and *then* to pints. **Notice that we can do this all in one step using the ratio shortcut method.**

Conversion via a Proportion	Conversion via the Ratio Shortcut	Conversion via Mental Math
$\dfrac{1 \text{ gal}}{4 \text{ qt}} = \dfrac{2 \text{ gal}}{?}$	$2 \cancel{\text{ gal}} \cdot \dfrac{4 \cancel{\text{ qt}}}{1 \cancel{\text{ gal}}} \cdot \dfrac{2 \text{ pt}}{1 \cancel{\text{ qt}}} = 16 \text{ pt}$	$2 \cdot 4 \cdot 2 = 16$
? = 8 qt	Answer: 16 pt	Answer: 16 pt
Now to convert from quarts to pints:	Notice how much easier this was!	The more steps to a conversion, the harder it is to keep track of mentally. Be careful when converting mentally!
$\dfrac{2 \text{ pt}}{1 \text{ qt}} = \dfrac{?}{8 \text{ qt}}$		
Answer: 16 pt		

Keeping Perspective — Multiple Tools

We've now explored many different ways to measure a three-dimensional shape: by describing a space's dimensions or volume, by describing the liquid a space could contain (i.e., its **liquid capacity**), and by describing the dry goods a space could contain (i.e., its **dry capacity**). And we've used a variety of different units in the process.

As you continue studying in both math and science, you'll learn about many other measurement units. Don't let all the new names and all the different units of measurement confuse you — each one is a tool that proves useful in different situations! God gave man the task of subduing the earth (Genesis 1:28), and math — including measuring systems — helps us in that process.

PRINCIPLES OF MATHEMATICS 1

18.5 Conversion to and from Cubic Units

What happens if you know the dimensions of a swimming pool in feet, and want to know how many *gallons* of water it will hold? Or if you want to find a container to hold a *bushel* of apples?

It's sometimes necessary to convert between liquid, dry, and cubic units. The chart shows a few conversion ratios. Knowing these ratios, you can convert between the units as you have between other units.[4]

Liquid Customary to Cubic

1 pint = 28.875 in^3
1 quart = 57.75 in^3
1 gallon = 231 in^3

Dry Customary to Cubic

1 quart = 67.2006 in^3
1 bushel = 2,150.42 in^3

Example: Convert 90 in^3 to gallons.

Conversion via a Proportion	Conversion via the Ratio Shortcut	Conversion via Mental Math
$\dfrac{90 \text{ in}^3}{?} = \dfrac{231 \text{ in}^3}{1 \text{ gal}}$	$90 \text{ in}^3 \cdot \dfrac{1 \text{ gal}}{231 \text{ in}^3} = 0.39 \text{ gal}$	$90 \div 231 = 0.39$
Answer: 0.39 gal	Answer: 0.39 gal	Answer: 0.39 gal

Example: Convert 15 in^3 to cups.

Since we don't know the conversion ratio between cups and cubic inches, we have to first convert to a ratio we do know (or figure out the conversion ratio). We'll convert to pints and *then* to cups.

Conversion via a Proportion	Conversion via the Ratio Shortcut	Conversion via Mental Math
$\dfrac{15 \text{ in}^3}{?} = \dfrac{28.875 \text{ in}^3}{1 \text{ pt}}$	$15 \text{ in}^3 \cdot \dfrac{1 \text{ pt}}{28.875 \text{ in}^3} \cdot \dfrac{2 \text{ c}}{1 \text{ pt}} = 1.04 \text{ c}$	$15 \div 28.875 = 0.52 \text{ pt}$
? = 0.52 pt	Answer: 1.04 c	
Now to convert from pints to cups:	Notice how this method allowed us to find the answer in one step!	Now convert from pints to cups:
$\dfrac{0.52 \text{ pt}}{? \text{ c}} = \dfrac{1 \text{ pt}}{2 \text{ c}}$		$0.52 \cdot 2 = 1.04$
Answer: 1.04 c		Answer: 1.04 c

Notice that you might need paper or a calculator to solve using the Conversion via Mental Math method on these examples.

Keeping Perspective

Notice how we keep applying the same principles over and over again to new situations. You're now using the same concepts of unit conversion you have been all along to convert between cubic and liquid/dry good units. Conversion skills will help you convert between any comparable units you encounter!

18.6 Measuring Capacity in the Metric System

Before we move on, let's also take a look at how to measure capacity using the Metric System. Remember, much of the world uses the Metric System, so it's important to become familiar with it.

Metric Liquid Capacity Units

As was the case with measuring distance in the Metric System, liquid metric units are all based on 10 — each liquid unit in the Metric System is worth 10 of the previous unit. And just as the metric distance units centered around the meter, metric liquid units center around the **liter**. Notice that liters can be abbreviated with either a lowercase l or a capital L.

> The common units are in bold — they are only the units you will be asked to know in this course.

10 milliliters (ml *or* mL) = 1 centiliter (cl *or* cL)
10 centiliters/100 milliliters = 1 deciliter (dl *or* dL)
10 deciliters/1,000 **milliliters** = 1 **liter** (*l or* L)
10 liters = 1 dekaliter (dal *or* daL)
10 dekaliters = 1 hectoliter (hl *or* hL)
10 hectoliters/1,000 liters = 1 kiloliter (kl *or* kL)

Metric Dry Capacity Units

The Metric System doesn't have a specific dry measurement system like the U.S. Customary System does; instead, dried goods are measured using distance units (meters, etc.), by mass (while it's technically different, for our purposes, you can think of mass as weight), or even using liters. For example, in a metric recipe, you might see the flour specified in grams.

We'll explore units of mass in the next lesson.

How the Systems Compare

> Even in the U.S. Customary System, food is often sold by weight rather than by capacity. Produce, packaged foods, etc., are often sold by the pound. Remember, we can use whatever unit makes sense in a situation — units are tools to help us describe a quantity in a way others will understand.

So how do the units in the Metric and the U.S. Customary Systems compare? A gallon is equal to almost 4 liters, so the liter is significantly smaller than the gallon. That's why if you ever head to France and try to fill a car up with gas, you might initially think the price of gas is pretty cheap…until you realize it's a price per *liter* instead of per *gallon*.

362 | PRINCIPLES OF MATHEMATICS 1

Customary to Metric

1 teaspoon ≈ 5 milliliters (ml)
1 gallon = 3.78541 liters

*Note: Again, the squiggly equal sign means the comparison is approximate —
1 teaspoon doesn't equal exactly 5 milliliters.*

Unit Conversion Between Systems

Once again, you can use the same skills you already have to convert between units in the Metric System, as well as to convert between the U.S. Customary System and the Metric System! As we've seen in the past, when we don't know the direct conversion between two units, it's easiest to use the conversion ratio shortcut, as the second example shows.

Example: Convert 789 mL to cL.

Conversion via a Proportion	Conversion via the Ratio Shortcut	Conversion via Mental Math
$\dfrac{789 \text{ mL}}{?} = \dfrac{10 \text{ mL}}{1 \text{ cL}}$	$789 \text{ mL} \cdot \dfrac{1 \text{ cL}}{10 \text{ mL}} = 78.9 \text{ cL}$	$789 \div 10 = 78.9$
Answer: 78.9 cL	Answer: 78.9 cL	Answer: 78.9 cL

Notice that you might need paper or a calculator to solve using the Conversion via Mental Math method on these examples.

Example: Convert 2 gallons to milliliters.

Since we don't know the conversion ratio between gallons and milliliters, we'll have to first convert to liters, and then to milliliters. **Notice that, once again, we can do it all in one step using the conversion ratio shortcut**, which is much easier than setting up separate proportions.

Conversion via a Proportion	Conversion via the Ratio Shortcut	Conversion via Mental Math
$\dfrac{1 \text{ gal}}{.78541 \, l} = \dfrac{2 \text{ gal}}{?}$	$2 \text{ gal} \cdot \dfrac{3.78541 \, l}{1 \text{ gal}} \cdot \dfrac{1,000 \text{ ml}}{1 \, l} = 7,570.82 \text{ ml}$	$2 \cdot 3.78541 \cdot 1,000 = 7,570.82$
$? = 7.57082 \, l$	Answer: 7,570.82 ml	Answer: 7,570.82 ml
Now to convert from liters to milliliters:	Notice how much easier this was!	
$\dfrac{1,000 \text{ ml}}{1 \, l} = \dfrac{?}{7.570.82 \, l}$		
Answer: 7,570.82 ml		

Keeping Perspective

Remember that each measuring system is but a different agreed-upon unit, or standard, for measuring volume. Measuring systems give us a way of describing the amount of milk that's for sale in a carton . . . or pricing gas so we can compare the prices between different pumps. Each measuring system is a useful tool.

18.7 Measuring Weight and Mass

Sometimes, rather than measuring dimensions or liquid/dry good capacity, it's helpful to measure an object's **weight** or, if using the metric system, its **mass**. Let's take a look at a few key units and at how they compare.

U.S. Customary Weight Units

Because of the consistent way God holds gravity in place, we can use it to measure objects based on what we call their weight. Without getting too technical, **weight** is a measurement of the gravitational pull on an object.

In the U.S. Customary System, we commonly measure weight in ounces, pounds, and tons.

16 ounces = 1 pound
2,000 pounds = 1 ton (called a "short ton")

Note that these ounces are a measure of weight rather than capacity. They're different than the fluid ounces we looked at previously.

Metric Mass Units

Rather than measuring weight, the Metric System has units for what we call **mass**. While they're technically different (mass remains the same no matter whether on the earth or in outer space, while weight would be different in a different gravitational field), we use the words interchangeably in nontechnical fields.

Notice that just as liquid volume was based on the liter and distance was based on the meter, mass in the Metric System is based on the **gram**. As with the liter and meter, each unit is worth 10 of the previous unit, and the same prefixes are used. Once again, common units are in bold.

10 **milligrams** (mg) = 1 **centigram** (cg)
10 centigrams/100 milligrams = 1 decigram (dg)
10 decigrams/1,000 milligrams = 1 **gram**
10 grams = 1 dekagram (dag)
10 dekagrams = 1 hectogram (hg)
10 hectograms/1,000 grams = 1 **kilogram** (kg)

In order to measure very large masses, the Metric System also includes a **ton**. A metric ton is worth 1,000 kilograms, which is slightly larger than what we mean when we say "ton" in the U.S. Customary System.

How the Units Compare

It's helpful to also know how the metric mass units compare to the U.S. Customary weight units.

Customary to Metric

1 ounce = 28.3495 grams
1 pound = 453.592 grams
1 U.S. ton (called a short ton) = 0.907185 metric tons

Notice that a gram is a *very* small unit!

Working with Weight and Mass

You can convert between units of weight and mass the same ways you've been converting between other units! The skills you've learned in unit conversion will keep applying over and over again.

Note, though, that we can only convert between units if we know how they compare. We don't have a universal way to switch from pounds to dry quarts, for example, as the weight of dry goods varies (a box of raisins weighs more than the same capacity of potato chips).

Keeping Perspective

Cubic, liquid capacity, dry capacity, weight, mass — there are a lot of different ways to measure! Each one comes in handy in different settings. Becoming familiar with them — and with how to convert between different units — will serve you well.

18.8 Chapter Synopsis

This chapter, we applied what we've been learning about area to new levels, using it to find the surface area and volume of different shapes, as well as to explore capacity, weight, and mass units. Below are a few key points.

■ The **total surface area** is the area of all exterior sides of a figure added together.

■ **Volume** is what we call the measurement of the total space an object takes up. It is a three-dimensional measurement, expressed in *cubic units*.

■ **Volume Formulas** — We learned that, to find the volume of prisms, cylinders, and cubes, we find the area of the base and multiply that by the height. This formula can be further simplified/customized for specific shapes based on how we would find their area.

Volume of a Prism or Cylinder

Volume = area of base • height   or $V = B \cdot h$

V = Volume
B = area of base
h = height

18. SOLID OBJECTS
AND VOLUME   365

- **Other Volume Systems** — Rather than finding the volume using cubic measure, we also learned some other **capacity**-based measurement systems, as well as how to measure contents by weight and mass.

- **Unit Conversion** — We practiced various volume, capacity, and weight/mass unit conversions.

This chapter, we also had the opportunity to touch on different applications of area, both here and in your *Student Workbook*. As we did, we saw again how math works as a tool and reveals the wonderful design God placed all around us (such as the area of a bird's wing) — a design that reminds us that He knows what He's doing in our lives, too.

Problem-Solving Reminder

Keep remembering the principles of problem solving you've learned. If you come across a problem you don't know how to solve, stop and draw it out. See if you can break it out into steps. It may require multiple tools from your toolbox. Ask yourself what you know and what you need to know. Figure out a plan. Execute the plan. Then check your work.

If you get something wrong, don't panic! Learn from your mistake. The goal of studying math is to learn the skills of problem solving where you have an answer key . . . because in real life, there isn't an answer key to check.

[CHAPTER 19]

Angles

19.1 Measuring Angles

It's time now to move on from measuring shapes to exploring them in more depth. We'll start our next level of investigations by taking a look at an important part of many shapes: angles.

Remember, an **angle** is a fancy name in Euclidean geometry for the intersection of two straight lines.

We call the point at which the lines intersect the **vertex** of the angle.

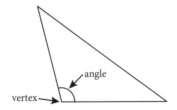

Up until now, we've used terms such as *right*, *obtuse*, and *acute* to describe the size of angles. But we need a way to more accurately describe an angle's size.

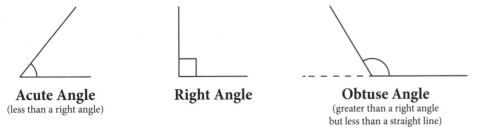

Acute Angle
(less than a right angle)

Right Angle

Obtuse Angle
(greater than a right angle
but less than a straight line)

One method for measuring angles is to describe the angle's location on a circle. This method breaks a circle into 360 small increments called **degrees**. The symbol to represent degrees is a small ° next to a number: 90° is read "ninety degrees."

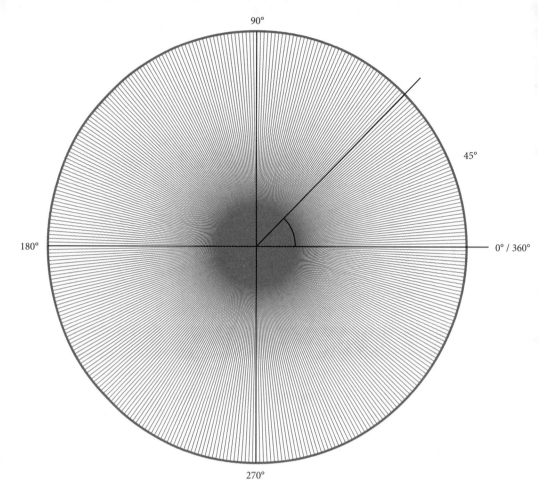

To measure an angle, we measure how many of the small increments — what we call "degrees" — the angle spans. The angle shown spans 45 degrees, so it has a measurement of 45°.

Exploring Degrees

The word **degree** comes from the Latin *de* meaning "down" and *gradus* meaning "step or grade."[1] You can think of degrees as little steps making up a circle.

We could have chosen any number of these little steps (degrees) to fit within the circle. In fact, rather than using 360 degrees, we'll sometimes think of a circle as consisting of 400 units called *gradians*, or, in the case of a clock, as consisting of 60 *minutes*. But however we choose to break up a circle, we need to be consistent. It wouldn't help us communicate if I broke a circle into 70 parts and you broke it into 360. So when working with degrees, we break the circle into 360.

Why 360? Perhaps 360 has remained the standard throughout the centuries because 360 has so many factors, making it easy to subdivide and work with.[2]

No one knows for sure why 360 was originally chosen to use in subdividing a circle. The Babylonian number system (which was based on 60, a factor of 360) or calendar (which was 360 days long) may have had something to do with it, although, according to one source, we don't find any evidence that the Babylonians themselves divided a circle by 360.[3]

368 | PRINCIPLES OF MATHEMATICS 1

We do know, though, that 360 was utilized extensively by the Greeks to break down circles.[4] The Greek mathematician Ptolemy (c. A.D. 90), in his work *Syntaxis*, broke each degree further down into 60 parts, or minutes, and each of those into *another* 60 parts, or seconds.[5]

> ### 180- and 360-Degree Turns
>
> Have you ever heard people talk about "making a 360"? Or that someone "made a 180"? Making a 360° turn takes you back to 0°, while turning 180° puts you facing the opposite direction. So when you "make a 360," you come back to your original position or viewpoint, while when you "make a 180," you change your position or viewpoint to the opposite one.

Down to the Mechanics

We use a device called a **protractor** to measure the degrees of angles. The protractor has half of a circle with the degrees marked on it. To use a protractor, simply place the **center of the bottom of the arc** at the place where the two lines meet (the vertex), with one line matching the straight edge. Then look to see at what degree the other line intersects the arc.

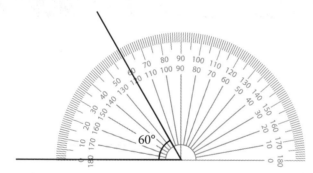

Note that this protractor has two different 0° — one on the left and one on the right. This allows us to start our measurement from either direction. We just have to be careful to read from the *same scale* as the 0° we're starting from.

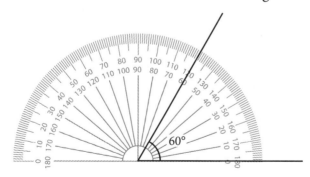

19. ANGLES | 369

If you need to, you can always extend a line so it reaches the protractor.

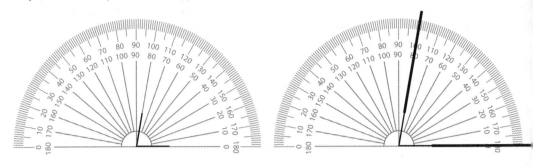

> # Keeping Perspective
>
> As you practice measuring angles today in your *Student Workbook*, keep in mind that angles can describe the intersection of any two lines. The wall intersects the floor at an angle, the oven door opens at an angle, the car window tilts at an angle — the list could go on and on. Degrees help us describe what angle to slope a roof, angle a road, sail a boat, etc.

19.2 More with Angles

Now that you know how to measure angles, let's take a look at how many degrees some common angles are, how to draw angles a specific size, and how to add angles together.

Naming Angles

While you might not think of a **straight line** as an angle, we could really think of a vertex as occurring anywhere on the line. If it did, the angle formed would always be 180°. We call this a **straight angle**.

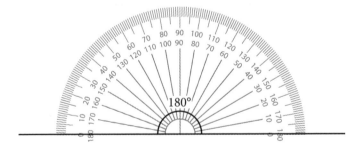

Notice that a **right angle** is 90°. Like a straight line, all right angles have only one measurement.

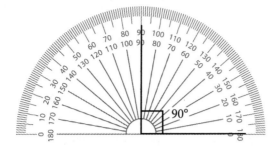

An **acute angle**, however, is any angle less than 90° but greater than 0 degrees. The one pictured as an example is 40°.

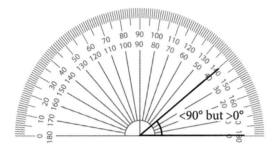

An **obtuse angle** is any angle greater than 90° but less than 180°. The one pictured as an example is 125°.

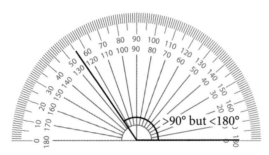

Drawing Angles

Drawing angles that are a specific degree is easy. Keep that protractor out and give it a try yourself.

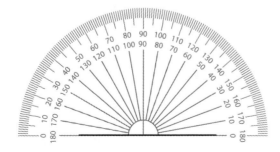

1. Use the straight edge (such as the bottom of your protractor or ruler) to draw a straight line.

19. ANGLES | 371

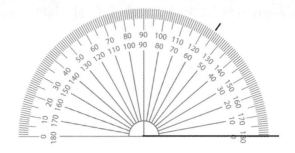

2. Rest a protractor on the line (align the middle of the protractor with where you want the vertex) and put a mark at the appropriate number of degrees.

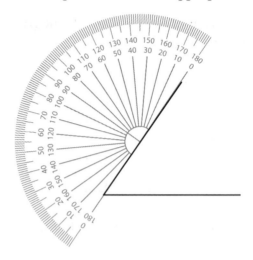

3. Use a straight edge to connect your lines, thus forming an angle.

And just when might you need to draw angles? Well, up until now you've drawn shapes without worrying about exact angles. But if you were trying to draw a scale drawing or blueprint, it would be important to get the angles accurate. And, using your protractor, you can!

For example, if you know you're drawing a rectangle (a parallelogram with right angles), you can begin by drawing a straight line to represent one side.

Then, since you know a rectangle has four right angles (it's part of the definition of a rectangle — a parallelogram with right angles), you know you need to make the next side intersect at a right angle.

90°

By making all the angles in the rectangle right angles, you'll also end up with two pairs of parallel lines.

[rectangle with 90° at each corner]

Adding Angles

Not only is it helpful to name and draw angles, it's also helpful to add them to see their total. We can add them numerically, as we would any other number.

20° + 30° = 50°

Notice that if we actually had angles those sizes and put them together, the larger angle formed would indeed be 50°.

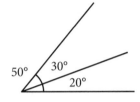

Along with other things, knowing how to add angles can help us draw or measure an angle greater than 180°. While a protractor stops at 180°, we can view larger angles as the addition of two smaller angles and can draw or measure the larger angles by drawing or measuring two smaller angles that sum to the desired larger angle.

Example: Draw a 200° angle.

To draw this, we'll draw two smaller angles whose sum equals 200°.

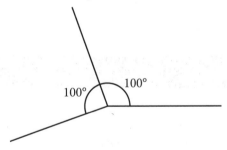

100° + 100° = 200°

Note: It doesn't matter what two smaller angles we draw, so long as the sum equals 200°.

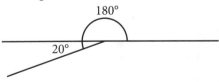

180° + 20° = 200°

Example: Measure the angle shown.

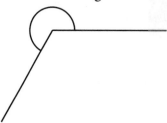

Since this angle is greater than 180°, we'll measure it by dividing it into two smaller angles, measuring the angle of those, and then adding them together.

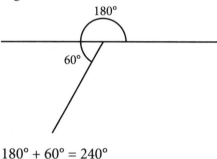

180° + 60° = 240°

The angle is 240°.

Keeping Perspective

As you practice the mechanics of naming, drawing, and adding angles, remember that angles are useful in describing God's creation and serving Him! We could even use angles to explore something as apparently commonplace as leaves. In fact, you'll get a chance to do just that in your *Student Workbook*.

19.3 Angles in Pie Graphs

Remember back to 12.3 when we explored using pie graphs to represent portions of a whole? Well, now that you know how to draw and add angles, you can draw these useful graphs.

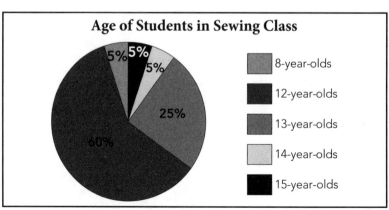

Since the pie graph is a circle, we now know it can be thought of as having 360°. We want to draw different sections within the circle so that they accurately reflect the percent of the total. For example, if we know a category represents 60% of the whole, we want to draw a section representing it that represents 60% of 360° . . . which would be 0.60 • 360° = 216°.

Notice that if you measure the angle the 60% section forms in the pie graph shown, you'll find that it does, indeed, measure 216°.

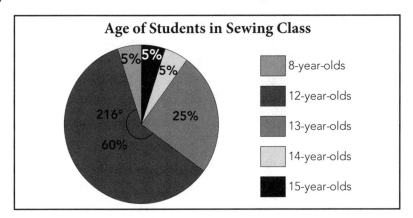

Let's move on to the 25% section. Its angle should be 25% of 360°, or 90° . . . and it is!

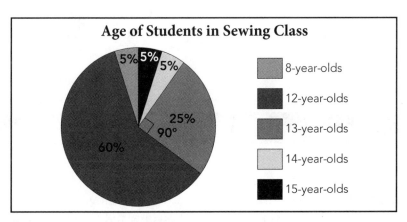

And our 5% angles should be 0.05 • 360°, or 18°, which they are.

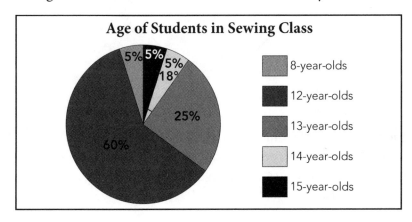

19. ANGLES | 375

Notice that the sum, or total, of all the angles in the circle equals 360° — this will always be the case.

$$216° + 90° + 18° + 18° + 18° = 360°$$

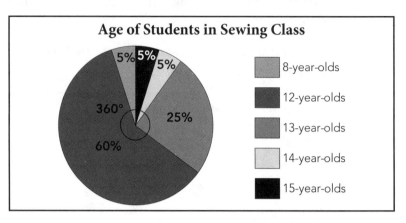

Keeping Perspective

Notice how drawing a pie graph uses angles! In fact, it requires combining a number of different concepts (percents, multiplication, angles, etc.). It's another example of how the different concepts you learn can be combined to accomplish a task. In the case of a pie graph, our "tools" helped us pictorially represent data.

19.4 Expanding Beyond

As you move on to high school geometry in a few years, you'll likely encounter a lot of apparently meaningless drawings such as the one shown. And you might be asked to spend a lot of time exploring those drawings and the angles and lines in them.

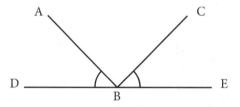

Always remember those drawings could be used to describe God's creation . . . and that goes beyond shapes! Up until now, our explorations in geometry have focused on exploring actual shapes, but the skills you've been learning help us describe many other aspects of God's creation too.

Representing Light

The apparently meaningless drawing shown could actually represent how light reflects off surfaces! Did you realize that God designed light so that when it reflects off a surface, light both hits the surface and reflects off it at the same angle? (Angles in action!)

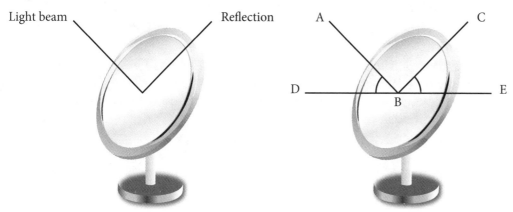

Not all surfaces reflect light, but when light does reflect, it follows this property.

Knowing this consistency has played a role in designing different technologies, such as CD and DVD players, devices that work based on shining a light and having it reflect at the same angle. That's right — it's the amazing consistency God created and sustains within the very light waves around us that makes DVDs possible.

Have you ever wondered why a glass prism makes rainbows of color? When light strikes something transparent, most of it shines through the object rather than reflecting off of it. But the transparent object changes the angle of the light. In some objects, such as glass triangular prisms, diamonds, and the crystals on most chandeliers, the angles in the transparent object cause the different colors within light to come out the other side at different angles. We then see them as separate colors instead of as one. God has hidden a kaleidoscope of color in every beam of sunlight.

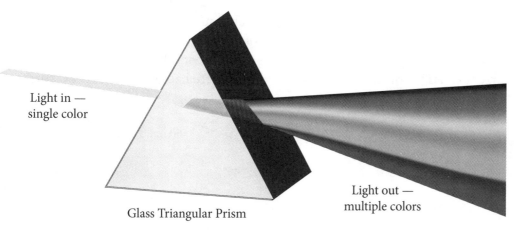

19. ANGLES

Application Thought

God created and sustains the light all around us. He understands all the intricacies of its workings. He also understands the intricacies and workings in our lives. We can take comfort knowing God is infinitely wiser than we are. His thoughts and ways are better than ours.

> He hath made the earth by his power, he hath established the world by his wisdom, and hath stretched out the heavens by his discretion (Jeremiah 10:12).

> Where is the way where light dwelleth? and as for darkness, where is the place thereof, That thou shouldest take it to the bound thereof, and that thou shouldest know the paths to the house thereof? (Job 38:19–20).

> For as the heavens are higher than the earth, so are my ways higher than your ways, and my thoughts than your thoughts (Isaiah 55:9).

Navigating a Ship

Compass

Have you ever wondered how sailors navigated across the ocean before the days of the GPS? How did they know their location and where to steer the ship? They used math . . . including angles! Angles are used extensively in navigation.

For example, notice the degrees on a compass — a common navigational tool. Degrees are used to refer to directions.

One method of navigation involves measuring the angle between the horizon and a celestial body (such as the sun or a star) and using that information, coupled with charts based off historical data about the sun's or a star's position, to figure out one's location.

Another useful technique known as "dead reckoning" involves angles as well. While we won't go into the details of this technique, the basic concept isn't hard to understand. If a ship starts at an angle of 40° and travels 80 miles at that angle before making a turn and traveling at 20° for another 80 miles, the ship's current location (or rather what it would be if there had been no "leeway, current, helmsman error,"[6] etc.) can be determined.

80 miles at 20°

80 miles at 40°

Starting point

> # Keeping Perspective
>
> The point? Angles help us describe many different aspects of God's creation. And, as we saw briefly when we looked at the angles in light, as we explore His creation, we end up seeing to a deeper level the wisdom and care with which God created this universe.

19.5 Chapter Synopsis and Faulty Assumptions

We've covered a lot of ground in this chapter! Hopefully you caught a glimpse of how angles apply in many more places than you might think.

Here's a quick recap of some of the mechanics:

- We can use units called **degrees** to measure angles. A degree is $\frac{1}{360^{th}}$ of a circle.

- **Protractors** are tools with degree markings we can use to easily measure and draw angles.

- A **right angle** is 90°, an **obtuse angle** is greater than 90° but less than 180°, an **acute angle** is less than 90° but greater then 0°, and a **straight line** can be thought of as an 180° angle, also called a **straight angle**.

- Exploring the angles within shapes and between lines proves quite useful and helps us explore real-life shapes . . . as well as other aspects of God's creation that we might not at first think of using geometry to describe (such as a beam of light).

As we continue our study of geometry, we'll keep discovering ways that we can use knowledge of angles, lines, and shapes to help us find other pieces of information. And you'll continue to encounter angles and the need to measure and draw them!

Ptolemy, the Sun, and the Medieval Ages

Earlier in this chapter, we briefly mentioned that the Greek mathematician Ptolemy broke down the circle into degrees, minutes, and seconds in his work *Syntaxis*. What we didn't have time to discuss then was that *Syntaxis* also laid out a mathematical theory that the sun circled the earth.[7] This theory then became accepted as fact for centuries — so much so that the Medieval Church, not wanting to contradict math and science, tried to add this teaching to the Bible.

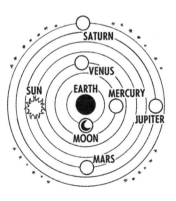

Why didn't people question that the sun circled the earth? Ptolemy's math was accurate. Math and reasoning were seen as absolute. Thus, how could the conclusion be wrong?

It wasn't until men like Copernicus (famous for proposing that the earth circled the sun) and Kepler (who discovered the laws of planetary motion) began to approach math as a *tool* (rather than as an unquestionable source of truth) that old ideas were shown to be false. Using math while observing the world around them, these men questioned the age-old teaching and discovered the truth about planetary orbits — a truth that accurately placed the sun, not the earth, at the center.

Always remember that sound math does not ensure a sound conclusion. If you start from an incorrect assumption, you can make a logical case for a very incorrect conclusion. Math and science are *tools*; they are *not* sources of truth.

The Medieval Church made the mistake of taking man's reasoning as true and finding a way to add it to the Bible.[8] Their mistake caused them lots of embarrassment when it was proven wrong.

Sadly, this backward way of approaching Scripture still goes on today in different forms. For example, many have tried to reinterpret Genesis in order to match Darwinian evolution rather than starting with Scripture and using it to interpret what we find in creation. Don't ever feel that you have to reinterpret Scripture to match what others say. Scripture is true and trustworthy, while man's opinion and even mathematical proofs are subject to error and revising.

For more information on how the literal Genesis account does indeed match the world in which we live, see AnswersinGenesis.org.

[CHAPTER 20]

Congruent and Similar

20.1 Congruent

Back in 13.5, we briefly played with what we called **congruent shapes** — that is, shapes that are "identical in form,"[1] even if their location is different.

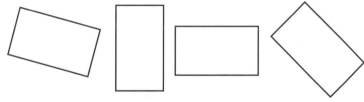

Congruent Rectangles

It's time to explore congruency in more depth, coupling it with some of the other tools you've learned.

Congruent Lines and Angles

Not only can shapes be congruent, but so can lines and angles! Notice that the two lines below are the exact same length. If we were to move them on top of each other, they would be identical. In other words, they're congruent!

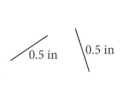 

Likewise, these angles are exactly the same degrees. Even though one is facing a different direction, if we were to move them on top of each other, we'd find that they were exactly the same. They, too, are congruent.

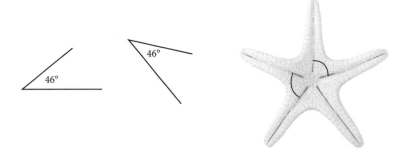

As you already know, a square has four equal sides and equal angles. We could also say that those sides and angles are congruent, since if we were to put them on top of each other, they would be identical.

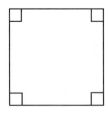

Term Subtleties

While *congruent* is very similar to *equal*, their meanings are subtly different. *Congruency* refers to the *forms* being the same — that they would be identical if put on top of each other. On the other hand, equal means "being the same in quantity, size, degree, or value."[2] *Equality* refers, not to the *form* of something, but to a *specific measurement* about it. However, congruent lines also have equal sizes (or they wouldn't be the same form!) . . . and equal-length lines are also congruent (if they're the same length, they have to be the same form).

Representing Congruency

It's often helpful to visually show congruent sides and angles. There are many different ways we could do this.

We could use color to show congruency, making lines that are the same length and angles that are the same degrees the same color. For example, in the figure shown, each different shade of gray could be a different color to show congruent sides and angles. But coloring lines and angles can get frustrating fast. And what if we don't have coloring materials available?

Another option would be to actually put the dimensions of each side or angle next to it. Notice how cluttered it becomes and how difficult it is to take in at a glance! Plus, there may be times when we don't know the exact measure of a line or angle — we just know they're congruent.

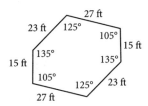

To better show congruency, it's common to use tick marks on the sides of shapes and arcs on the angles. Sides with one tick mark are congruent to all other sides with one tick mark, those with two tick marks are congruent to all other sides with two tick marks, etc. Likewise, angles with one arc are congruent, those with two are congruent, etc.

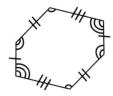

Note: There's an exception with angles. We still use the L-shaped marker for right angles (angles that measure 90°). Notice in the triangles shown, the right angles are marked with the L-shaped markers, and the other four angles have a single arc to show that they're congruent with each other.

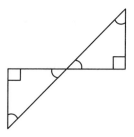

Keeping Perspective

In the next lesson, we'll apply congruency to help us find missing measurements — such as the distance across a stream. For now, though, just get used to the term and familiarize yourself with how congruency is shown. Tick marks and arcs are a convention to help us easily show that certain sides or angles are congruent.

20.2 Corresponding Parts — Applying Congruency

In the last lesson, we discussed that the word **congruent** means "identical in form."[3] As we've mentioned, both shapes and parts of shapes can be congruent.

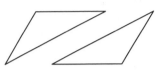

Congruent Triangles   Congruent Sides and Angles

Obviously, if two shapes are congruent, then **their corresponding sides and angles — that is, the sides and angles that represent the *same part* of the shape — have to be congruent.**

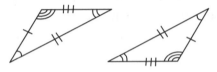

It's important to understand that a corresponding side or angle may move locations if a congruent shape is a rotated, reflected, etc., version of the original, but that its size and dimension don't change.

Notice that the 2.9 cm side stays 2.9 cm no matter how we move the triangle.

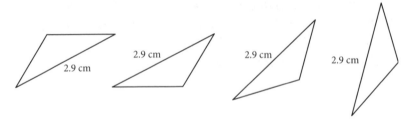

Likewise, the 125° degree angle stays 125° degrees no matter how we move the triangle.

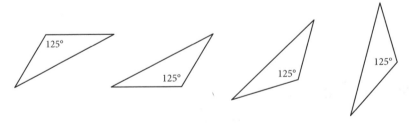

In other words, no matter how we translate, rotate, or reflect this triangle, the side that's 2.9 cm long is always 2.9 cm long, and the angle that's 125° is always 125°.

Using the congruency notation we looked at in the last lesson, we can easily show that *all* the corresponding sides and angles are congruent, despite how the triangle is moved.

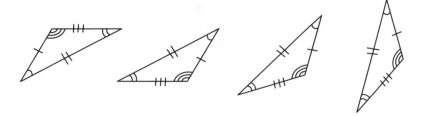

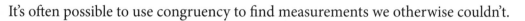

The idea that the same parts of congruent triangles are congruent is often summarized like this: **Corresponding Parts of Congruent Triangles Are Congruent.** This same idea applies beyond triangles — if two or more shapes are congruent, their corresponding parts will be as well.

It's often possible to use congruency to find measurements we otherwise couldn't.

For example, we know the dimensions of △ABC.

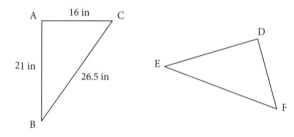

If we know that △DEF is congruent to △ABC, then we automatically know the measurements of all the corresponding sides and angles in △DEF — that is, the sides and angles that represent the *same part* of the triangle. They will be the same as those in △ABC.

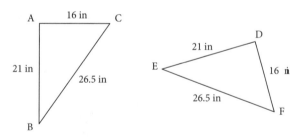

Let's apply this to a nontextbook scenario. Say we were trying to build a bridge across a stream, and we needed to know the width of the stream. There's just one problem: it's not practical to measure the stream's width with a measuring tape.

Don't worry; congruency can help us. The first step is to mark off two congruent triangles. We can do this by picking an object across the stream as close to the stream as possible to serve as a vertex (i.e., corner) of one triangle. For illustration purposes, let's say we use a tree growing by the stream's bank.

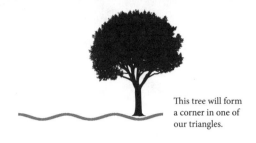

This tree will form a corner in one of our triangles.

We'd then need to mark a spot on our side of the stream that's directly across from the tree (but as close to the bank as possible), thus forming an imaginary line about the width of the stream. We'll call that spot Marker 1.

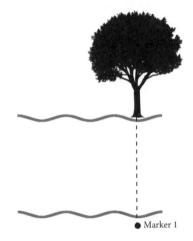

Standing at Marker 1 and looking at the tree, we'll then turn 90° and walk in a straight line a certain distance — say 4 ft — and put another marker there. We'll call it Marker 2.

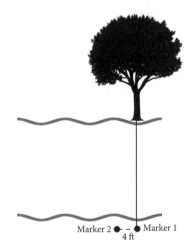

We've now just marked out a **triangle**, with our markers as the corners and a **right angle** at Marker 1 (after all, we turned **90°** from looking at the tree at Marker 1 to get to Marker 2).

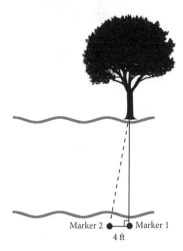

Now we need to form a congruent triangle to the triangle we just formed. To do so, we would now need to start at Marker 2 and walk the same distance we did before (4 ft), putting our third marker there.

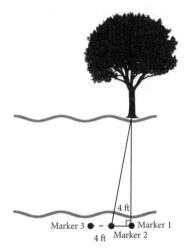

Now, starting at Marker 3, we need to turn our back to the stream (this will form another 90°, or right angle) and walk in a straight line until our path would cross a line going through Marker 2 to the tree. We'll put Marker 4 here.

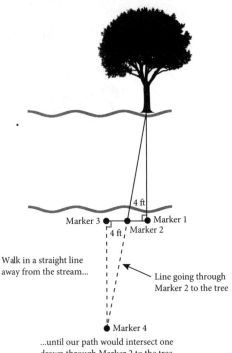

We've now formed two congruent triangles! How do we know the two triangles are congruent?

Basically, we made sure we laid out the triangles with a touching angle, side, and angle that had equal measurements. If you take a full geometry course, you'll learn that the only way a touching angle, side, and angle of a triangle can all be equal in measure is if the entire triangle is congruent.

And how do we know we have a touching angle, side, and angle that have equal measure? Well, the two 90° angles are the same because we turned 90° when we made them, the 4-foot sides are the same because we measured them, and the two angles where the diagonal and horizontal lines intersect have the same measure because of a property of intersecting lines.

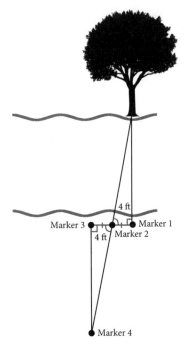

Since we now have two congruent triangles, we know that *all* of our corresponding sides will be equal. Provided we lined up our angles and measured correctly, the distance between Marker 1 and the tree will be the same as the distance between Marker 3 and Marker 4 — a distance we could measure!

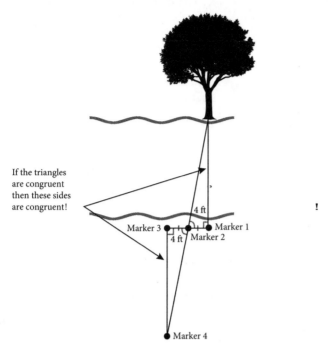

So if the distance between Marker 3 and Marker 4 turns out to be 20 feet, for example, we know the distance from Marker 1 to our tree is 20 feet. We now know the stream is approximately 20 feet wide. (I say approximately because the tree and our marker are not *exactly* at the edge of the stream and because we have estimated our angles.)

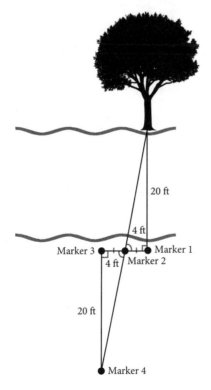

Illustration of Thales from a book published in 1875.

Distance to a Ship?

Legend has it that the Greek mathematician/philosopher Thales may have found the distance to a ship using congruent triangles. Whether or not the legend is true, know that the things you're learning about triangles, angles, and congruency can help find various distances!

Keeping Perspective

While you may never need to find the distance across a stream, it's an example of geometry helping us find information we otherwise couldn't. We used the pieces of information we knew (in this case, the measurements we could make on one side of a stream) to find what we wanted to know (the approximate distance across a stream).

As you learn more geometry, it's easy to get lost in the details and processes. Don't forget that geometry helps us measure the earth — including measurements we can't just use a measuring tape to find. The reason why we look so in depth at shapes, angles, etc., is *so that* we'll be able to figure out information we couldn't otherwise.

20.3 Exploring Similar Shapes

Sometimes, shapes are not only moved around, but they're also scaled — that is, made bigger or smaller. As we briefly mentioned in 13.5, we call shapes that are the same except for their size **similar shapes**. For example, every time you've been drawing a scale drawing, you've been drawing a similar shape — a shape that was just a scaled-down version of the other.

Similar Rectangles — *The one on the left is a scaled-down version of the one on the right.*

Notice that it wouldn't matter if the two rectangles were also rotated, reflected, or translated — so long as one is a scaled-down version of the other, the shapes are similar.

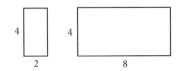

Similar Rectangles — *These rectangles are still similar — the one on the left is still a scaled-down version of the one on the right.*

It's time to explore similarity in a little more depth. By exploring it mathematically, we'll be able to use the knowledge that shapes are similar to help us measure and explore God's creation. Before we jump to that point, though, let's dig a little deeper into what similarity means and look at how we can apply it.

Exploring Similarity

If one shape is truly a scaled-down version of the other, then the **corresponding angles** (angles that represent the same part of the shapes) **will be congruent**, just as they were in congruent shapes. Unlike in congruent shapes, though, in similar shapes, the dimensions of corresponding sides only have to be **proportional**, not congruent. Proportional just means that their sides form equal ratios.

When you think about it, both of these conditions have to be true for one shape to truly be a scaled-down version of the other. After all, if the angles were different or if one side wasn't proportional, we'd no longer have a true scaled-down version. So it may seem like we're stating the obvious. However, by understanding the attributes of similar shapes, we can use those attributes to find missing dimensions in similar shapes.

For example, let's say we have the two rectangles shown and we know they're similar.

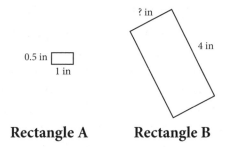

Rectangle A   **Rectangle B**

Because we know these rectangles are similar, we can use math to find the missing side — the side marked with *? in*.

We'll start by writing a proportion to describe the length and width of rectangle A and rectangle B. We know the ratio between the corresponding sides of each of the rectangles will be the same because the rectangles are similar.

$$\frac{\text{length of A}}{\text{width of A}} = \frac{\text{length of B}}{\text{width of B}}$$

$$\frac{1 \text{ in}}{0.5 \text{ in}} = \frac{4 \text{ in}}{? \text{ in}}$$

Here we can see that the length of Rectangle B is 4 times the length of Rectangle A (1 in • 4 = 4 in), so, in order to form an equal ratio (and thus form a similar shape), the width must have also increased by 4. Thus, the width of the large rectangle is **0.5 in • 4**, or **2 in**.

$$\frac{1 \text{ in}}{0.5 \text{ in}} = \frac{4 \text{ in}}{\textbf{2 in}}$$

The rectangle on the right is *4 times* the size of the rectangle on the left, so *every* corresponding side will be 4 times larger. Another way to think about it is that every corresponding side in the rectangle on the left is $\frac{1}{4}$ the dimension of the rectangle on the right.

> If two or more shapes are truly similar, then
>
> - **their corresponding angles** (angles that represent the same part of the shape) **will be congruent** and
>
> - the **dimensions of their corresponding sides** (sides that represent the same part of the shape) **will be proportional**. (In other words, their sides will form equal ratios.)
>
> Important note: **If these two conditions are not met, no matter how similar two or more shapes** *look*, **they are not similar in mathematical terms.**

Keeping Perspective

In this lesson, we explored a little deeper what we mean by similar shapes — a term used to describe shapes that are just enlarged or reduced versions of each other. We saw that no matter which angles we look at in similar shapes, the corresponding angles will be congruent. And no matter which sides we choose to compare, the *ratio* between the corresponding sides of similar shapes will be the same — that is, they will be proportional. For example, if one side of a shape is 8 times the corresponding side of a similar shape, then *every* side will be 8 times the corresponding side of the similar shape.

In the next couple of lessons, we'll apply our knowledge of similar shapes to find different measurements (including real-life measurements, such as the height of a tree). Don't forget that we're only able to categorize, name, and explore shapes (including similar shapes) because God gave us this ability.

20.4 Angles in Triangles and AA Similarity

When you take a full geometry course, you'll learn a lot of ways to tell if shapes — especially triangles — are congruent or similar. If we can tell that two shapes are congruent or similar, we'll be able to use that knowledge to help us solve problems.

In this lesson, we're going to take a look at just *one* way you can tell if two triangles are similar. In the next lesson, we'll apply what we learn to find the height of a tree (without climbing a ladder).

Angles in a Triangle

To begin with, we need to look at the angles in triangles. Rather than just telling you about the angles in a triangle, I want you to stop, grab a protractor, and measure the angles in each of these triangles. Add up all the angles in each triangle, as is done on the first triangle.

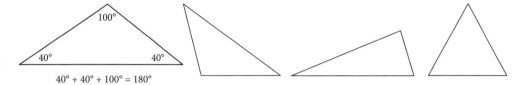

If you measure and add the angles correctly, you will find that the sum of the angles in *every* triangle equals 180°. This holds true for all triangles, not just the ones shown. The sum *has* to equal 180° in order to form a three-sided polygon (otherwise, we'd end up with more sides . . . or wouldn't be able to form a closed, straight-sided figure).

Finding the Missing Angle

Since the sum of the measure of all the angles in a triangle always equals 180°, then if we know the measure of two of the angles, we can figure out the third!

For example, in this triangle, the two angles we know total 44°, as 22° + 22° equals 44°. Thus the remaining angle must be 136°, since the sum of all the angles in a triangle equals 180°.

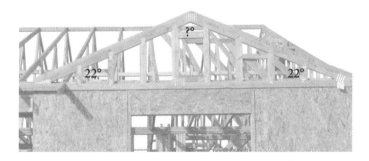

$$22° + 22° + \underline{} = 180°$$
$$44° + \underline{} = 180°$$
$$44° + \mathbf{136°} = 180°$$

AA Similarity Theorem

Are these two triangles similar? Remember, to be similar, all the angles have to be congruent and the sides have to be proportional. We don't know the measure of the sides or of one set of angles, so how can we tell for sure if the triangles are similar?

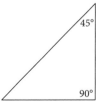

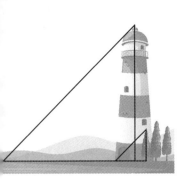

While we could just pick up a protractor and ruler and find the lengths of the sides and measure of the angles in the drawing to see if they're similar, remember that this drawing could stand for something in real life. It's not really the drawing we care about — it's the piece of real life these triangles are representing. And those are often impossible or difficult to physically measure. For example, these triangles might represent the heights of a lighthouse and a tree, along with the lengths of their shadows.

It turns out that it doesn't even really matter if we know the measurements of any of the angles or sides in the triangles. If we know that **two sets of their corresponding angles are congruent, two triangles *have* to be similar**. This is often referred to as **AA (for "angle angle") Similarity Theorem**.

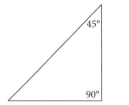

These triangles are similar, as two sets of corresponding angles are congruent.

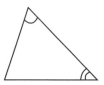

These triangles are similar, as two sets of corresponding angles are congruent.

The AA Similarity Theorem makes sense when you think about it. After all, since we know the sum of the angles in a triangle total 180°, we know that if two angle measurements are the same, the third must be too, or else they both wouldn't total 180°. And if all the angles in a triangle are congruent, the corresponding sides have to be proportional — otherwise, they'd form different angles.

Keeping Perspective

When you get to a full geometry course, you'll learn many other ways to tell if triangles are similar or congruent. As you do, keep in mind that knowing different ways to tell how triangles relate to each other helps us because, in real life, we can't always measure every side of a triangle. Triangles can be used to represent the heights of buildings, the distance to a ship, the height of a mountain, and so much more. Geometry helps us use what we know about angles and shapes to measure aspects of God's creation we couldn't otherwise!

20.5 Similar Triangles in Action — Finding the Height of a Tree

Say I wanted to figure out the height of a tree before chopping it down (to make sure it won't hit the house!), but I can't just extend a measuring tape to the top of the tree. I could find the height using similar triangles!

Notice that we could think of a tree and its shadow as a triangle.

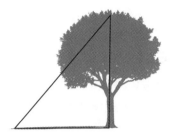

If we were to put a stick in the ground near the tree, we could think of it and its shadow as forming another triangle. We know the sun is striking the stick at the same angle it strikes the tree. So we know these two angles — whatever their measure — are the same in both triangles.

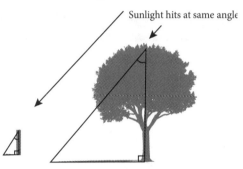

Since both the stick and the tree are basically perpendicular to the ground, they can be thought of as forming right angles (90° angles). So we know that these two angles are equal in measure.

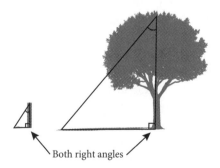

We saw in the last lesson that if two corresponding angles of two triangles are congruent, then the triangles have to be similar. The triangle formed with our stick forms a similar triangle with the triangle formed with the tree.

It follows, then, that if we measure the stick and its shadow, as well as the tree's shadow, we can set up a proportion to find the height of the tree!

Disclaimer: Before you head outside and chop down a tree near your home, know that this isn't meant to be a lesson on when it's safe to chop down a tree. It's only meant to show you that math can help you find real-life missing heights.

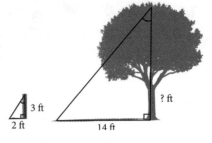

$$\frac{3 \text{ ft}}{2 \text{ ft}} = \frac{? \text{ ft}}{14 \text{ ft}}$$

The height of the tree is 21 ft. $\frac{3 \text{ ft}}{2 \text{ ft}} = \frac{21 \text{ ft}}{14 \text{ ft}}$

> ## Keeping Perspective
>
> Do you see what we just did? We built on our knowledge of triangles, angles, and similarity to figure out a measurement we couldn't easily measure — in this case, the height of a tree, which might prove useful if chopping a tree down.

20.6 Chapter Synopsis and a Peek Ahead

This chapter, you began learning how to use information about shapes to find unknown information. You . . .

- **learned about congruency** (i.e., "identical in form"[4]) and at how both shapes and parts of shapes could be congruent. We discussed using tick marks and arcs to easily represent congruent lines and angles, and looked at how corresponding parts of congruent shapes are congruent.

- **learned about similar shapes** (i.e., shapes where one is a scaled up or down version of the other). We saw that if shapes are truly similar, their corresponding angles will be congruent and their corresponding sides will be proportional.

- **learned about the angles inside triangles and the AA Similarity Theorem**. We saw that the angles in a triangle always total 180° and that if two sets of corresponding angles in two triangles are congruent, the triangles have to be similar.

Above all, I hope you caught sight of how, as we learn more about shapes and their properties and relationships to other shapes, we can **use that knowledge to find missing information**, such as the distance across a stream . . . or the height of a tree.

Looking Ahead

When you take an actual geometry course, you'll learn more ways to find missing information as well as how to do it more formally through a mathematical proof. For example, you might learn how to find the measure of a certain angle by first figuring out that two triangles are congruent. You might then show that two other angles are congruent and continue using information you know to find information you don't know (many geometry proofs are quite extensive) until you've found the measure of another angle you're missing.

Formal mathematical proofs are incredibly helpful in learning to think through problems logically — and logical reasoning helps us solve real-life problems, such as finding the height of a tree. Logical reasoning can also help us outside of math, such as in giving a reason for what we believe or in thinking through a decision.

Just be careful. Our culture today uses terms in a confusing way. Some scientists like to claim that they've "mathematically proven" things about the past (such as that the world evolved from nonlife). Yet we can't *prove* the past. Since we live in the present, we have to make assumptions about what happened in the past and what conditions were like. And if our assumptions are wrong, our conclusion will be too. As creation author and speaker Ken Ham often points out, both evolutionists and creationists have the same data — they just interpret the data through different starting assumptions.[5]

Ironically, science and math don't make sense in an evolutionary worldview. If life evolved and wasn't created, we'd have no reason to even explain our ability to reason. As Dr. Lisle points out, if our thoughts were "the inevitable result of mindless chemical reactions," why should we even trust our thinking abilities?[6] On the other hand, it would make sense that we can reason if we live in a consistent universe and God gave man the ability to think logically. In other words, to even trust that we can reason logically through a problem, we have to borrow from the biblical worldview.

Reasoning all starts with assumptions, and a proof holds true only if those starting assumptions are true. Let's make sure we start our reasoning with the truth of God's unchanging Word.

> *Thy word is true from the beginning: and every one of thy righteous judgments endureth for ever (Psalm 119:160).*

> "If evolution were true, science would not make sense because there would be no reason to accept the uniformity of nature upon which all science and technology depend. Nor would there be any reason to think that rational analysis would be possible since the thoughts of our mind would be nothing more than the inevitable result of mindless chemical reactions." — Dr. Jason Lisle[7]

[CHAPTER 21]

Review

21.1 Arithmetic Synopsis

We started our explorations of math at the very beginning — understanding that the symbols and words in math were ultimately a way of describing the quantities and consistencies God created and sustains around us. We explored standard conventions that, like a language, help us clearly communicate basic properties (characteristics), and the four basic operations: addition, subtraction, multiplication, and division.

Once we had the basic skills, we kept expanding on those skills by applying them to additional notations (fractions, decimals, ratios, and percents). We next looked at expanding numbers to include negative numbers and at organizing numbers into categories, or sets. We then touched on statistics and looked at how we can use math to help us make sense out of data.

Conclusion

All the rest of math continues to build on the foundational concepts with which we began. For example, in geometry, we continually used the basic operations (addition, subtraction, multiplication, and division), along with the various notations: fractions ($\frac{1}{16}$ of an inch, etc.), decimals (a side that's 4.5 in long), ratios (unit conversion!), percents (finding percents of shapes or in problems related to shapes). Even sets proved themselves handy (*polygons* can be thought of as a set).

As you review the foundational concepts, remember that they all ultimately rest on God. If He were not holding this universe together with unerring consistency, math would be a meaningless paper exercise. Instead, it's a useful tool that applies in every field. You may be surprised at how even nonmathematical occupations use math in one way or another.

21.2 Geometry Synopsis

Let's take a quick look back at what we learned about geometry, the branch of math that helps us "measure the earth," so to speak.

- **Naming Shapes** —We started by exploring names and terms we use to categories shapes and parts of shapes (i.e., refer to them as sets). We later used these terms to help us measure shapes with different properties.

- **Measuring Shapes** — We next moved on to learning about units — standards against which we could measure distances, areas, volumes, weights, and more. We explored different ways to convert between units.

  We also measured the perimeter, area, and volume of a variety of shapes. Along the way, we learned about exponents and square roots and saw how exponents greatly help us in describing area and volume, among other things. We also learned that we could represent relationships as formulas — and learned quite a few of them for perimeter, area, and volume. And we discovered that irrational numbers (such as π) defy our understanding and just keep going and going and going. God has truly created a complex universe.

- **Exploring Shapes** — In the last few chapters, we explored angles, triangles, congruent shapes, and similar shapes. We touched on how we can use the information we know to find information we don't know, such as the height of a tree or the distance across a stream.

We've only touched the surface of geometry. There's much, much more we could have discussed! But hopefully what we have covered has equipped you to go out and begin "measuring the earth" using math. The most important thing to remember is that math — including shapes and measurements — help us describe God's creation.

21.3 Life of Johannes Kepler

Today we are going to look at the life of a famous mathematician and scientist whose work changed history. This mathematician's life both provides an example of math in action and is resplendent with lessons we can apply to our own lives.

Known as the discoverer of the laws of planetary motion, Johannes Kepler proposed that the planets circle the sun in elliptical orbits (basically, in ovals[1] 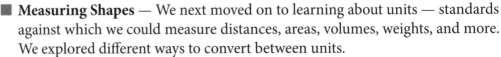) rather than in circular orbits as previously thought. Although often thought of as a scientist, Kepler was also a mathematician. In his study of planetary motion, Kepler used extensive math, including geometry. And while we can now quickly read about his discovery in the pages of history, the discovery took him years . . . and a long journey.

The Beginnings

Johannes Kepler did not plan on becoming a mathematician — he set out to become a minister. But toward the end of his university studies, his professors recommended him for a math position.

The young minister-to-be didn't like the idea of giving up his divinity studies. Although he eventually agreed to take the math position, Kepler still planned on becoming a minister one day. But God had something very different in mind for Kepler, as Kepler himself later recognized.

Kepler had always been interested in the movement of the heavens and had admired Copernicus and his sun-centered theory. As a professor, Kepler now had more time to investigate these matters. He spent years developing a theory to explain the movements of the heavens, only to later discover his theory was insufficient. Undaunted, Kepler kept trying. His belief in the universe as an orderly creation of God made him certain the movement of the heavens could be explained by geometry.

In 1600, Kepler's teaching career at the school came to an abrupt halt. Along with others who refused to convert to Catholicism, Kepler was told to leave the country! Yet, although Kepler probably could not see it at the time, God had a plan to transform persecution and exile into a tremendous blessing.

Exiled from his own country, Kepler soon found himself assisting (and depending on the generosity of) the famous astronomer Tycho Brahe — that is, until Brahe died in 1601. After Brahe's death, Kepler inherited Brahe's position and records. Because he had been exiled from his own land and forced to take shelter under Tycho Brahe, Kepler now had the records he needed to discover the laws by which God caused the planets to orbit the sun. Who would have thought God would use a persecution and forced exile to help Kepler accomplish his life work?

Johannes Kepler

Max Caspar, one of Kepler's biographers, says, "Looking back later when, through the discovery of his planet laws, he had become aware of his ability, he recognized the voice of God in the call which had come to him. It is God who by a combination of circumstances secretly guides man to the various arts and sciences and endows him with the sure consciousness that he is not only a part of the creation but also partakes in the divine providence."[3]

The Obstacles Mount

Discovering the planetary laws did not prove an easy task. From a collection of numbers Brahe had made over a period of many years chronicling where in the sky Mars had appeared, Kepler tried to find some sort of orderly law that could express the way God caused Mars to orbit the sun.

How hard was this task? According to Robert Wilson, "It took Kepler eight years and nearly a thousand pages of closely written calculations before he cracked the problem and discovered his first two laws of planetary motion (the third was to wait another nine years)."[2] Can you imagine spending eight years on a geometry problem you are not even sure can be solved and then another nine years to finish the task?

Kepler's Beliefs

No one reading through Kepler's *Harmonies of the World* — Kepler's explanation of his discoveries — can doubt Kepler's belief in God. He often paused in the middle of an explanation to mention his Creator and sometimes even broke off into a hymn of praise. It seems almost as if Kepler still viewed himself as a minister, trying to uncover the glory of God throughout creation. His book on planetary motion ends with this tribute to God:

> *Crying out with the royal Psalmist: Great is our Lord and great His virtue and of His wisdom there is no number: praise Him, ye heavens, praise Him, ye sun, moon, and planets, use every sense for perceiving, every tongue for declaring your Creator. . . . To Him be praise, honour, and glory, world without end. Amen.*[4]

In fact, Kepler's willingness to persevere amidst hardships stemmed from his deep faith that God had created an orderly universe. He was unwilling to accept the "close" results obtained from the lingering Greek idea that celestial bodies orbited in circles, albeit Copernicus had shown they orbited the sun instead of the earth. Instead, he searched for a better model.

Questioning the Greeks was a huge step. For centuries, the Greek philosophers' teachings had been taught as fact. To question them was equivalent to questioning proven fact. Kepler could only be so daring because he believed in God as the source of truth, not the Greeks' human reasoning.

At the same time, though, Kepler's theology and outlook on math were far from perfect. He carried over a lot of Greek mysticism into his beliefs about God and the universe. Forgetting that creation and our minds are both fallen, Kepler often drew unbiblical spiritual parallels and inferences about God. Kepler also dabbled in astrology (although he admitted it held no weight) and brought a good deal of mystical thinking into his astronomy.

Conclusion

Within Kepler's life, we see God's sovereignty at work, using even an exile to accomplish His purposes. We also find a challenge to persevere, to view the universe as God's handiwork, and to worship Him while using math to explore it. At the same time, we find a fallen man who allowed many falsehoods into his thinking — a human tendency we all need to be careful about.

21.4 Course Review

Throughout all our explorations, I hope you've seen two overarching principles:

1. **Math describes real life.** There's a purpose to what you're learning.

2. **Math points us to the Lord.** The very fact that math does indeed work outside of a textbook is because this universe is created and held together by a consistent, covenant-keeping God we can trust. This same God created man, giving us the ability to, in a very limited way, "think His thoughts after Him" (Johannes Kepler). As we use math to explore His creation, we find evidences of God's care and wisdom everywhere.

The Main Elements

As we've journeyed, we've encountered specific elements of math over and over again. Let's do a quick review.

1. **Notations, Symbols, Terms, and Conventions** — From place value to fractions and decimals to abbreviations (such as *ft* for *feet*) to exponents to terms (such as *rectangle*) to letters in formulas, notations and terms aid in effective communication. This portion of math is like a language in that it is one way of describing the quantities, relationships, ideas, etc., all around us. While it's easy to get lost in the language portion of math, keep in mind that we need so many different notations and terms *because God created a complex universe*. Math's very complexity should cause us to stand in awe of the Creator!

2. **Operations** — Throughout this course, we've encountered the basic operations — addition, subtraction, multiplication, and division — in different forms. We started learning them straightforward (i.e., $3 + 4 = 7$), progressing on to working with partial quantities (fractions, decimals, and percents).

   While the applications have gotten more involved as we've applied these operations to different notations and used more than one operation at a time, the operations have stayed the same. If we want a math that works, we have to make it so that, no matter what notation we use to describe it, addition, subtraction, multiplication, and division mirror the consistencies God created and sustains.

   Whether working with a basic problem or a complicated one, every time we successfully perform a math operation, we know it is because God is still consistently governing all things. Math is shouting out at us that we serve a covenant-keeping God whom we can trust. Just as He is faithful to the day and the night and the "ordinances," or laws, He has placed all around us (including addition and subtraction), He will be faithful to everything else He says in His Word.

21. REVIEW | 403

Thus saith the LORD; If my covenant be not with day and night, and if I have not appointed the ordinances of heaven and earth; Then will I cast away the seed of Jacob and David my servant, so that I will not take any of his seed to be rulers over the seed of Abraham, Isaac, and Jacob: for I will cause their captivity to return, and have mercy on them (Jeremiah 33:25–26).

[referring to Jesus] Who being the brightness of his glory, and the express image of his person, and upholding all things by the word of his power, when he had by himself purged our sins, sat down on the right hand of the Majesty on high (Hebrews 1:3).

For by him [Jesus] were all things created, that are in heaven, and that are in earth, visible and invisible, whether they be thrones, or dominions, or principalities, or powers: all things were created by him, and for him: And he is before all things, and by him all things consist (Colossians 1:16-17).

3. **Algorithms** — From adding multi-digit numbers to dividing fractions to converting units of measure to converting numbers into scientific notation, algorithms (step-by-step methods) take much of the thought process out of a mathematical problem. However, it's always important to make sure you really understand the algorithm as a way of describing a consistency God created.

Moving Forward

As you move into your next math course, you'll find that the same main elements and principles apply over and over again, just in more and more involved scenarios. The biblical worldview we explored in these math concepts applies to future concepts you'll encounter too.

Most Important Takeaway — It's Absolutely Absolute

The most important lesson I hope you've taken from our time together is that God is in charge of everything and is entirely trustworthy. The only reason math works outside a textbook is because we live in a consistent universe. And why do we live in a consistent universe? Because the consistent God of the Bible created and sustains all things!

Truth is not relative, nor does it change to match the whims of the times. No matter what symbols or words we use to describe it, the universe operates according to the "ordinances" that *God* created and sustains. If we want a math that works, we have to make it match what He ordained.

We live in a culture that wants to believe each person can find his or her own way to God. The problem is, though, that *God* is the One who decides how He can be approached, not us. If we want to know Him and escape the punishment our sins deserve, we have to come to Him His way.

One day, each one of us will stand before God. At that moment, it won't matter how good we thought we were, or how we compared with other people. God sees the depths of our hearts, and before Him, no one is righteous.

As it is written, There is none righteous, no, not one (Romans 3:10).

If He judges us based on our own merit, each one of us is worthy of hell. If you think you're righteous, just try to be perfectly good for one day. You'll quickly discover how impossible it is.

But, as we've mentioned before, there's very, very good news. God sent His Son, Jesus, to take the punishment we deserve. If we will repent (turn away) from our sin and believe (put our faith and hope) in Jesus for salvation, then God will credit our sin as belonging to Jesus and credit Jesus' righteousness as our own. Now that's an absolutely amazing exchange!

As you finish your math for the year, my challenge to you is to make sure you know and are reconciled with math's Creator. And if you are, go tell someone the good news of what Jesus has done.

APPENDIX A

Light, Distance, and Measuring Time in Space

A light-year is a useful measurement unit based on the distance we observe light travel in a second here on earth. Just because a star is 4.3 light years from earth does not mean the light took 4.3 years to get here. Light may very well travel differently in the deep recesses of space, or God may have used some other way to get the light here much faster than normal. Although we can see stars billions of light years away, Genesis makes it clear that the earth is only a few thousand years old, so we know that light has not been traveling for billions of years.

A question that often arises when discussing creation versus evolution is, "How can light from distant stars be seen in a 'young' universe?" The truth is that we do not really know how light travels those vast distances. But neither do secular scientists know how light travels. In fact, light-travel time is also one of the big problems for the old-earth big-bang theory (Robert Newton, "Light-Travel Time: A Problem for the Big Bang," *Creation*, September 2003, http://www.answersingenesis.org/creation/v25/i4/lighttravel.asp, accessed 11/24/14). The fact is, no one fully understands how light, gravity, and other physical forces operate on a cosmic scale.

That said, there are many ways to explain our ability to see the distant starlight, and there are now several creationist cosmologies (including one by Dr. John Hartnett, author of *Starlight, Time, and the New Physics*) based on "gravitational time dilation." Dr. Jason Lisle, author of *Taking Back Astronomy*, has proposed the "Alternate Synchronization Model," suggesting the universe is divided into "time zones" similar to what we have on earth (Robert Newton, "Distant Starlight and Genesis: Conventions of Time Measurement," *Creation*, April 2001, http://www.answersingenesis.org/tj/v15/i1/starlight.asp, accessed 11/24/14).

Dr. Lisle also mentions that many "people mistakenly think that Einstein's theory of relativity demands that the speed of light has not changed in time. In reality, this is not so. Relativity only requires that two different observers would measure the same velocity for a beam of light, even if they are moving relative to each other" (Ham, Ken. *The New Answers Book 1: Over 25 Questions on Creation/Evolution and the Bible*. Green Forest, Ark.: Master Books, 2006, p. 247).

While we do not know exactly how God did it, we do know God created the stars to give light upon the earth. Getting the light here was no problem for our God! Remember, He holds all things together with the power of His Word — let's give Him the credit for knowing how to do what He said He did, even if we don't know exactly how He did it. "And God set them in the firmament of the heaven to give light upon the earth" (Genesis 1:17).

For more information on this important topic, please check out the resources on www.answersingenesis.org. (Thanks to Zak Klein for his help with this information.)

Chapter 1. Introduction and Place Value

1. *Webster's New Collegiate Dictionary*, 1974 ed., s.v. "Neutral."

2. David A. Noebel, *Understanding the Times*, rev. 2nd ed. (Manitou Springs, CO: Summit Press, 2006), p. 16.

3. Herbert Westren Turnbull, *The Great Mathematicians* (New York: New York Univ. Press, 1962), 141. Quoted in Larry L. Zimmerman, *Truth & the Transcendent: The Origin, Nature, & Purpose of Mathematics* (Florence, KY: Answers in Genesis, 2000), p. 13.

4. Walter W. Sawyer, *Mathematician's Delight* (Harmondsworth Middlesex: Penguin, 1943), 10, quoted in James D. Nickel, rev. ed., *Mathematics: Is God Silent?* (Vallecito, CA: Ross House Books, 2001), p. 290.

5. Freemon J. Dyson, "Mathematics in the Physical Sciences," National Research Council's Committee on Support of Research in the Mathematical Sciences (COSRIMS), *The Mathematical Sciences: A Collection of Essays* (Cambridge, MA: Massachusetts Institute of Technology, 1969), p. 99.

6. Denis Guedj, *Numbers: Universal Language* (New York: Harry N. Abrams, 1996), p. 124.

7. Alberit Einstein, *Letters to Solovine* (Paris: Gauthier-Villars, 1956), 114–115, cited in James D. Nickel, *Mathematics: Is God Silent?* 2001 rev. ed. (Vallecito, CA: Ross House Books, 2001), p. 210.

8. Rousas J. Rushdoony, *The Philosophy of the Christian Curriculum* (Vellecito, CA: Ross House, 1981), p. 58. Found in James D. Nickel, *Mathematics: Is God Silent?* rev. ed. (Vallecito, CA: Ross House Books, 2001), p. 204.

9. Dyson, *The Mathematical Sciences: A Collection of Essays*, pp. 101–102.

10. *New Oxford American Dictionary*, 3rd edition (Oxford University Press, 2012), found in Dictionary, Version 2.2.1 (156) (Apple, 2011), s.v. "cryptography."

11. *The World Almanac and Book of Facts*, 2012 (New York: World Almanac Books, 2012), p. 732.

12. Top Row: One of several possible symbols the Egyptians may have used as a form of an equals sign (this particular hieroglyphic means "together" and could have been used to symbolize the results of addition); symbol used by Diophantus (200s); form of modern symbol presented by Recorde (1557); symbol used by Buteo (1559); symbol used by Holzman, better known as Xylander, that several other mathematicians adopted.

    Second Row: Symbol used by Hérigone (1634); another symbol used by Hérigone; one of the symbols used by Leonard and Thomas Digges (1590); symbol popularized by Descartes (1637) — it was this symbol that proved our modern symbol's main competition; a way our modern sign was sometimes printed.

    Symbols based on Florian Cajori, *A History of Mathematical Notations: Two Volumes Bound As One* (Mineola, NY: Dover Publications, 1993), 1:297–309, and on David Eugene Smith, *History of Mathematics*, vol. 2, *Special Topics of Elementary Mathematics* (New York: Dover Publications, 1958), pp. 410–411.

13. Pronunciation from *The American Heritage Dictionary of the English Language*, 1980 New College Edition, s.v. "quipu."

14. For an in-depth look at quipus, see Marcia and Robert Ascher, *Mathematics of the Incas: Code of the Quipu* (Mineola, NY: Dover Publications, 1981).

15. The hieroglyphic style shown here was the oldest, more decorative style, and the one typically associated with the Egyptians. However, the Egyptians also developed two cursive styles (the hieratic and the demotic) that were faster to write. Notation within the styles also varied. Florian Cajori, *A History of Mathematical Notations: Two Volumes Bound As One* (Mineola, NY: Dover Publications, 1993), 1:11, and David Eugene Smith, *History of Mathematics*, vol. 2, *Special Topics of Elementary Mathematics* (New York: Dover Publications, 1958), p. 47.

16. Florian Cajori, *A History of Mathematical Notations: Two Volumes Bound As One* (Mineola, NY: Dover Publications, 1993), p. 1:13.

Chapter 2. Operations, Algorithms, and Problem Solving

1. "Hebrew Lexicon :: H2708 (KJV)." (Blue Letter Bible, accessed 9/15/14, http://www.blueletterbible.org/lang/lexicon/lexicon.cfm?Strongs=H2708&t=KJV).

2. Daniel Adams, *The Scholars Arithmetic; or, Federal Accountant. Containing . . . the Whole in a Form and Method Altogether New, for the Ease of the Master and the Greater Progress of the Scholar*, steriotype ed. Rev. and corrected, with additions (Leominster, MA: Adams & Wilder, 1830), p. 11.

3. Ahmes papyrus, problem 28. Quoted in Cajori, *History of Mathematical Notations*, 1:230, as from T.E. Peet, *The Rhind Mathematical Papyrus*, Plate J, no. 28. As is often the case with math notation, the use of this symbol was not universal; Cajori notes on page 229 that a different Egyptian papyrus uses a pair of legs to represent squaring a number instead of adding. David Eugene Smith notes that the actual Ahmes papyrus was written right to left, meaning the symbol would have been the reverse of what is pictured, appearing as a pair of legs walking backwards when read from left to right. See David Eugene Smith, *History of Mathematics*, vol. 2, *Special Topics of Elementary Mathematics* (New York: Dover Publications, 1958), p. 396.

4. Cajori, *History of Mathematical Notations*, 1:229–230. Cajori also notes that Diophantus, a Greek writer from the third century A.D., used juxtaposition (placing numbers next to each other) instead of a line.

5. For example, p´ was one of the symbols English mathematician Oughtred used to represent addition. See Cajori, *History of Mathematical Notations*, 1:190, in a table of Oughtred's Mathematical Symbols "first published, with notes, in the *University of California Publications in Mathematics*, vol. 1, no. 8 (1920), p. 171–86." Italian mathematicians in the 1500s also used this symbol (p. 230).

6. Cajori, *History of Mathematical Notations*, 1:230. See also Midonick, ed., *Treasury of Mathematics*, 93, and Smith, *History of Mathematics*, p. 396.

7. Cajori, *History of Mathematical Notations*, 1:230, and Smith, *History of Mathematics*, p. 398.

8. Cajori, *History of Mathematical Notations*, 1:236–239.

9   While *Mathematics: The Loss of Certainty* does not approach the topic from a biblical perspective, it nonetheless clearly shows the complete loss of certainty man is forced to adopt when he abandons a biblical basis for mathematics. [Morris Kline, *Mathematics: The Loss of Certainty* (New York: Oxford University Press, 1980)].

10   Vivian Shaw Groza, *A Survey of Mathematics: Elementary Concepts and Their Historical Development* (New York: Holt, Rinehart, and Winston, 1968), p. 215. The footnote adds, "The Hindus used the dot, . , as the symbol for zero." A zero was used here instead of a dot for simplicity.

11   Based on Leonardo Pisano, *Fibonacci's Liber Abaci: A Translation into Modern English of Leonardo Pisano's Book of Calculation*, translated by L.E. Sigler (New York: Springer-Verlag, 2002), pp. 20, 46.

12   Pisano, *Fibonacci's Liber Abaci*, p. 15.

13   Ibid.

14   Ibid., pp. 15–16.

15   Walter W. Sawyer, *Mathematician's Delight* (Harmondsworth Middlesex: Penguin, 1943), 10, quoted in Nickel, *Mathematics: Is God Silent?* p. 290.

16   For more information on problem-solving steps, see James Robert Overman, *Principles and Methods of Teaching Arithmetic* (New York: Lyons and Carnahan, 1920), http://books.google.com/books?id=6gcCAAAAYAAJ&ots=dYbiCX-qHS&dq=Principles%20and%20 Methods%20of%20Teaching%20Arithmetic.&pg=PP1#v=onepage&q&f=false, accessed 11/24/14, 256–274, and Samuel Chester Parker, *Methods of Teaching in High Schools*, rev. ed. (Boston, MA: Ginn and Co., 1920, http://books.google.com/books?id=pdtEAAAAIAAJ &ots=JIokmNSOfc&dq=methods%20of%20teaching%20in%20high%20schools&pg=PP1#v=onepage&q&f=false, accessed 11/24/14, 185–205. Steps here were inspired by those given in *Principles and Methods of Teaching Arithmetic*.

17   While many mathematicians have tried to explain math's ability to work without acknowledging God, sooner or later something has come along to dislodge their theory. For more details of the philosophy behind math and the errors in non-biblical worldviews, see *Mathematics: Is God Silent?* by James Nickel.

     Morris Kline's *Mathematics: The Loss of Certainty* is another resource on the philosophy behind math. While *Mathematics: The Loss of Certainty* does not approach the topic from a biblical perspective, it nonetheless clearly shows the complete loss of certainty man is forced to adopt when he abandons a biblical basis for mathematics (New York: Oxford University Press, 1980).

     You may also wish to look at a presuppositional book on logic, such as *The Ultimate Proof* by Dr. Jason Lisle. In this book, Dr. Lisle makes the case that only the biblical worldview makes sense of the laws of logic and that, apart from the biblical God, there is no way to really prove or know anything. He also makes the case that it is the biblical God, not any other God, who makes sense out of logic. (I would add that this is true of math too. Only the biblical God has the characteristics necessary to account for what we see in logic and math.) [Jason Lisle, *The Ultimate Proof of Creation: Resolving the Origins Debate* (Green Forest, AR: Master Books, 2009.]

Chapter 3. Mental Math and More Operations

1   *New Oxford American Dictionary*, 3rd edition (Oxford University Press, 2012), Version 2.2.1 (156) (Apple, 2011), s.v. "complement."

2   *The American Heritage Dictionary of the English Language*, 1980 New College Ed., s.v. "multiplicand."

3   Ibid., s.v. "multiplier."

4   *Merriam-Webster Online Dictionary*, s.v. "divide" (2009), http://www.merriam-webster.com/dictionary/divide, accessed 4/01/09.

5   Cajori, *History of Mathematical Notations*, 1:268–271.

     Top row, from left to right: Diophantus; Bakhshālī arithmetic; common Hindu method (using modern notation); several mathematicians used this symbol; American/English symbol used today (although many mathematicians have actually used it to represent subtraction!).

     Second row, from left to right: A variation on our current symbol used in a Buenos Aires text; symbol used by Gallimard; symbol used by Da Cunha; symbol used by Leibniz; another symbol used by Leibniz. Note: This list is by no means exhaustive!

6   According to Wolfram MathWorld, "An identity is a mathematical relationship equating one quantity to another (which may initially appear to be different)." Eric W. Weisstein, "Identity," from *MathWorld* — A Wolfram Web Resource. http://mathworld.wolfram.com/ Identity.html.

7   John H. Saxon Jr., *Algebra 2: An Incremental Development*, 2nd ed. (Norman, OK: Saxon Publishers 1997), p. 8.

Chapter 4. Multi-digit Multiplication and Division

1   Pronunciation from *The American Heritage Dictionary of the English Language*, 1980 New College Edition, s.v. "distribute."

2   Based on a problem in Eugene Henry Barker, *Applied Mathematics for Junior High Schools and High Schools* (Boston, MA: Allyn and Bacon, 1920), p. 32.

3   See "Bit by Bit: A Course Resource for the History of Mechanized Thought," "1.3 Napier's logs and Napier's rods," http://ds.haverford .edu/bitbybit/bit-by-bit-contents/chapter-one/3-napiers-logs-and-napiers-rods/, accessed 11/24/14, and Georgi Dalakov, "Biography of John Napier (1550–1617)," http://history-computer.com/People/NapiersBio.html, accessed 11/24/2014, for more information about John Napier and the development of counting machines.

Chapter 5. Fractions and Factoring

1   Pronunciation from *The American Heritage Dictionary of the English Language*, 1980 New College Edition, s.v. "fraction."

2   *New Oxford American Dictionary*, 3rd edition (Oxford University Press, 2012), Version 2.2.1 (156) (Apple, 2011), s.v. "notation."

3   *New Oxford American Dictionary*, 3rd edition (Oxford University Press, 2012), Version 2.2.1 (156) (Apple, 2011), s.v. "ordinal number."

4   Florian Cajori, *A History of Mathematical Notations: Two Volumes Bound As One* (Mineola, NY: Dover Publications, 1993), 1:11.

5   Cajori, *History of Mathematical Notations*, 1:11–15. See also David Eugene Smith, *History of Mathematics*, vol. 2, *Special Topics of Elementary Mathematics* (New York: Dover Publications, 1958), pp. 45–47.

6   Cajori, *History of Mathematical Notations*, 1:26.

7   Ibid., 1:36. See also Smith, *History of Mathematics*, 2:208–209.

8   Ron Aharoni, *Arithmetic for Parents: A Book for Grownups about Children's Mathematics*, translated from Hebrew by Danna Reisner (El Cerrito, CA: Sumizdat, 2007), p. 9.

9   *New Oxford American Dictionary*, 3rd edition (Oxford University Press, 2012), Version 2.2.1 (156) (Apple, 2011), s.v. "cryptography."

10 Edward Britton, Mary Ann Huntley, Gloria Jacobs, and Amy Shulman Weinberg, *Connecting Mathematics and Science to Workplace Contexts: A GUIDE to Curriculum Materials* (Thousand Oaks, CA: Corwin Press, 1999), p. 2.

11 *New Oxford American Dictionary*, 3rd edition (Oxford University Press, 2012), Version 2.2.1 (156) (Apple, 2011), s.v. "multiple."

Chapter 6. More with Fractions

1   *Merriam-Webster Online Dictionary*, http://www.merriam-webster.com/dictionary/divide, accessed 4/01/09, s.v. "divide."

2   *Merriam-Webster Online Dictionary*, http://www.merriam-webster.com/dictionary/division, accessed 11/20/14, s.v. "division."

Chapter 7. Decimals

1   *The American Heritage Dictionary of the English Language*, 1980 New College Edition, s.v. "decimal."

2   Symbols based on those presented in Cajori, *History of Mathematical Notations*, 1:314–335. Top row, left to right: Belgian Simon Stevin, 1585; Wilhelm von Kalcheim, 1629; J.H. Beyer, 1603; John Napier, 1617; Richard Balam, 1653. Second row, left to right: William Oughtred, 1631; Johann Caramuel, 1670; "other writers," such as A.F. Vallin, 1889; one of many methods referenced by Samuel Jeake (1696) as being in use; another one referenced by Samuel Jeake.

Chapter 8. Ratios and Proportions

1   *The American Heritage Dictionary of the English Language*, 1980 New College Edition, s.v. "ratio."

2   See http://mathforum.org/library/drmath/view/58042.html for an exploration of some of the different definitions for "rate" verses "ratio." The author concludes with this: "A rate generally involves a 'something else,' either two different kinds of units (such as distance per time), or just two distinct things measured with the same unit (such as interest money per loaned money)." Doctor Peterson, The Math Forum, "Rate vs. Ratio" The Math Forum @ Drexel, http://mathforum.org/library/drmath/view/58042.html, accessed 10/1/14.

3   *New Oxford American Dictionary*, 3rd edition (Oxford University Press, 2012), Version 2.2.1 (156) (Apple, 2011), s.v. "proportional."

Chapter 9. Percents

1   *The American Heritage Dictionary of the English Language*, 1980 New College Edition, s.v. "per cent."

2   Ibid.

3   David Eugene Smith, *History of Mathematics*, vol. 2, *Special Topics of Elementary Mathematics* (New York: Dover Publications, 1958), pp. 247–250.

4   Names based on those given at Michael Kellogg's www.wordreference.com, s.v. "bird."

5   To learn more about how stats can be misrepresented, see Darrell Huff, *How to Lie with Statistics* (New York: W. W. Norton, 1993).

Chapter 10. Negative Numbers

1   Quote and most information from *The Encyclopaedia Britannica: A Dictionary of Arts, Sciences, and General Literature*, 9th ed., vol. 23 (Philadelphia, PA: Maxwell Sommerville, 1894), http://books.google.com/books?id=KGlJAAAAYAAJ&pg=PA308&dq=The+Encyclopaedia +Britannica:+A+Dictionary+of+Arts,+Sciences,+and+General+Literature+%E2%80%9Cprecursor+of+accurate+thermometers %E2%80%9D&hl=en&sa=X&ei=IvpsVPmpHe3gsAT-h4LwAw&ved=0CC4Q6AEwAQ#v=onepage&q&f=false, accessed 11/19/14, p. 308.

2   Ibid., p. 308.

3   *The Encyclopaedia Britannica: Arts, Sciences, Literature, and General Information*, 11th ed., Vol. 11 (New York: The Encyclopaedia Britannica, 1910), http://books.google.com/books?id=dT0OAQAAMAAJ&pg=PA407&dq=The+Encyclopaedia+Britannica+galileo &hl=en&sa=X&ei=nYsrURiPvvYEi-6BmAQ&ved=0CEoQ6AEwBQ#v=onepage&q&f=false, accessed 2/25/13, p. 411.

4   *The New International Encyclopaedia*, 2nd ed. Vol. 9 (New York: Dodd, Mead, & Co., 1915), http://books.google.com/books?id =QRwoAAAAYAAJ&pg=PA410&dq=The+New+International+Encyclopaedia+galileo&hl=en&sa=X&ei=DY0rUaqJCovC9gSI4ICoCw &ved=0CEoQ6AEwBQ#v=onepage&q=The%20New%20International%20Encyclopaedia%20galileo&f=false, accessed 11/19/14, p. 410.

5   David Halliday, Robert Resnick, and Jearl Walker, *Fundamentals of Physics*, 7th ed. (England: John Wiley & Sons, 2005), p. 88.

Chapter 11. Sets

1   John Venn, *Symbolic Logic* 2nd ed., revised and rewritten (New York: Macmillan, 1894), pp. 509–510.

2   Ibid., p. 7.

3   *The American Heritage Dictionary of the English Language*, 1980 New College Ed., s.v. "sequence."

Chapter 12. Statistics and Graphing

1   *The American Heritage Dictionary of the English Language*, 1980 New College Edition, 1980, s.v. "statistics."

2   Eric W. Weisstein, "Sampling," *MathWorld* — A Wolfram Web Resource, http://mathworld.wolfram.com/Sampling.html, accessed 9/27/14.

3   Gallup, Gallup Editors, "Romney 49% Obama 48% in Gallup's Final Election Survey," http://www.gallup.com/poll/158519/romney-obama-gallup-final-election-survey.aspx, accessed 11/18/14.

4   *New Oxford American Dictionary*, 3rd edition (Oxford University Press, 2012), Version 2.2.1 (156) (Apple, 2011), s.v. "coordinate."

5   Ibid., s.v. "axis."

6   The Weather Channel, "Monthly Planner for Moscow, Russia," http://www.weather.com/weather/monthly/RSXX0063, accessed 10/25/14.

7   Ibid.

8   National Center for Education Statistics, Kids' Zone, "Graphing Tutorial," http://nces.ed.gov/nceskids/pdf/graph_tutorial.pdf, accessed 10/14/14, p. 5.

9   Centers for Disease Control and Prevention (CDC), Department of Health and Human Services, "Using Graphs and Charts to Illustrate Quantitative Data," *Evaluation ETA: Evaluation Briefs*, No. 12 (July 2008): 2, http://www.cdc.gov/healthyyouth/evaluation/pdf/brief12.pdf, accessed 10/25/14.

10 The Weather Channel, "Daily Averages for Phoenix, AZ," http://www.weather.com/weather/climatology/daily/USAZ0166, accessed 10/10/13.

11 Babe Ruth — Standard Batting Chart, Baseball-Reference.Com, http://www.baseball-reference.com/players/r/ruthba01.shtml, accessed 10/1/14.

12 Based on definition in *The American Heritage Dictionary of the English Language*, 1980 New College Edition, s.v. "statistics."

Chapter 13. Naming Shapes: Introducing Geometry

1   Euclid, "Euclid's Elements: Book 1, Definitions, Definition 2." Found in David E. Joyce, *Euclid's Elements*, Department of Mathematics and Computer Science, Clark University, http://aleph0.clarku.edu/~djoyce/java/elements/bookI/bookI.html, accessed 10/04/14.

2   Pronunciation from *New Oxford American Dictionary*, Apple Version 2.2.1, s.v. "plane."

3   Cajori, *A History of Mathematical Notations*, 1:405,406. Note: The left-most sign was placed on top of ABC to indicate angle ABC.

4   Based on definitions given in *New Oxford American Dictionary*, Apple Version 2.2.1, s.v. "polygon," "poly," and "gon."

5   Eric W. Weisstein, "Circle." From *MathWorld*--A Wolfram Web Resource. http://mathworld.wolfram.com/Circle.html, accessed 10/15/14.

6   Prism and cylinder definitions were based on *Ray's New Higher Arithmetic*, rev. (Cincinnati: Van Antwerp, Bragg & Co., 1880), p. 390.

7   Ibid., p. 391.

8   *New Oxford American Dictionary*, 3rd edition (Oxford University Press, 2012), Version 2.2.1 (156) (Apple, 2011), s.v. "operation."

9   Euclid, "Euclid's Elements: Book 1, Definitions, Definition 2."

Chapter 14. Measuring Distance

1   Tina Butcher, Linda Crown, Rick Harshman, and Juana Williams, eds. *NIST Handbook 44: 97th National Conference on Weights and Measures 2012*, 2013 ed. (Washington: U. S. Department of Commerce, 2012), B-3, found on http://www.nist.gov/pml/wmd/pubs/h44-13.cfm, accessed 10/6/2014.

2   Ibid., B-10.

3   Ibid., B-3.

4   Ibid., B-6.

5   *The American Heritage Dictionary of the English Language*, 1980 New College Edition, s.v. "millennium."

6   "one thousand watts," *The American Heritage Dictionary of the English Language*, s.v. "kilowatt."

7   See *NIST Handbook 44*, B-5.

8   Ibid., B-3.

9   Ibid., B-4.

10 *The American Heritage Dictionary of the English Language*, s.v. "ratio."

Chapter 15. Perimeter and Area of Polygons

1   *New Oxford American Dictionary*, found in Apple's Dictionary Program (Apple, 2005–2007), s.v. "formula."

2   *The American Heritage Dictionary of the English Language*, s.v. "a priori."

3   *The American Heritage Dictionary of the English Language*, s.v. "a posteriori."

Chapter 17. More Measuring: Triangles, Irregular Polygons, and Circles

1   See http://exzuberant.blogspot.com/2011/07/really-really-understanding-area-of.html for some additional explanations of why the area formula works. Enzuber, "Really, Really Understanding the Area of a Triangle, Part 1," Exuberant, http://exzuberant.blogspot .com/2011/07/really-really-understanding-area-of.html, accessed 10/7/14.

2   Eric W. Weisstein, "Circle," From *MathWorld* — A Wolfram Web Resource, http://mathworld.wolfram.com/Circle.html, accessed 10/15/14.

3   Archimedes, *The Measurement of a Circle*, found in T.L. Heath, *The Works of Archimedes Edited in Modern Notation with Introductory Chapters* (Cambridge: J. and C.F. Clay, 1897), p. 96, found on Google Books, accessed 01/03/13.

4   Pi, symbolized π, is "A transcendental number, approximately 3.14159, representing the ratio of the circumference to the diameter of a circle and appearing as a constant in a wide range of mathematical problems," *The American Heritage Dictionary of the English Language*, 1980 New College Edition, s.v. "pi."

5   Although used beforehand by a few mathematicians, D.E. Smith credits Euler's adoption of the symbol in 1737 as what brought it into general use. David Eugene Smith, *History of Mathematics* vol. 2, *Special Topics of Elementary Mathematics* (New York: Dover Publications, 1958), p. 312; Florian Cajori *A History of Mathematical Notations: Two Volumes Bound As One* (Mineola, NY: Dover Publications, 1993), 2:8–15; this offers a more detailed look at its gradual adoption.

6   If you'd like to learn more about the process, see Michael D. Huberty, Ko Hayashi and Chia Vang, "A Slice of Pie," http://www.geom.uiuc .edu/~huberty/math5337/groupe/welcome.html, accessed 10/8/14.

7   Alfred S. Posamentier and Lehmann Ingmar, π: *A Biography of the World's Most Mysterious Number* (Amherst, NY: Prometheus Books, 2004), p. 13.

8   Ibid., p. 28.

9   For more details, see Dr. Jason Lisle, "Contradictions: As Easy as Pi," Answers in Genesis, http://www.answersingenesis.org/articles /2009/06/08/contradictions-as-easy-as-pi, accessed 01/03/13.

Chapter 18. Solid Objects and Volume

1   *New Oxford American Dictionary*, 3rd edition (Oxford University Press, 2012), Version 2.2.1 (156) (Apple, 2011), s.v. "capacity."

2   Units in chapter were based on various online sources as well as the official standards given in Tina Butcher, Linda Crown, Rick Harshman, and Juana Williams, eds. *NIST Handbook 44: 97th National Conference on Weights and Measures* 2012, 2013 ed. (Washington, DC: U.S. Department of Commerce, 2012), Appendix C. Found on http://www.nist.gov/pml/wmd/pubs/h44-13.cfm, accessed 10/6/2014.

3   Butcher, et al., *NIST Handbook 44: 97th National Conference on Weights and Measures 2012*, B-8, http://www.nist.gov/pml/wmd/pubs /h44-13.cfm, accessed 10/6/2014.

4   Based on the official standards given in Butcher, et al., *NIST Handbook 44: 97th National Conference on Weights and Measures 2012*, Appendix C.

Chapter 19. Angles

1   *New Oxford American Dictionary*, 3rd edition (Oxford University Press, 2012), Version 2.2.1 (156) (Apple, 2011), s.v. "degree."

2   See David Eugene Smith, *History of Mathematics*, vol. 2. *Special Topics of Elementary Mathematics* (New York: Dover Publications, 1958), p. 231.

3   Ibid., p. 230.

4   Florian Cajori references Hypsicles and Hipparchus as two prior Greek mathematicians who broke a circle down into 360. Florian Cajori, *A History of Mathematical Notations: Two Volumes Bound As One* (Mineola, NY: Dover Publications, 1993), 1:28.

5   These parts also had other names as well, such as "first sixtieths" or "first parts." See Smith, *History of Mathematics*, p. 232, and Sir Thomas Little Heath, *A History of Greek Mathematics*, vol. 1 (Oxford: Clarendon Press, 1921), found on Google Ebooks, accessed 10/27/14, http://books.google.com/books?id=h4JsAAAAMAAJ&dq=Sir%20Thomas%20Little%20Heath%2C%20A%20History%20of%20 Greek%20Mathematics%2C%20Vol.%201&pg=PA45#v=onepage&q&f=false, p. 45.

6   Nathaniel Bowditch, *The American Practical Navigator: An Epitome of Navigation*, 2002 Bicentennial Ed. (Bethesda, MD: National Imagery and Mapping Agency, 2002), p. 99, see also p. 749, found on http://msi.nga.mil/NGAPortal/MSI.portal?_nfpb=true&_st =&_pageLabel=msi_portal_page_62&pubCode=0002, accessed 10/27/14.

7   See James D. Nickel, *Mathematics: Is God Silent?* rev. ed. (Vallecito, CA: Ross House Books, 2001), pp. 57–58.

8   See Ibid., p. 68.

Chapter 20. Congruent and Similar

1   *New Oxford American Dictionary*, s.v. "congruent."

2   *New Oxford American Dictionary*, s.v. "equal."

3   *New Oxford American Dictionary*, s.v. "congruent."

4   Ibid.

5   See Ken Ham, *The Lie, Evolution/Millions of Years*, rev. (Green Forest, AR: Master Books, 2012).

6   Dr. Jason Lisle, *The Ultimate Proof of Creation: Resolving the Origins Debate* (Green Forest, AR: Master Books, 2009) p. 62.

7   Ibid.

Chapter 21. Review

1   Merriam-Webster.com defines "Ellipse" as "oval," Merriam-Webster.com, s.v. "ellipse," accessed October 24, 2014.

2   Robert Wilson, *Astronomy Through the Ages: The Story of the Human Attempt to Understand the Universe* (Princeton, NJ: Princeton University Press, 1997), p. 69. For more details about the obstacles Kepler faced, see Max Caspar's *Kepler*, translated/edited by C. Doris Hellman (New York: Dover Publications, 1993) or Wilson's *Astronomy Through the Ages*.

3   Max Caspar, *Kepler*, trans./ed. by C. Doris Hellman (New York: Dover Publications, 1993), p. 51.

4   Johannes Kepler, *Harmonies of the World*, found in *On the Shoulders of Giants: The Great Works of Physics and Astronomy*, Stephen Hawking, ed. (Philadelphia, PA: Running Press Book Publishers, 2002), p. 723.

Additional Endnotes

Over the course of writing this curriculum, hundreds of websites and books were consulted, along with various learning standards, tests, and other curriculums. Below are a few additional resources not yet included in the endnotes or text that I want to acknowledge.

Bennett, Jeffrey O., and William L. Briggs. *Using and Understanding Mathematics: A Quantitative Reasoning Approach.* 2nd ed. Boston: Addison Wesley, 2002.

COMAP. *For All Practical Purposes: Mathematical Literacy in Today's World.* 6th ed. New York: W. H. Freeman, 2003.

Eugene Henry Barker, *Applied Mathematics for Junior High Schools and High Schools* (Boston: Allyn and Bacon, 1920). Available on Google Books, http://books.google.com/books?id=-t5EAAAAIAAJ&vq=3427&pg=PR2#v=onepage&q&f=false

Groza, Vivian Shaw. *A Survey of Mathematics: Elementary Concepts and Their Historical Development.* New York: Holt, Rinehart, and Winston, 1968.

Halliday, David, Robert Resnick, and Jearl Walker. *Fundamentals of Physics.* 7th ed. John Wilely & Sons, 2005.

John C. Stone and James F. Millis, *A Secondary Arithmetic: Commercial and Industrial for High, Industrial, Commercial, Normal Schools, and Academies* (Boston: Benj. H. Sanborn & Co., 1908). Available on Google Books, http://books.google.com/books?id=R-tYGAAAAYAAJ&pg=PP1#v=onepage&q&f=false

Johnson, David B., and Thomas A. Mowry. *Mathematics: A Practical Odyssey.* 2nd ed. Boston: PWS Publishing, 1995.

Joseph Victor Collins, *Practical Algebra: First Year Course* (New York: American Book Co., 1910). Available on Google Books, http://google.com/books?id=hNdHAAAAIAAJ&pg=PP1#v=onepage&q&f=false

For links to helpful resources, curriculum reviews, and other information to guide further study, please see my website, www.Christian-Perspective.net.

[INDEX]

Bolded references are to definitions.

24–hour clock 50–51
a posteriori **312–313**
a priori 19, **312–313**
AA similarity theorem **392–394**
abacus **27**, 46–50, 159–161
absolute value **206–208**, 214
acute **257–258**, 264, 367, 371, 379
Adam 22–23, 99, 132, 260, 263
addend **41**
Addition **37–42**, 403
    on an abacus 46–47, 159–160
    of angles 373–374, 393
    associative property and **77–81**, 83, 270
    Bhāskara method 56
    and carrying 47
    commutative property and ... **76**, 79–81, 83, 270
    of decimals 158–161
    of fractions 125–130, 138–139
    identity property of **78–81**, 101
    multi-digit 46–47
    of negative numbers 197–200
    of percents 189–191
    solving problems with 41–42, 60–62
    symbols **40**
    terms (addend, sum) **41**
    with zero 101
    (*See also* checkbook, mental math, order of
        operations, time)
additive identity (*See also* identity property of
    addition) **78**, 84
additive inverse **198–199**, 213
algebraic symbols 26
algorithms **46**, 64, 85, 404
angle **256–260**, 377–379
    adding 373–376, 393
    congruent 382–385
    drawing 371–373
    labeling 259, 383
    and light 377–379
    measuring 367–370
    names for (acute, obtuse, right, and straight)
    .... **257–258**, 264, 367, **370–371**, 379
    and navigation 378
    in pie graphs 374–376
    in a triangle 264, 393–394
    vertex of **257**, 367
approximate (*See* rounding)
area **305–309**, 311, 316–317, 344–345
    of circles 342, 345
    of parallelograms 309–312
    of polygons 336–338
    of rectangles 306–308
    of squares **308, 317**, 319
    surface area 347–349
    of triangles .... **333–335**, 413n2 (chapter 15)
    (*See also* measuring, unit conversion)
Archimedes 340
associative property **77–81**, 83, 270
average **244–251**

axes (*See* axis)
axis **238–240**
Babel, Tower of 24, 253
bar graph **232–237**, 244
base
    area and **310**, 334–335
    percents and **187–189**
    place value and **32–34**
    volume and **351–355**
Bhāskara addition method 56
binary system **31–34**
borrowing 49
Cain 253
calculator 104–105, 107, 329
capacity **358–366**
    conversion between systems 361–363
    metric 362–363
    U.S. 358–360
carrying 47
Celsius **204–205**
Centigrade **204–205**
checkbook (*See also* checks) 52–56
checking work 41, 59–62, 103–104
checks 52, 54, 156
circle **263–264**, 338
    area of 342, 345
    and angles 367–369
    circumference of
    .... **339–341**, 345, 413n4 (chapter 17)
    drawing 263–264
    examples of 263–264, 271
    planetary orbits 400–402
    terms (diameter and radius) **339**, 345
    (*See also* pi, pie graphs)
circumference (*See also* circle)
    .... **339–341**, 345, 413n4 (chapter 17)
clock .... 43, 50–51, 368 (*See also* time)
column graph **233–237**, 244
common difference **225–226**
common ratio **225–226**
commutative property **76**, 79–81, 83, 270
compass (geometry) .... 263–264 (*See also* circle)
compass (navigation) 378
complement **66**, 83
computer 34–35, 223, 236, 239, 271, 329
confidence level (*See also* statistics) **231**
congruent **270**, 381–390, 396
conventions (*See also* order of operations)
    .... **81–84**, 403
conversion ratio (*See also* unit conversion) **282**
coordinate graphs (*See* coordinates)
coordinates **238–241**, 251, 268–270

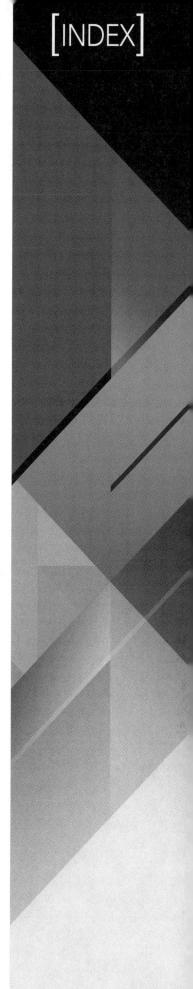

Copernicus 205, 379, 401–402

corresponding parts (*See also* congruent)
.....................**384–385**, 391–392, 394–396

counting (numbers, type of) 220–221

cryptography **23**, 121

cube 266, **352**

cubit **281**

cubic (units) (*See under* measuring: units)

Customary System, U.S. (*See under* measuring)

cylinder **266**, 350, 352, 354–355

data points **228**, 249–250

decagon **261**, 337

decimal system (*See also* place value, decimals)
.....................**27–32**, 57, 66, 153–158, 162–163

decimal point (*See also* decimals, scientific notation)
154, 156–157

decimals **153–168**, 399, 403
adding and subtracting 158–161
and fractions 155–157, 166–167
dividing 162–168
mental math with 179–180
multiplying 162–163
and negative numbers 210
and percents 184–186
in ratios 174–175
reading 157
representing (decimal point) **154**, 156–157
rounding **164–166**, 168, 180, 192–193
(*See also* scientific notation)

degree **367–369**

denominator (*See also* fractions, factoring)
.......................**110–111**, 128, 130–131

diamond **262**

diameter **339**, 345

difference **41**

distributive property **92–94**, 105

dividend **74**, 97

divisible **123**

division **73–76**, 403
by 10 or powers of 10, 162–163, 185
of decimals 162–168
Egyptian 73
as fractions (*See also* fractions, ratios)
.............110–112, 147, 169–171, 212–213
of fractions 148–152
multi-digit 95–100
of negative numbers 209–210, 214
of percents 188
with remainders 100
solving problems with 103–105
symbols 75
terms (dividend, divisor, quotient) **74**, 97
with zero 101
(*See also* order of operations)

divisor (*See also under* factoring) **74**, 97

Egyptian ....
26, 28–30, 40, 73, 113, 409n12 (chapter 1),
409n15 (chapter 1), 409n3(chapter 2)

Elements **254**, 256

equality (equal) (*See also* symbols: worldview of)
.....................................**26**, 42, 382

equilateral **264**

Euclid 254

even (numbers) **220–221**

evolution 20, 380, 397, 407

exponent (*See also* square roots, scientific notation)
.....................**315–318**, 331, 351–352, 403

extremes **171**

factor **72**, 104, 121–125
common **124–125**
finding a missing 75
greatest common **124–125**
and greatest common denominator **124**
and greatest common divisor **124**, 132
and least common multiple **130–132**
(*See also* factoring)

factor tree **122–125**, 320

factoring **121–123**, 143–144
factor tree **122–125**, 320
and square roots 319–321
(*See also* factor, fractions)

Fahrenheit **203–205**

Fibonacci Sequence (*See also* Pisano, Leonardo
"Fibonacci") **225–226**

flight 349

force 18, **211–212**

formulas **303–305**, 311, 317, 355, 365

fractions............................**109–152**, 403–404
adding and subtracting
............125–130, 138–139, 202, 210–213
and decimals........................155–157, 166–167
dividing 148–152
as division 110–112, 147, 169–171, 212–213
equivalent (*See also* proportions, unit
conversion) 117–120, 132
in history 113–114
improper **113**, 120
lowest terms and **120**, 124–125, 132
mixed numbers and **114–117**, 138–141, 151, 210–212
multiplying............................135–138, 140–147
negative numbers and 202, 210–213
and percents 185
pronunciation of 112
proper **113**
within ratios 174–175
simplifying 119–120, 142–145
terms of (numerator, denominator) **110–111**
whole numbers and **112**, 128–129,
136–137, 146–147, 151 (*See also* factor,
factoring, ratio)

fixed–value systems **28–31**

frequency...................228, 232–233, 251

Galilei, Galileo............................204–206

Gelosia multiplication method (*See also* Napier's
rods) **86–88**, 106–107

geometry
    examples of ..264, 379, 390, 392, 394, 401–402
    other types of....................................256, 273
    overview of **253–254**, 272–273, 397, 400
    graphs....................232–244, 251, 374–376
    bar **232–237**, 244
    column...................................**233–237**, 244
    computer–generated...............................236
    coordinate...................**238–241**, 251, 268–270
    histogram...**234**
    labeling.......................................234, 237
    line **241–244**
    pie **232–233**, 374–376
    scale (*See also* proportions: scale drawings and.
        See also scale) 236
    (*See also* statistics)

greater than................................**26–27**, 42, 223

greater than or equal to **223**

Greeks (*See also* Euclid, Ptolemy)
    26 (symbol), 40, 75 (symbol)
    254, 273, 369, 402, 409n12 (chapter 1),
    409n4 (chapter 2), 410n5 (chapter 3)

heptagon............................**261**, 299–300, 337

hexadecimal **34–35**

hexagon...............................**261**, 336–337

histogram...**234**

honeycomb......................................254, 345

humanism..19, 273

hummingbird...349

identity (*See also* additive identity, multiplicative
    identity)...........................**78**, 410n6 (chapter 3)

identity property of addition **78–81**, 101

identity property of multiplication
    ....**78–81**, 118–119, 152, 168, 323, 325

Incas 28

inequality...**26–27**, 42, 223

integer 19, **220–221**, 343

International system of units (SI)
    (*See* measuring: metric)

inverse (*See* multiplicative inverse, additive inverse)

irrational numbers (*See also* pi) 343–345, 400

isosceles **264–265**

Kepler, Johannes ... 16, 400–403, 414n2 (chapter 21)

length..............................275–281, **301–302**

less than.............................**26–27**, 42, 223

less than or equal to..............................**223**

letters (as math symbols) 60, 303–305

Liber Abaci...57–58

light (angles and)..............................377–379

line graph **241–244**

lines................................**255–260**, 376
    and line segments......................................**255**
    categorizing (straight, parallel, perpendicular)
    ....**255, 258**, 370–374
    labeling....................................259, 382–383
    congruent (*See also* congruent)...........**381–384**

lowest terms **120**, 124–125, 132

margin of error **231**

mass.......................324, 328–329, **362, 364–365**

math
    defined....................................**15–17**
    gospel and 21

mean...**244–251**

measuring........................... 400, 413n2 (chapter 18)
    capacity 358–363
    distance 275–281, 288–290, 297–298
    with congruency (*See also* congruent).........385
    Customary, U.S. (*See also* measuring: capacity;
        measuring: distance; measuring: weight)
    ....**275**, 281
    history of....................................276, 281, 358
    mass.......................................**362, 364–365**
    meter **275–278**
    metric (*See also* measuring: capacity;
        measuring: distance; measuring: mass)
    ....**275, 277–278**, 281
    with similarity (*See also* similar)................391
    temperature.....................203–206, 241–245
    time (*See also* time) 294–297, 298, 407
    units...**275**
    cubic.............................**350–352**, 361–362
    and exponents.............................316–317
    square **305–306**, 316–317
    weight..364–365
    (*See also* absolute value, angles, area, circles,
        geometry, perimeter, unit conversion,
        volume)

median...**249–251**

Medieval Ages 57, 379–380

mental math
    adding and..65–66
    complements and **66**, 83
    conversions via..................................285
    decimals and........................165–166, 179–180
    division and.......................................180–181
    money and.................165–166, 179–181
    multiplying and..........................94, 180–181
    percents and.................................192–193
    rounding and 67–68, 165–166
    subtracting and..65–66
    (*See also* rounding)

metric system (*See also* measuring: capacity;
    measuring: distance; measuring: mass)
    ....**275, 277–278**, 281

minuend..**41**

miracles...38

mixed number (*See also* fractions)
    ....**114–117**, 138–141, 151, 210–212

mode..**249–251**

models, scale **176–179**

money
    conversions with 286–287
    mental math with.................165–166, 179–181
    negative numbers and debt....................195–197
    writing with fractions and decimals.............157

multiple, least common (*See under* factor)

multiplicand **71–72**

INDEX 417

multiplication 68–73, 85–91
    associative property and............**77–81**, 83, 270
    commutative property and ...**76**, 79–81, 83, 270
    of decimals 162–163
    distributive property and.................**92–94**, 105
    of fractions
       ......135–138, 140–141, 142–145, 146–147
    Gelosia method **86–88**, 106–107
    identity property of
       ....**78–81**, 118–119, 152, 168, 323, 325
    multi-digit 85–91
    Napier's rods and **70–72**, 106–107
    of negative numbers...................208–210, 214
    of percents 188
    solving problems with........................102–105
    symbols of.................................71, 83
    table..70–71
    terms (multiplier, multiplicand, factor, product)
       ....**71–72**
    with zero.....................................101
    (*See also* multiplicative inverse, multiplicative
       identity)
    (*See also* mental math)

multiplication table 69–71

multiplicative identity 78, 83–84

multiplicative inverse..........................**146–147**, 152

multiplier.....................................**71–72**

music 137–138, 225

Napier, John ....70–71, 106–107, 157 (symbol),
    411n2 (chapter 7)

Napier's rods..... **69–72**, 106–107, 410n3 (chapter 4)

natural (numbers) 220–221

naturalism................................14, 18, 273

navigation...........................378, 414n6 (chapter 19)

negative numbers
    ....**195–214**, 219–220, 239, 243, 321, 399

neutral 11, **13–14**, 19–22

nonagon...............................**261**, 337

notation (*See also* fractions, decimals, percents,
    exponents, roots) **110**

numerals (*See* numbers)

number line **196**

numbers
    different approaches to writing 27–35
    reading (*See also* fractions, decimals, ratios) ...25
    types of.....................................219–222
       negative and positive (*See also* negative
          numbers) **219–220**
       whole............................112, **220**
       integer 19, **220–221**, 343
       natural **220–221**
       counting........................**220–221**
       even and odd...................**220–221**
       prime **121**, 221, 226
       irrational and rational (*See also* pi)
          ....**343–345**, 400
      writing................................22–27
    (*See also* decimal system, negative numbers)

numerator **110–111**

obtuse **258**, 264, 367, 371, 379

octagon................................**261**, 337

odd (numbers) **220–221**

operations (*See also* addition, subtraction,
    multiplication, division, order of operations)
    ....64, 403–404

order of operations **81–83**, 318

ordered fixed–value systems **30–31**

ordinal (numbers).............................**112**

parallel lines **258**

parallelogram **262–263**, 309–312, 372

parentheses.......................................
    coordinates and 238
    distributive property and 92–94
    exponents and.............................317–318
    multiplication and 71
    negative numbers and 198
    order of operations and 77, 82–84

pentagon................................**261**, 263, 304

percentage **187–189**, 191–194

percents **183–184**, 186, 399, 403
    adding and subtracting 189–191
    applying...............................184, 186–193
    decimals and.............................184–186
    division of 188
    fractions and...................................185
    mental.....................................192–193
    multiplication of..................................188
    relative frequency and.................228, 232–233
    (*See also* pie graph)

perimeter (*See also* circumference)
    ....**299–304**, 311–312, 321, 400

perpendicular lines **258**

polygon **260–263**, 299–313, 336–338

pi ....**341–346**, **413n4 (chapter 17)**,
    .... 413n9 (chapter 17)

pie graph..............................**232–233**, 374–376

Pisano, Leonardo "Fibonacci," 57–58, 225

place value (*See also* decimals, scientific notation)
    ....23–25, **27–28**, 31–35

point **255**

population **230–231**

positive (numbers) **219–220**

power.....................................**316**

prime factors (*See also* factors, factoring)
    ....**121–124**, 130–132, 221, 320

prime numbers (*See also* factor, factoring, sets)
    ....**121**, 132, 221, 226

prism (shape)...............................**265–266**, 350–356

prism (glass)..................................377

problem-solving
    ....58–60, 102–105, 366, 410n16 (chapter 2)

product **72**

properties
    associative property **77–81**, 83, 270
    commutative property **76**, 79–81, 83, 270
    distributive property........................**92–94**, 105

identity property of addition **78–81**, 101

identity property of multiplication **78–81**, 118–119, 152, 168, 323, 325

    of numbers 219–222

    of shapes....................................263–264, 301, 396

    (*See also* multiplicative identity, additive identity)

proportions **171–173**, 181

    conversions and.............................281–283

    decimals inside...............................174–175

    fractions inside 174–175

    models and 176–179

    scale drawings and **176–179**, 269–272, 390

    scale ratio and **176**

    similar shapes and..........................269–270

    terms in (extremes and means) **171**

    (*See also* proportional, similar, unit conversion)

proportional..........................**176–177**, 270, 391–394

protractor.....................................369–374

Ptolemy (*See also* Greeks) 369, 379–380

pyramid 253, 355

quadrilateral **261–263**, 299–300

quipu .... **28**, 409n14 (chapter 1)

quotient **74**, 97

radius.....................................**339**, 345

raising a number....................................**316**

rates (*See also* ratios, proportions)
    .... **169–171**, 181, 184, 411n2 (chapter 8)

rational numbers (*See also* irrational numbers) ...**343**

ratios...**169–179**, 181

    in circles (*See also* pi) 340–341

    common..**225–226**

    conversion **282**

    decimals inside....................................174–175

    fractions inside 174–175

    and percents 183–184, 186–187, 192

    reading...170

    shortcut, conversion

    ....284–286, 291–293, 323, 357

    (*See also* fractions, proportions, rates, unit conversion)

reciprocal..............................**146–147**, 152

rectangles 258, **262**

    area of...306–308

    drawing 372

    perimeter of..301

    similar shapes and........................174, 391–392

reflect (reflection).................**268**, 270–272, 377, 384

relative frequency....................**228**, 232–233

remainder **100**, 114–115, 164, 166

rename (renaming).............................49, 159

rhombus.......................................**262**

right (angle/line)...........**257–258**, 264, 367, 371, 379

right (triangle) **264–265**

Roman 30–31, 114, 281

rotate **269**, 270–272, 384

rounding ...**67–68**, 164–166, 168, 180, 192–193, 359

sales tax.......................................**183–184**, 191

sample 227, **229–231**

scale

    drawings/shapes **176–179**, **269–272**, 390

    choosing a 204–205, 236–237, 240

scale ratio **176**

scalene....................................**264–265**

scientific notation....................................**324–331**

sequence (*See also* sets)...........................**224–226**

sets **215–226**

    division and..74

    inequalities with 223

    multiplication and 71

    random 222

    sequences and.......................................224–226

    shapes and..258, 262–263

    statistics and 230

    Venn diagrams and.......**216–218**, 220–221, 258

    (*See also* place value, subsets)

    (*See* numbers: types of)

SI (International system of units) (*See also* measuring: metric) 277, 297–298

signs (mathematical) (*See also* symbols)

similar (*See also* proportional).....**269–270**, 392–396

square (units). (*See also* measuring: units)

squares..................................**262**, 300, 308

square roots **319–321**, 331

stars (distance to) 329–330, 407–408

statistics (*See also* graphs)
    .... 191–192, **227–251**, 399, 411n5 (chapter 9)

straight (angle/line) **255, 370**, 379

stream (distance across) 385–390

subset..........................**215**, 217, 220–221, 230, 262

subtraction....................................**37–42**, 75 (symbol), 399, 403–404, 410n5 (chapter 3)

    on an abacus.......................46, 48–51,159–161

    and borrowing 49

    of decimals...158–161

    of fractions 127–130, 139, 202, 210–213

    Leonardo's method...............................56–58

    multi-digit 46, 48–51

    of negative numbers
    ....197–200, 201–203, 210–213

    of percents 189–191

    repeated 73

    solving problems with..................41–42, 58–63

    symbol..39

    terms (minuend, subtrahend, difference)**41**, 225

    with zero..101

    (*See also* checkbook, mental math, order of operations, time)

subtrahend..**41**

suffering 132–133

sum **41**

surface area **347–349**

symbols (mathematical)
  comparing with 42
  decimal point (*See also* decimals, scientific
    notation) **154**, 156–157
  division...75
  equality...26–27, 42
  fractions...113–114
  inequality...26–27, 42, 223
  letters as 60, 303–305
  minus...39
  multiplication 71, 83
  plus 39–40
  worldview of...........16, 19, 21–24, 63–64, 403
  (*See also* negative numbers, notation)

Syntaxis 369, 379–380

temperature 203–206, 241–244, 245

theorem (AA similarity)..............**393–394**, 395, 396

three-dimensional (*See also* cube, cylinder, prism,
  volume) 255–256, 265–266, 350–351

time 42–45, 50–51, 407

time zones 44–45, 407

translate...............................**267–268**, 270–272, 384

triangle 261, **264–265**
  categories of (isosceles, equilateral, scalene,
    right, acute, obtuse).....................**264–265**
  angles in 264, 393–394
  area of.................**333–335**, 413n2 (chapter 15)
  congruent (*See also* congruent)
    ....270–271, 384–390
  perimeter of...**299–300**
  similar 269–270, 392–396
  using to find other areas.......................336–338

truth (basis for and nature of)..............312–313, 397

two-dimensional (*See also* circles, polygons)
  ....**255–256**

undefined term **254–255**

unit of measure (*See* measuring, unit conversions)

unit conversions
  capacity 360–361, 363
  cubic...356–358
  currency..286–287
  between customary and metric.....292–294, 298
  mass...365
  methods
    via mental math............**285–286**, 291–292
    via proportion.......**281–283**, 286, 291–292
    via the ratio shortcut
      ....**284–286**, 291–293, 323, 357
  multi-step **291–292**, 323, 357
  metric 288–290
  square 322–324
  time 294–297
  weight..365
  (*See also* conversion ratio)

Venn diagram **216–218**, 220–221, 258

Venn, John..218

vertex...**257, 367**

volume (*See also* capacity) **350–356**

vulture 349

whole (number) (*See also* fractions)
  ....**110, 219–222**

width 301–302

worldview (overview of)
  ....**13–14**, 15–22, 63–64, 312–313, 403–405

Author Bio

Katherine A. (Loop) Hannon, is a homeschool graduate who had her own view of math transformed. Understanding the biblical worldview in math made a tremendous difference in her life and started her on a journey of researching and sharing on the topic. For more than a decade, she has been researching, writing, and speaking on math, as well as other topics. Her previous books on math and a biblical worldview have been used by various Christian colleges, homeschool groups, and individuals.

 Website and Blog: www.ChristianPerspective.net
(Check out the Additional Resources page and optional eCourse available there.)

 Facebook: https://www.facebook.com/KatherineALoop
(Join the conversation!)

 @MathNotNeutral

 YouTube: https://www.youtube.com/user/mathisnotneutral

THE WORLD'S STORY

BE SURE TO EXPLORE THE ENTIRE SERIES

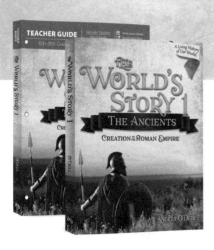  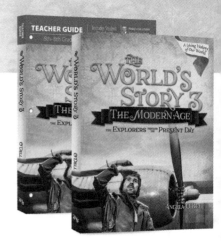

THE WORLD'S STORY 1
THE ANCIENTS
GRADE 6-8

Students will start with God's Creation and learn all about Biblical history and ancient civilizations up to the time of the Early Church.

2 Book Set 9781683441359

THE WORLD'S STORY 2
THE MIDDLE AGES
GRADE 6-8

Students will cover the Middle Ages through the Renaissance, beginning with the fall of Rome and surveying history from around the world.

2 Book Set 9781683441397

THE WORLD'S STORY 3
THE MODERN AGE
GRADE 6-8

Students will study the Age of Explorers through the modern day and learn all about the wars, revolutions, and culture changes that defined these times.

2 Book Set 9781683441403

VISIT MASTERBOOKS.COM *Where Faith Grows!* TO SEE OUR FULL LINE OF FAITH-BUILDING CURRICULUM OR CALL 800-999-3777

Mathematical Curriculum that Adds a Biblical Worldview!

Discover this easy to use text written in a conversational style directly to the student. You will find that all math boils down to a way of describing God's creation and is a useful means we can use to serve and worship Him.

Book 2 offers understanding and focus on:
> Essential principles of algebra
> Coordinate graphing
> Probability and statistics
> Functions, and many more important areas of mathematics!

This curriculum firms up the foundational concepts and prepares students for upper-level math in a logical, step-by-step way. It is aimed at grades 6-8, consists of the *Student Textbook* and *Teacher Guide*, which contains all the worksheets, quizzes, and tests, along with an answer key and suggested schedule.

Understanding the core principles of arithmetic and geometry helps students learn even more skills that will allow them to explore and understand the many aspects of God's creation!

Principles of Math Book 2
Package (Student & Workbook)
978-0-89051-915-8

Student
978-0-89051-906-6
Paperback 420 pages

Teacher Guide
978-0-89051-907-3
Paperback 440 pages

Katherine Loop is the owner of Christian Perspective, a family ministry dedicated to encouraging homeschooling families through a variety of methods, including resources, free e-mail newsletters, e-mail/phone support, and speaking engagements. She is a freelance writer, editor, marketer, and video editor with a mission to help families "Seek the LORD, and His Strength" (Psalm 105:4) in every aspect of their lives.

masterbooks.com   800-999-3777

JACOBS' MATH

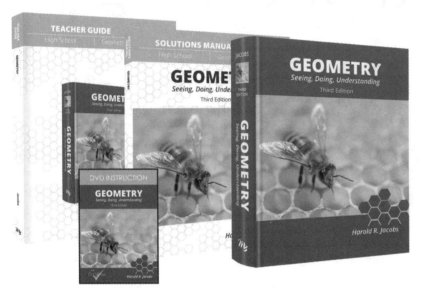

JACOBS' GEOMETRY

An authoritative standard for years, with nearly one million students having learned geometry principles through the text.

Jacobs' Geometry	978-1-68344-020-8
Solutions Manual	978-1-68344-021-5
Teacher Guide	978-1-68344-022-2
3-BOOK SET	**978-1-68344-036-9**
Geometry DVD	713438-10236-8
3-BOOK / 1-DVD SET	**978-1-68344-037-6**

JACOBS' ALGEBRA

This provides a full year of math in a clearly written format with guidance for teachers as well as for students who are self-directed.

Elementary Algebra	978-0-89051-985-1
Solutions Manual	978-0-89051-987-5
Teacher Guide	978-0-89051-986-8
3-BOOK SET	**978-0-89051-988-2**
Elementary Algebra DVD	713438-10237-5
3-BOOK / 1-DVD SET	**978-1-68344-038-3**

AVAILABLE AT MasterBooks.COM 800.999.3777
& OTHER PLACES WHERE FINE BOOKS ARE SOLD.

CREATION BASED
HIGH SCHOOL SCIENCE
WITH LABS

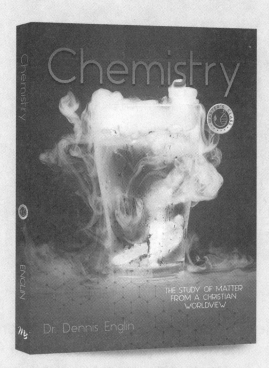

MASTER'S CLASS: CHEMISTRY
GRADE 10-12 | 978-1-68344-134-2

MASTER'S CLASS: BIOLOGY
GRADE 9-12 | 978-1-68344-152-6

TO SEE OUR FULL LINE
OF FAITH-BUILDING
CURRICULUM